Building Repair and Maintenance Management

Building Repair and Maintenance Management

PS Gahlot ME (Str. Engg.)

Ex-Professor
National Institute of Technical Teachers' Training and Research
Chandigarh

Principal
Yagyavalkya Institute of Technology
Sitapura, Jaipur (Rajasthan)

Sanjay Sharma ME (Hyd. and Irrigation)

Assistant Professor
National Institute of Technical Teachers' Training and Research
Chandigarh

CBSPD

CBS Publishers & Distributors Pvt Ltd

New Delhi • Bengaluru • Chennai • Kochi • Kolkata • Lucknow • Mumbai
Hyderabad • Jharkhand • Nagpur • Patna • Pune • Uttarakhand

Building Repair
and
Maintenance
Management

ISBN: 81-239-1243-9

Copyright © Authors and Publisher

First Edition: 2006

Reprint: 2008, 2010, 2011, 2012, 2013, 2014,
2015, 2016, 2017, 2019, 2025

Published by Satish Kumar Jain and produced by Varun Jain for

CBS Publishers & Distributors Pvt Ltd

4819/XI Prahlad Street, 24 Ansari Road, Daryaganj, New Delhi 110 002, India
Ph: 011-23289259, 23266838   Website: www.cbspd.com
                                e-mail: delhi@cbspd.com

Corporate Office: 204 FIE, Industrial Area, Patparganj, Delhi 110 092
Ph: 011-4934 4934   Fax: 011-4934 4935   e-mail: publishing@cbspd.com; publicity@cbspd.com

Branches

• **Bengaluru:** Seema House 2975, 17th Cross, K.R. Road, Banasankari 2nd Stage, Bengaluru 560 070, Karnataka, India
  Ph: +91-80-26771678/79   Fax: +91-80-26771680   e-mail: bangalore@cbspd.com

• **Chennai:** 18/8B, Subbarayan Street, Shenoy Nagar, Chennai 600 030, Tamil Nadu, India
  Ph: +91-44-42032115, 26681266   e-mail: chennai@cbspd.com

• **Kochi:** 42/1325, 1326, Power House Road, Opp KSEB, Power House, Ernakulam 682 018, Kerala, India
  Ph: +91-484-4059061-65   Fax: +91-484-4059065   e-mail: kochi@cbspd.com

• **Kolkata:** 147, Hind Ceramics Compound, 1st Floor, Nilgunj Road, Belghoria, Kolkata-700056, West Bengal, India
  Ph: 033-25633055, 033-25633056   e-mail: kolkata@cbspd.com

• **Lucknow:** Basement, Khushnuma Complex, 7-Meerabai Marg (Behind Jawahar Bhawan), Lucknow 226001, India
  Ph: 0522-4000032   e-mail: tiwari.lucknow@cbspd.com

• **Mumbai:** PWD Shed. Gala no. 25/26, Ramchandra Bhatt Marg, Next to JJ Hospital Gate no. 2 Opp. Union Bank of India Noorbaug
  Mumbai-400009, Maharashtra, India
  Ph: 022-66661880/89   e-mail: mumbai@cbspd.com

Representatives

• **Hyderabad** 0-9885175004 • **Jharkhand** 0-9811541605 • **Nagpur** 0-8692091830
• **Patna** 0-9334159340 • **Pune** 0-9664372571 • **Uttarakhand** 0-9716462459

Printed at: SRK Graphics, Delhi, India

FOREWORD

The book on "Building Repair and Maintenance Management" is of great significance to all Civil and Construction Engineers. This publication will facilitate quality construction followed by repair and maintenance management of buildings.

The book contains details of the latest techniques of construction and new building and repair materials. The book is well structured for easy comprehension of various techniques of repair and maintenance management.

I congratulate the authors for their commendable job in bringing out such a comprehensive book on building repair & maintenance management. I hope that both the engineering students and field engineers will benefit a great deal from this publication. Those engaged in the work of building repair and maintenance management, must possess a copy of this book for easy reference.

Dr. Brahm Datt
Vice Chairman
Som Datt Builders Pvt. Ltd.
Engineering Contractors,
New Delhi

FOREWORD

Maintenance and house keeping of buildings and other civil engineering structures in the country have been the areas of great neglect and not given the importance they deserve in both, the education and training of civil engineers as well as in the practicing field. The subject is of vital importance to the practicing engineers on the field to ensure that the buildings designed and constructed at huge expense of time and money are maintained properly so as to ensure their optimal functionality, durability and longevity. To realize this, there is an urgent need to strengthen the curricula of courses of study in civil engineering as well as offer specialized extensive training programs to practicing engineers on the subject of maintenance of buildings and other civil engineering structures.

Concrete and masonry structures are susceptible to cracking, while steel gets corroded. Steel structures manifest loss of serviceability by excessive deformation and buckling. Other than that, many a times' deterioration occurs due to lack of adequate care and coordination at the time of construction to enforce primary objectives of safety, durability and serviceability of structures.

Maintenance of buildings is required to ensure that all the elements in the buildings function as per the requirements of users and do not fail at inconvenient moments. It is the responsibility of the maintenance cells/departments in an organization that the buildings are restored and maintained as per standards, prescribed according to the functional requirements, rules, regulations and acts of the land.

There are many a factors such as ageing, destructive action of natural forces, poor selection of materials, occupational and human factors that could contribute to the deterioration and malfunctioning of the various sub-systems of building structures. A thorough understanding of the principal ways in which deterioration of buildings employing different types of materials occur, is a pre-requisite for investigating the exact causes of deterioration, so as to undertake appropriate remedial measures for the repair and maintenance of buildings.

The book on Building Repairs and Maintenance Management written by the authors who have a wide and rich experience on the subject, provides the latest information on methods/procedures, techniques and tools used for carrying out investigations for making assessments of deterioration in buildings and systematic planning for their repair and maintenance. The book also provides an extensive treatment on the use of the latest available materials and techniques of repair and maintenance of different part/sub-systems of buildings.

Generally it is the lack of awareness that leads to neglect of timely remedial measures for the repair and maintenance of buildings. As a consequence, the same problems manifest themselves again and again. Hence timely investigation will facilitate taking remedial measures to prevent further deterioration.

The book meets the long standing need of a comprehensive treatment on the subject of repair and maintenance of buildings. I hope that it will prove very useful for both the students of civil engineering at Degree and Diploma levels and practising engineers.

Dr. M. M. Malhotra
Former Consultant to CPSC,
Manila (Philipines)
Ex-Principal, NITTTR,
Chandigarh

FOREWORD

It gives me great pleasure to write this foreword for 'Building Repair and Maintenance Management' as I foresee this book to fill a long-standing gap in this neglected area. I appreciate the authors' insight into the subject as I discover, on going through the script, that they have very well measured the length and breadth of the subject before attempting this book.

The building has been viewed by authors as a complete unit, including water supply and sanitary systems besides structural components. Between identifying causes which lead to damage of building and providing an entire cross-section of maintenance and protection techniques, suitability and availability of latest materials to be used, and the remedial measures to adopt for longevity and smooth function, the book is comprehensive and complete.

The comprehensive nature of the book is expected to make it equally relevant to professional engineers, academicians engaged in research and teaching and the students. I heartily congratulate the authors for their path-breaking efforts to produce this pioneering work of scholarship.

Dr. Baljeet S. Kapoor
Principal, CCET,
Chandigarh

PREFACE

Considering the importance and knowing it well that no good books are available in the market on the subject of "Repair and Maintenance Management", the authors have tried to bring out a comprehensive book on the subject generally neglected in the curricula of civil engineering. The major purpose of undertaking this venture has been to make available a text book in *"Repair and Maintenance Management"* for engineering students. The competence of understanding, investigating and providing suitable approach to repair and maintenance problems is most essential for all level of civil engineering professionals. Many Universities and Boards are introducing this subject at all levels of programmes (M.Tech., B.Tech., Diploma, etc.). While writing this book authors have considered and focused on the needs of building construction industry as a whole to suit all levels of engineers and technologists. The book is written considering the civil engineering background and is even suitable for working professionals associated with building industry as contractors, consultants, engineers, and supervisors. The book provides description of repair and maintenance approaches using innovative techniques and new materials.

Each chapter of the book starts with the broad expected learning outcomes followed by inputs of related introduction, description of materials and techniques of repair, and illustrative applications in specific situations. The purpose of providing learning objectives at the beginning of the chapter is to raise expectations in the learner of what the learner would achieve at the end of studying the chapter. At the end summary of various techniques are given for quick reference by students and practising engineers. The illustrations are followed by repair and maintenance problem related questions for practice to develop engineering competence. Suggestive questions are given for practice to develop professional competence of solving repair and maintenance related problems. Problems and solutions have been explained with sketches. Chapters of Unit-I, mostly deal with concepts, principles and fundamental aspects of diagnosing the problem correctly and then planning the repair and maintenance. Unit-II is based on the problem analysis and common materials and techniques of repair while Unit-III deals with specific problems and techniques. Major emphasis, while presenting the subject matter, had been laid on developing diagnostic analysis of the problems and preparing suitable plan of action.

The sample cases of repair and maintenance are based on real life situations drawn from the field. Many of the solutions are based on **chemical stability** and providing barriers to moisture movement in materials of building components within the available environmental conditions.

System Internationale (SI) Units of measurement have been used wherever necessary. Teachers of engineering colleges and polytechnics and working engineers have provided information on the repair and maintenance problems. This has helped the authors to write this book incorporating such common and specific problems and solutions.

The authors wish to express their sincere thanks to Mr. Devinder Chawla, Managing Director of Samriti Engineers Pvt. Ltd., Chandigarh and our post-graduate student for providing information on many problems and solutions carried out. The authors are also thankful to Lt. Col. P. M. Meena for providing useful information collected from various engineering organizations during his ME Thesis. The authors are also thankful to Mr. Yogesh Kaushal for initial typing of the manuscript. We have all praise for M/S CBS Publishers and Distributors for the quality and speed they have displayed in bringing out this book.

Authors

CONTENTS

UNIT-I

BASIC PRINCIPLES OF MAINTENANCE

Chapter 1

IMPORTANCE OF BUILDING REPAIR AND MAINTENANCE

IMPORTANCE OF BUILDING REPAIR AND MAINTENANCE

LEARNING OBJECTIVES

After studying this chapter, the learner knows the importance of repair and maintenance of building and will be able to:

- **Describe** the importance of repair and maintenance of buildings;
- **State** significance of repair and maintenance of buildings;
- **State** objectives of maintenance;
- **Describe** factors influencing the repair and maintenance of buildings;
- **Describe** economic aspects of repair and maintenance

1.1 INTRODUCTION

On the inception of civilization when man decided to settle and abandon a wandering life style, his immediate requirement was a suitable shelter. He used to live in caves and shifted to man made huts constructed with materials available in nature. This very art of construction slowly developed in to a science of construction technology over the centuries. The skill of man developed further and he could make so many complex, beautiful monuments and intricate structures in addition to magnificent residential accommodations. The examples are old Forts, Havelies, Taj Mahal, Jal Mahal, Lake Palace, Kutab Minar, Jama Masjid, Golden Temple, Planetariums, Vatican city, White House, Pyramids, Sanchi Stupa, Stadia and many other structures.

Available materials in nature were used along with artificial materials for construction of buildings. Further the use of these materials not only enhanced the aesthetic look but also increased the life of the structures. Over a period of time some of these buildings have retained their original glamour and serviceability while others do not. Considering this aspect, few questions arise, viz. ! Are these buildings, structures, monuments, memorials and havelies still in the same serviceable condition as at the time of its construction ? If yes, then what special methods and materials have been used ? If not, then why not? What all could have been done to keep them intact in original shape and serviceable condition? If there was a solution to keep these buildings intact in original shape and serviceable condition then why this has not been undertaken? Are there any constraints and hurdles to maintain these buildings in original condition? These questions make one to think about repair and maintenance seriously.

Obviously there are some restrictions and factors because of which, all these forts, havelies and monuments are not in the original shape and serviceable condition in which they were envisaged. Therefore, certain maintenance system needs to be adopted. An appropriate maintenance can be planned by addressing the following questions:

   (a) Is special material for repair and protection available ?
   (b) Is skilled manpower for maintenance available ?
   (c) Is know-how to use such materials available ?
   (d) Is required funding for the same available ?
   (e) Is access for repair available?

The answer to above questions will facilitate planning of repair and maintenance for the affected buildings. Therefore, for long serviceable life the need for repair and maintenance of such buildings is important and starts at the design stage itself.

1.2 HISTORICAL BACKGROUND

When man used to live in huts, the repair and maintenance was simple and easy to carry out. At the same time the required materials for the purpose were available in abundance in nature. There was no worry about availability of repair and maintenance material in nature. During ancient times, the huts were constructed at locations which were least affected by natural calamities. With advancement, construction activity acquired the gigantic scale and industrial status. Factors like man, material, machine and money started playing an important role in construction industry. Only some of the buildings could be managed and maintained in excellent condition. The excellent condition of such structures indicates the appropriate design, quality of construction, workmanship and good maintenance practices. Thus, it can be said that repair and maintenance of building structures was one of the vital aspect right from the beginning.

1.3 SIGNIFICANCE

It is a well known fact that every one of us want things of proven standards and same pertains to buildings and its services too. The objectives of building construction are mainly concerned with the **stability** of buildings, the **weather tightness, internal comfort** level, the optimum **use of the building** and its **longer serviceable life**. In order to achieve these objectives, it is necessary to carry out periodic and planned maintenance, at regular intervals.

The maintenance should start immediately after the building has been constructed for its intended use. All the users and maintenance engineer should be concerned with the performance of the materials used and the standard of workmanship in construction. If these are unsatisfactory, deterioration begins and defects become apparent, requiring urgent remedial measures. After completion, all buildings require regular, planned and timely maintenance to preserve and enhance their serviceability.

The importance and need of regular and planned maintenance can be seen from the following examples:

(a) If electricity fails due to non-maintenance in a Hospital, particularly when operations are being conducted, it becomes matter of life and death.

(b) In present day, life can get disrupted due to lack of reliable maintenance of services. In multistoreyed buildings if lifts are out of order then the entire life of the occupants is affected seriously.

(c) In precision Laboratory where moisture is detrimental to working of equipment, the leakage, seepage and dampness can create total chaos in results.

Mr. Adam Neville in his key note on Maintenance and Durability of Concrete Structure said "In order to ensure durability it is not enough to make durable concrete, but it is also necessary to put into place a system of regular and planned maintenance. It is therefore, necessary to have regular and planned maintenance of buildings and its services.

1.4 WHAT IS MAINTENANCE ?

The term maintenance comes from the French verb "Maintenir", which connotes to hold. It means to hold, keep, sustain or preserve equipment, building or structure to an acceptable standard of serviceability.

Building maintenance is therefore the act of maintaining the building in its serviceable condition. The act of maintaining may require repair or replacement but the primary objective of maintenance is to avoid as far as practicable the need for repair or replacement of the structural elements, fittings, services, equipment or finishing which collectively make up the building and its environment.

An eminent Engineer Mr. White gives an apt definition of Maintenance, **"Maintenance is synonymous with controlling the condition of a building including services so that the serviceability remains within specific region of acceptability"**.

Maintenance is the work undertaken to keep, restore or improve every facility in every part of a building, its services and surroundings to a accepted standard. Maintenance is essential to sustain utility value of the building and its facilities.

1.5 OBJECTIVES OF MAINTENANCE

The objectives of maintenance are to ensure that building and its services can perform its designed functions for the desired period of time with a high degree of reliability. Due regard should be paid to overall economics of the maintenance operations and to safety of persons living and working in it. The main objectives of maintenance are :

(a) To **preserve buildings** and its services in good serviceable condition.

(b) To **restore buildings** and its services in its original standard, when deterioration occurs due to any reason.

(c) To make **improvements in serviceability** whenever required.

(d) To **sustain the utility** value.

(e) To **prevent and slow down the rate of deterioration** of structures.

(f) To **enhance serviceability** of the structures.

(g) To avoid crisis maintenance by **regular and planned maintenance programme**.

There are two types of basic methods of maintenance by which the above objectives of maintenance can be achieved. These are:

(a) **Corrective** maintenance, and

(b) **Preventive** maintenance.

The first method is the common or usual method to carry out repair and rehabilitation when an item fails or when it falls below the level of an acceptable standard. This method is known as **corrective maintenance** *e.g.* corrosion maintenance, repair of cracks, etc.

The second method is to intervene in the life cycle of each item immediately before it is expected to deteriorate in its health and to restore it to an acceptable standard of the health. This is known as **preventive maintenance** *e.g.* protective coatings, water proofing membranes, etc.

Corrective maintenance often warrants no action until failure occurs and then repair or replacement becomes a matter of urgency. This of course is not a very satisfactory state of affairs from the user's point of view as well as from an economic aspect. Thus, the need for preventive maintenance gains importance.

1.6 FACTORS INFLUENCING MAINTENANCE

There are various factors which influence the decision to carry out preventive or corrective maintenance. Therefore, it is necessary to consider these factors for effective maintenance of building structures. These important factors are:

    (a) Cost

    (b) Age of building

    (c) Availability of physical resources

    (d) Urgency of maintenance

    (e) Future use

    (f) Social consideration.

(a) Cost

The cost of maintenance may at first sight seems to be a simple matter of how much to spend on the material and labour. The cost of maintenance comprises of **direct and indirect** cost. The maintenance materials vary to a great extent and its cost also varies dramatically. The direct cost in maintenance operations ranges generally from 70% to 90% of the total cost. Before coming to a decision to implement a particular item of maintenance, indirect cost factors like restricted access, production stoppage, safety aspects, availability of time, overhead expenses etc. must also be considered along with the direct cost.

(b) Age of Building

All buildings and structures consist of materials and components linked together to form the desired unit of accommodation. All such materials and components will start 'aging' from the moment these are used in the building construction.

    Any building structure constructed will have certain life expectancy since the materials and components wear out. This wearing out will reduce the overall serviceability of the building and also affect its remaining useful life.

    To obtain the maximum life out of materials, components, services, equipment and the building itself, a planned programme of inspections and maintenance should be established as soon as the building has been constructed.

(c) Availability of Physical Resources

Physical resources in the context of maintenance of buildings can be defined as all the materials, components, services and equipment which are necessary for maintenance. Therefore, when an item of maintenance is being planned, the availability of all these physical resources must be considered and ensured.

(d) Urgency

The matter of urgency may outplay other factors when decision is to be taken to carry out a specific maintenance job. An urgent maintenance task may be required for a number of reasons such as the repair of services which, unless rectified immediately, would render them unserviceable, causing lot of inconvenience. When such a problem arises, the paramount question which must be posed is, how urgent is the urgency? Urgency is a relative term and therefore it must be established whether the repairs need to be carried cut immediately, within hours or within days. Accordingly, action for maintenance must be undertaken.

(e) Future Use

The future use of a building as a whole must be considered while deciding when and how much maintenance to carry out at any given period of time. If the lease is for a short period and changed occupancy is expected then maintenance of the building in question must be accordingly planned. If required some efforts be made to carry out the maintenance in the context of the proposed future use.

(f) Social Considerations

Agencies engaged in maintenance works cause influence on social environment also. The results of good endeavours of maintenance agencies are left behind as an asset to the owner if no inconvenience is caused to the society and the environment is also maintained clean and safe.

The agencies carrying out maintenance activities can create disturbances such as noise, safety, dust, smells, and temporary interruption of services. It must, therefore, be one of the objective to recognize this social responsibility. Plan the maintenance in such a manner that the disturbance will be kept to a minimum level, particularly when working within the occupied building. Pleasing environment should be created by regular and planned maintenance of building structures.

1.7 MAINTENANCE AND THE GROSS NATIONAL PRODUCT

The gross national product (GNP) is a term in economics to indicate the total market value of the goods and services produced by the nation's economy during a specific period of one year. The GNP does not take into account or allow for any depreciation or consumption of the goods or services used to calculate the figure. If these are taken into account the result is known as the Net National product (NNP).

It has been found that during a given period, a fair amount of money spent in terms of maintenance, repairs, conservation, rehabilitation and adaptation of buildings contribute significantly to the GNP.

The immediate benefits of good maintenance operations are:

(a) Maintaining the **value** of the property;

(b) Maintaining the building or structure in a condition which will enable it to fulfil its **intended function;** and

(c) Presenting a **good appearance** to the general public.

At a glance it appears that good maintenance only results in benefit to the user, but a well-maintained structure or building will also achieve a **high degree of efficiency** in its use which in turn will **increase morale** of those living or working within its environment. It may also lead to an **increase in productivity**. The three benefits mentioned above are interrelated.

Human morale also plays an important role in GNP. If the moral is high, the quality and efficiency of a person usually increases which in turn result in increased production without extra effort or abnormal increase in costs. Efficiency can be related to human, equipment and services. If services and equipment are maintained to a high standard this again will mobilise high morale in the people using these equipment and services. Thus completing the cycle of better efficiency and increased production which affects the GNP. Figure 1.1 explains the GNP cycle and the contribution of maintenance towards higher GNP.

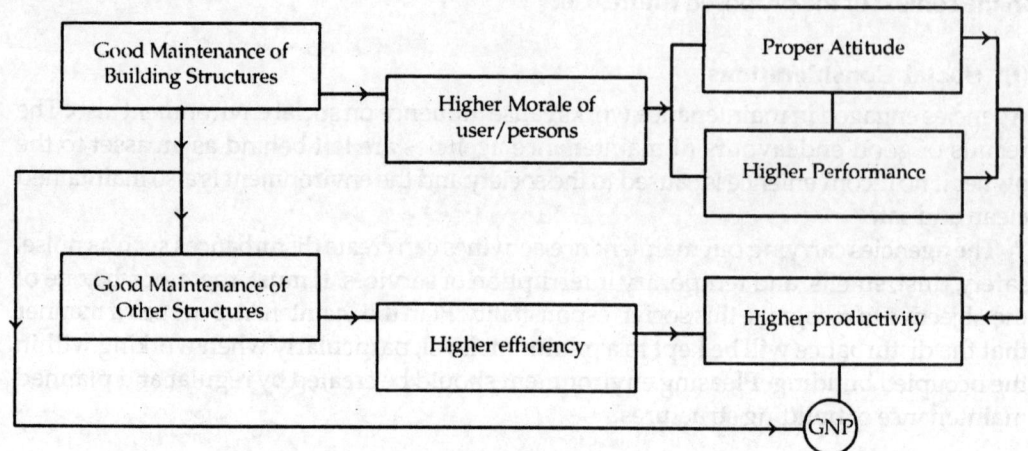

Fig. 1.1: GNP Cycle

1.8 ECONOMIC CONSIDERATIONS

There is a wrong notion in the minds of people that regular maintenance makes the building usage costly and uneconomical. If the real cost of building for the serviceable life is considered then the regularly maintained buildings are much economical and cost effective due to enhanced serviceable life.

The serviceable life of a building is governed by a number of factors like :-

(a) The extent to which maintenance was considered and included in the design.

(b) The degree to which the users carry out maintenance work during the life of the building.

(c) Economic considerations.

Life of Buildings

A building can be considered as having three distinct lives, namely

(a) **Physical Life:** assuming the property is structurally sound, this life can be extended almost indefinitely by careful, regular and planned maintenance.

(b) **Functional Life:** A building may no longer fulfil its original intended function due to social and technological changes. However, it may be possible to adapt the building to cater for these changes or alternatively convert it suitably for a different usage.

(c) **Economic Life:** A comparison of costs of maintaining a property against replacement is usually the best indicator of economic viability, but in some cases the economic life may be compared to the value of the site on which the building is situated.

The most important decision to be taken is, to carry out building work, adaptation work, or maintenance work, based on the life and cost of building work. The technique which can be employed to assess the economic life of a building is discounting. **Discounting** is method of calculating the present value of a sum of money due in the future. Discounting is basically the **reciprocal of compounding**. Discounting brings all the moneys involved in the project or proposal to a common base which is usually taken at its current or prevailing **value**.

Thus the appraisal of all the proposals can be made and a decision in regard to adoptation of a particular proposal, based on economic soundness, can be made.

1.9 CONCLUSION

To conclude it can be said that regular and planned maintenance of buildings is of vital importance. Planned maintenance goes a long way in preserving building's longevity, durability and serviceability. Maintenance is a continuous process which start from commissioning of the building and continues till the building acquires a state beyond economical repairs.

Premature loss of property occurs due to **ignorance, negligence** or **abuse of building**. Deteriorated buildings create unhealthy environment and unsound health of occupants which ultimately result in loss of productivity of people. Lack of repair and maintenance of buildings would further entail a huge cost on replacement at a later stage. Like longevity of human life, building life longevity also depends on timely and appropriate treatment of ailment and maintenance of building health and serviceability. Serviceability and health of a building can be enhanced by suitable repair and preventive maintenance and treatment commensurate with attacking environment. Building structures should not be abused by indiscriminate **over loading** and **over stressing** without suitable rehabilitation works.

The topic of maintenance of serviceability of buildings is of universal importance and gigantic in nature as the cost of maintenance, repair and rehabilitation runs into many thousands crores of rupees. Enhancement of **durability, longevity** and **serviceabilty** of building structures depend upon **careful design**, sound **construction**, good **workmanship** and continuous maintenance.

Invention of new protective materials have made maintenance a highly controlled and scientific process. Well planned maintenance goes a long way in preserving our coveted structures and buildings.

1.10 SUMMARY

Since the inception of civilization, many beautiful monuments and intricate structures have been constructed. Few of them have however retained their glamour and serviceability. In order to keep these structures in good condition, one has to think about the steps needed for their repair and maintenance.

Appropriate maintenance can be planned by having know-how about new repair and protection materials, training manpower and making enough funds available for this purpose. To achieve weather tightness, internal comfort level and longer service life, regular repair and maintenance of buildings is necessary.

Building maintenance is the act of maintaining the building in **serviceable condition**. It is essential to sustain **utility values** of the buildings. Corrective and preventive maintenance should be carried out with objectives of **preserving buildings** in good serviceable condition, **slow down rate of deterioration** and avoid crisis maintenance. The various factors which influence effective maintenance are **cost, age of building, availability of physical resources, urgency, future use**, social and economic considerations.

A fair amount of money in terms of maintenance, repairs, conservation, rehabilitation and adaptation of buildings contribute to the GNP. Immediate benefits of good maintenance includes maintaining the **value of the property**, presentation of **good appearance** and use of building for its **intended function**.

Regular maintenance proves to be much more economical in the long run due to enhanced serviceable life. Thus, well planned maintenance goes a long way in preserving structures and buildings.

QUESTIONS

1.1 Describe the importance of repair and maintenance of buildings.

1.2 Define maintenance with reference to buildings.

1.3 List down the main objectives of maintenance of buildings.

1.4 Differentiate between corrective maintenance and preventive maintenance.

1.5 Explain the various factors which influence the decision to carry out maintenance.

1.6 Explain "how good maintenance contributes to GNP".

1.7 Define the following:
   (i) Physical life
   (ii) Functional life
   (iii) Economic life.

1.8 Describe briefly the importance of economic consideration in maintenance.

Chapter 2

PRINCIPLES OF MAINTENANCE MANAGEMENT AND QUALITY ASSURANCE

PRINCIPLES OF MAINTENANCE MANAGEMENT AND QUALITY ASSURANCE

LEARNING OBJECTIVES

After studying this chapter, the learner understands maintenance management of buildings and will be able to:

- **Know** principles of **maintenance management and quality assurance;**
- **Explain organisational structure** for maintenance;
- **Describe** type of **maintenance work force;**
- **Explain** the **information communication** system for effective maintenance;
- **Describe building inspections** and reports;
- **Explain** the type of **maintenance budget** estimates;
- **Explain specifications** for maintenance jobs;
- **Explain health and safety** requirements in maintenance;
- **Explain** the concept of **quality assurance in maintenance;**
- **Explain** life **expectancy** of building elements;
- **Explain preventive maintenance** in buildings;
- **Describe** importance of maintenance manual;
- **Explain** economic considerations in maintenance jobs

2.1 INTRODUCTION

Maintenance management involves two distinct activities-Management and Maintenance. Management, in its simplest meaning is an act or manner of dealing, planning, controlling, directing and guiding other people in the organisation or team for achieving the desired goals. **Management, is therefore, an art of solving a problem or seeing a job carried out efficiently by people**.

Maintenance refers to maintaining the plant (equipment) and buildings in efficient operational conditions so as to accomplish goals of production without interruptions and breakdown. Maintenance is of utmost importantance in every field in light of highly competitive world market. In this modern competitive era no part of any organisation can afford to use its resources (man, machine, materials, method and environment) inefficiently. **Maintenance, is therefore, an act of keeping the components of production process in good working condition for its efficient and effective use in** production.

Maintenance management, is therefore, considered as **an act of dealing, planning, controlling, directing and guiding maintenance team** for keeping the **production process components in good working condition**. Maintenance management involves following basic functions:

(a) **Planning and Designing** of maintenance system

(b) **Coordinating, communicating and controlling** of maintenance jobs

(c) **Motivating, Directing and Guiding** maintenance team.

(a) Planning and Designing of Maintenance System

This would include organizing the maintenance department by ensuring availability of all the maintenance process components (men, machines, materials, and work instructions) of adequate quality and quantity at the required time. Another aspect of planning and designing would be **setting of targets and standards** of maintenance. These targets and organisational vision, mission and goals, practicability and profitability should also be considered in planning and designing of maintenance job.

(b) Coordinating, Communicating and Controlling of Maintenance Jobs

Controlling is a **continuous process of inspection, measuring**, and **comparing** with the planned targets and standards. Controlling includes recording communication and feeding back the information for future designing and modifications in design and standards. **Coordinating** refers to making cross-**functional teams to work harmoniously** in achieving the set goals and objectives using different resources. For optimum results specially from human resources, an appropriate balance and coordination amongst different teams is necessary. This balance is obtained by proper communication, feedback of information and coordination amongst cross-functional teams. Controlling and coordinating functions facilitate adoption and management of the required changes in maintenance system to suit organisational goals.

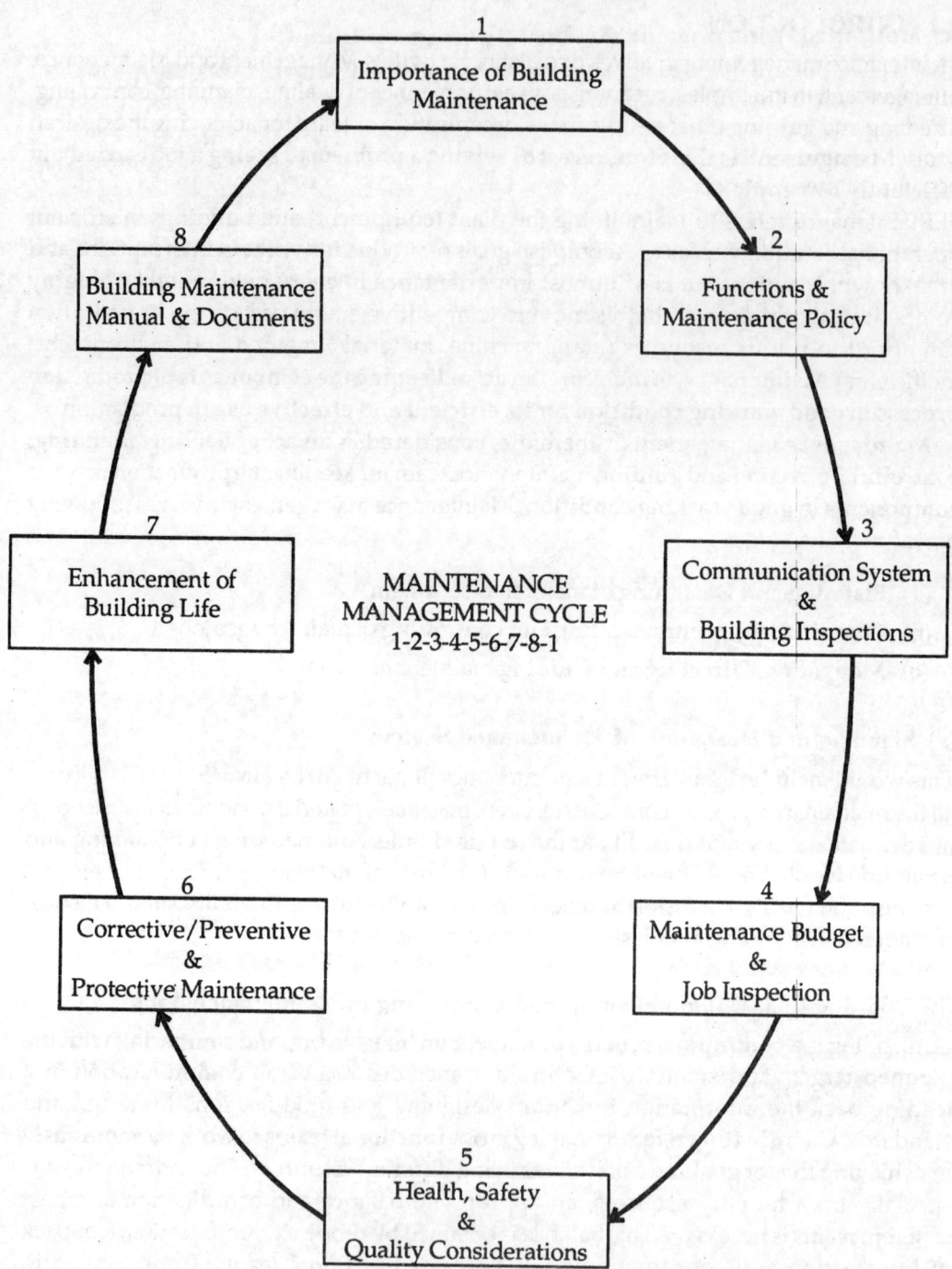

Fig. 2.1: Approach for Effective Management of Maintenance

(c) Motivating, Directing and Guiding Maintenance Team

This is an important management function of maintenance team leader to **inspire loyalty, effectiveness, quality, efficiency,** pride and sense of **commitment** to the goals of the organisation. The team members can be motivated through education and training and developing proper **communication system** to create **mutual trust,** and commitment in respect of accomplishment of goals. Clear directions and guidelines need to be provided regarding approaches, targets and responsibilities in achieving the desired goals. **Performance measurement** tools and procedures need to be clearly specified for every one to know and use. Motivation and guidance help maintenance team in realising the organisational goals of maintaining good working conditions at the optimum level of efficiency and economical in terms of cost. Motivation and guidance facilitate in development of mutual trust amongst members and concern for quality of maintenance in the organisation.

Organisation structure of maintenance team/department will depend on maintenance policy, approaches, size and other factors related to the organisation itself.

2.2 FUNCTIONS AND ORGANISATIONAL STRUCTURE

Every project or organisation needs to have certain organisational structure, depicting the flow of authority and responsibility for various tasks and achieving goals. Building maintenance activities in an organisation can be carried out by employing **direct labour** by the owner of the premises. Alternatively maintenance can be carried out through a **labour contractor.** Before arriving at an organisational structure, the decision has to be taken as to whether **direct labour** or **contract labour** or **mixed pattern** will be employed for maintenance jobs. In case mixed pattern of direct and contract labour for maintenance job is adopted, there has to be an appropriate balance between them, depending on the nature, size, frequency and specificity of various jobs.

The actual structure of a maintenance department will be governed by several factors, such as, the **size of organisation, the anticipated functions, work load, budget allocation and the overall policy towards maintenance and employment of direct labour.** In large organisations, the employment of separate direct labour force to cover the major and regular jobs of maintenance (such as cleaning, environment related services and house keeping), is usually considered necessary. While in smaller organisations, it may be possible to allocate some of the maintenance tasks as part of or in addition to other regular duties of workforce. In developing organisational structure of any organisation, the functions of a building maintenance department vested with the responsibility of maintaining the production process in good condition, must be **ascertained** first.

The maintenance department plays the following important roles:

(a) Organisational
(b) Planning & Controlling
(c) Advisory, and
(d) General

(a) Organisational Functions

These functions include work allocation, path responsibility and accountability, setting standards and procedures, establishing lines of communication, providing the necessary accommodation and equipment, employing maintenance staff, purchasing maintenance material and tools, and preparing and administering maintenance contracts whenever outside contractors are engaged.

(b) Planning and Controlling Function

These functions include inspection, measurement, and reporting on the need for maintenance works, planning of the work, both in the short and long term contexts; preparing cost estimates and budget estimates; controlling quality and quantity of output with reference to standard norms; and developing a system of recording and reporting for future planning.

(c) Advisory and Guiding Functions

These functions include informing and advising top management on maintenance policy, standards, cost requirements, suitable type of maintenance system, and special maintenance aspects at the design stage of proposed buildings, extensions, alterations or adaptations.

(d) General Functions

Maintenance department may be required to look after fire protection and safety, general security and safety, campus maintenance, and collection and disposal of waste.

When deciding about the organisational structure of any company, maintenance department must find an appropriate slot considering various functions, it is expected to carryout. For effective maintenance the organisational structure plays a critical role in terms of responsibilities and authority enjoyed by the maintenance department.

2.3 MAINTENANCE WORK FORCE

Maintenance workforce comprises of **direct, contract,** or **mixed labour** force. We should study the functional needs of the organisation, and advantages and disadvantages of direct workforce to decide most suitable maintenance workforce arrangements. It may be noted that the disadvantages of direct labour force becomes advantages of contractual (indirect) labour. It may also be kept in mind that sometimes occasions may arise when it becomes necessary and desirable to employ a **specialist agency** from outside the company, for certain specific maintenance problem.

Advantages of Direct Labour Force

(i) **Flexibility:** Direct labour force will provide greater flexibility and quicker response to any emergency maintenance problem due to full control over the movement and work allocation of a direct labour force. Direct workforce is available with the organisation at all the times and can be deputed straight away to attend the problem. Direct work force has intimate knowledge of locations, and type of problems and solutions required. Lack of quick response by building maintenance work force may result in interruption or stoppage of production process and loss of goodwill. Control over direct workforce can go a long way in limiting the need for emergency maintenance by establishing and supervising a system of regular planned and preventive maintenance.

(ii) **Intimate Knowledge:** The members of direct labour force acquire an intimate knowledge and experience of the building, equipment, and its services over a period of time. This intimate knowledge and experience becomes a great asset of direct maintenance workforce specially in case of large volume of equipment and services. Although these services and items of equipments may be standard but some times they can develop their own peculiarities. Familiarity with these peculiarities facilitate fast diagnosis of the fault and the remedy required to put the equipment or service back into operation with the minimum delay. This will also help in promoting good relationships between production and maintenance staff and better overall environment in the organisation.

(iii) **Effective Control:** A direct labour force enables management to control the overall strategy of maintenance work more effectively because of well known abilities and limitations of individuals in direct workforce. Effective balance can also be obtained between skilled labour and engineering staff depending on the nature and type of maintenance problem. The planning of the workload is also under better control in the case of direct work force in comparison to the contract work force. The quality control is also better since the direct labour force can be motivated for doing an effective job so as to reduce or eliminate future maintenance problems to be set right again by them.

Disadvantages of Direct Labour Force

It is very difficult to be specific on disadvantages of a direct labour force, since these depends on the type of premises and/or organisation. Each type of premises and organisation has different type of maintenance problems. In small organisations the direct maintenance labour force may **remain under employed**. Insufficiently planned, or preventive maintenance, may not keep them fully occupied and wasteful and uneconomical labor lead them to **wait** for some thing to do wrong.

It may also be argued that a large direct labour force may be equipped to deal with problems which occur within their jurisdiction only. However, they do not always have within their own work force, the necessary qualified or specifically experienced staff (to take full advantage of the facilities available in the organisation) to deal with **complex problem** which might crop up at any time or suddenly.

Productivity is often quoted as being lower in case of direct labour force due to older persons and security of job in comparison to the contract workforce. For the contract work force salary payments and incentive bonus schemes are based on quantitative output. It is also argued that **older workforce** tends to have wide knowledge and experience which reduces overall need for maintenance due to adequate quality in their maintenance work.

The main disadvantage of direct labour maintenance system is **low productivity** (as low as 65 per cent). This low productivity may not necessarily be due to poor workmanship or lack of work but very often by its nature and locations. Engaging contract labour to carryout small isolated work, spread over a large area could be more expensive than direct labour. If the maintenance contract is on a long-term basis whereby the contractor is called upon to supply the necessary labour and materials as required at some specific place and specific time since only productive time would be chargeable. This type of contract enables the contractor to direct his labour force on to other jobs when the maintenance requirements are non-existent. Direct labour may not yield the desired results in case of a **non-committed, workforce** and **leadership**.

If any conclusion is to be made as to whether direct or contract labour is to be used for maintenance work, it must be based on the **workload, nature of maintenance**, and the **amount of unproductive time** anticipated because of travelling between the jobs. Importance of maintenance is gradually increasing in light of highly competitive world market environment and requirements of total quality in every walk of life. In successful organisations management and maintenance teams have started taking maintenance tasks more seriously to decide about the choice of direct, contract or mixed type of labour force. Apart from economic considerations, this selection will involve **nature of maintenance, size** and **policy of the company** and feedback information from **previous experience**. The option may be decided based on balancing various advantages, disadvantages and other considerations of leadership.

2.4 INFORMATION MANAGEMENT AND COMMUNICATION SYSTEM

Design and development of effective **data based maintenance** system will require accurate information from similar projects from the past experience. We shall be required to collect information from similar projects on capital costs, running costs, nature of maintenance problems and teams, user requirements, maintenance hinderances and evaluations. Feedback information shall also be required to modify and improve quality of existing maintenance system. The designer of the system will need specific information of maintenance requirements at various stages of design and development of the organisation. All the information and maintenance requirements are required to be communicated in very clear terms in respect of buildings and equipment, to achieve desired goals and targets. Different type of **information collected needs to be analysed and scrutinised scientifically before being made applicable** to achieve the required quality and maintenance targets. Various types of information required at different stages are given as under:

(i) **Initial Brief:** The designer will require an outline of the maintenance requirements and objectives proposed at this stage. The proposed new buildings, extensions and adaptations must be clearly specified. These outline information would consist of all the basic **site details**, approximate areas for various usages together with **occupation, densities, service requirements, standards** required, and **time and cost targets**. This information enables the designer to produce sketch designs of buildings with approximate **cost summaries** and maintenance requirements.

(ii) **Working Drawings and Specifications:** The sketch designs with approximate costs and maintenance requirements facilitate the management to approve suitable alternative designs for the preparation of **working drawings** and **detailed specifications** for the new or maintenance jobs. Working drawings and detailed specifications shall be prepared for the selected alternative(s). At this stage, detailed informations are collected and supplied with regards to all aspects of the project and in particular the information and provisions for maintenance. The information for maintenance would include any special features required in terms of **access to different parts,** or building or equipment in the context of the maintenance policy of the building owner. Any restrictions as to the choice of materials based on the proposed methods of cleaning or renewal should be communicated, especially in the context of floors. Floors form one of the biggest element for maintenance, **renovation, cleaning** and **safety** consideration in any building. Poor location of permanent elements such as **ducts/shaft, stairs,** and **lifts** can very often limit the possibility of changes, and modifications required in future. The location of these permanent elements will influence the further changes and maintenance requirements. The above information can be communicated to the designer in the form of detailed written instructions, sketch drawings of any specific requirements and manufacturers' detailed literature for special materials, equipment, or fixtures.

(iii) **Choice of Maintenance Contracts:** The method of selecting a maintenance contract will depend upon the **size and nature** of the maintenance job to be undertaken. Simple and relatively small buildings on clear isolated sites are usually straight forward contracts for which full documentation is available. Such maintenance contracts are therefore suitable for a **fixed price tender** obtained through competition from amongst a selected number of contractors. Large buildings on congested sites are, however, best served by competitive tenders from selected contractors based on a **Bill of Approximate Quantities (BAQs).** Complicated additions and alteration works are better carried out on a negotiated price with a contractor specializing in such type of works. In many specialized and complicated cases it is better to contract on a cost plus basis. This includes the cost of **actual labour and materials used plus** an agreed percentage to cover overheads and profit of the contractor. When the accurate tender figure can not be assessed before hand, generous allowance for unforeseen items are required to be included, maintenance contract with cost plus basis is quite suitable.

Management of any organisation must provide guidance and direction in the connection with the tasks so as to decide the type of contract and type of maintenance task involved. Different sections in any organisation should provide information and advice about the type of maintenance task to the management, who should in turn pass on the information to the maintenance team/section. The information may be communicated through discussion and/or feedback system. Projected trends must be considered for achieving maximum flexibility of the proposed building or adaptation in future.

The maintenance policy of the management, plays very important role for the successful maintenance and keeping the building in operational condition. Proper **information and communication system** plays a vital role in the successful maintenance of building premises. Short term and long-term maintenance must be planned and incorporated in the design information in the beginning so as to achieve objectives of optimum economy, through **clarity in maintenance policy.**

Maintenance manual, specifying various details and special requirements of complicated maintenance jobs, must be provided along with completion of building or adaptation works. Preparation of a Maintenance Manual will go a long way in ensuring proper maintenance of a building and keeping the same in operational condition, at the most economical cost as a result of clarity in the information and communication system right from the design stage.

2.5 PROPERTY INSPECTION AND REPORTS

Inspections of buildings and their fixtures are carried out for a number of reasons and purposes. The purpose of these inspections must be established before commencement for obtaining necessary information required to prepare an acceptable report. The inspection form should be designed to ensure that as far as practicable there are no omissions in the report. Different types of inspections may be listed, as below:

- Complete building inspections or survey
- Inspection to rectify
- Planned inspections
- Control inspections.

Inspections form the basis for deciding on the nature and type of maintenance. Building inspections also provide appropriate data for budget estimates, type of maintenance work force and urgency of maintenance.

2.5.1 Complete Building Inspection or Survey

This form of inspection is usually carried out to obtain a complete and accurate **record of the property,** its service and fixtures. This inspection would normally be carried out where such data does not exist, particularly at the beginning of a lease or prior to sale/purchase. The inspection needs to be carried out by an experienced surveyor, specially in the case of older buildings with good knowledge of similar type of construction. The information should be gathered in such a manner that when finally presented, errors and omissions are negligible.

The building should be measured with a tape and using **running dimensions** to reduce the risk of cumulative errors which can easily occur if the process of 'piece meal' measuring is adopted. All fittings, fixtures, services, and any other features must be noted and measured to obtain their position and size.

A written report should also be prepared to indicate the condition, need for repair or any other items in need of attention and protection. The report should be submitted with the required drawings. The drawings may be presented in the form of fully dimensioned and annotated working drawings or, alternatively, as measured drawings without dimensions or annotation but including a drawn scale. Complete building inspection facilitates preparation of **standard measurement book** with bill of quantities (**BOQ**) required again and again during maintenance.

2.5.2 Inspection to Rectify

This form of inspection is carried out by the operational field staff. The inspection may well be planned according to timing or carried out as the result of a request from the user.

The typical examples are the inspection for cleaning out of necessary rain water gutters and down pipes. Springs and self closing doors are also checked and adjusted. A dated report of the inspection and action taken should be submitted for record purposes to the maintenance management team. Alternatively this information could be extracted from work order sheet or time sheets. This inspection helps in resolving petty items of maintenance on a continuous basis.

2.5.3 Planned Inspections

All types of buildings need planned inspection. Planned inspection is carried out after defining exactly the purpose of the inspection and it has to be carried out by competent inspector or surveyor. As far as practicable appropriate survey sheet should be designed and used to enable the inspector to gather and record his findings without overlooking or ignoring any item in the report. Standard formats must be used to facilitate collection of appropriate information for maintenance planning and execution.

Planned inspections are undertaken for a number of reasons, such as:

(i) To prepare complete inventory or record of the premises as to its condition and contents in the form of services, fittings, and fixtures;

(ii) To ascertain the need for current, predicted or future maintenance for the purposes of planning the work load and/or the budget;

(iii) To investigate the cause and extent of any occurrence of defect necessitating maintenance, so that its priority and also the standard of maintenance required can be determined.

The need for recording the information and data obtained during planned inspection for the **feedback system** is most critical. The feedback system used to share the information and data should ensure easy filing and retrieval. Such filing system could range from one based on **location, item** or **type of service, inspection date**, to one based on a **trade department or craft classification**. Any or all of these headings could be used but if more than one is selected there should be an adequate system of cross referencing. Card index systems are probably better than box or lever arch filing systems, for both storage and easy retrieval.

With development of computers, now it has become very easy to store and retrieve the information under a variety of information classification. Use of computers has revolutionized storage and retrieval of very precise information at any time with suitable software programmes.

2.5.4 Control Inspections

As the name indicates, the aim of this form of inspection is to check that a particular maintenance work has been executed with standard workmanship and quality. Control inspections are carried out to check whether the instructions and guidelines in accordance with the organisations maintenance policy have been followed and the required standards achieved. Control inspections are carried out to measure the quantitative **progress, quality, time** and **cost** for comparison with the planned standards. Control inspections are based on the fundamental principle of management which states "**what gets measured gets done.**" Thus for achieving quality in maintenance works, we need to undertake control inspections on a regular basis.

Chain of activities/events occur for the successful conclusion from the moment any property inspection is requested or planned. The whole process should be recorded for historical record and feedback purposes to ensure effective maintenance management. Various type of inspections form the backbone of any maintenance management system. Maintenance management cannot succeed without appropriate inspections.

2.6 MAINTENANCE BUDGET ESTIMATES

Before approval of any maintenance or adaptation work, an **estimate** of likely **cost and time** is needed. The maintenance manager prepares an estimate of cost and time to assess the cash needs, labour requirements, and priorities of jobs. Management requires this information for the following purposes:

(i) To make **comparative evaluation** between competitive estimates.

(ii) To establish long and short-term **budget requirements**.

(iii) To **apportion** or allocate money from an existing budget.

(iv) To provide **evidence** for a request for extra money where necessary maintenance or adaptation is not covered by an existing budget.

The estimator will require both data and an agreed system by which an estimated cost may be calculated. The data available will govern the accuracy of the estimate prepared. Preparing budget estimate will require following information for greater accuracy:

(i) Nature of proposed work

(ii) Extent or scope of work

(iii) Method of operation

(iv) Restrictions of any kind

(v) Current labour costs and availability

(vi) Past data of performance for the similar jobs and conditions, usually obtained from feedback or historical records

(vii) Direct or contract labour considerations

(viii) Specialist services and/or consultancy fee

(ix) Standards and level of specifications.

These estimates can be detailed or approximate, depending on the purpose for which these are required. Approximate estimates are a quick and simple method of assessing likely costs. These should be based on past records and performance which need to be adjusted taking into account any current inflation trends. For budget purposes cost estimation can also be predicted by using projected trends. Adjustments also need to be made for variation in condition or nature of the work with regard to the past data. Following systems of cost estimation are commonly used for budget allocation for maintenance jobs:

(i) **Cost per Plinth Area Covered:** The plinth/floor area is usually measured gross inside the external walls where the maintenance jobs are carried out. This plinth area cost varies for variety of items of maintenance job and conditions under which the job is likely to be carried out.

(ii) **Cost per Volume:** This approach is based on cost per unit volume of building elements involved in maintenance job. This method also accounts for height. In a large size building, gross area of the external walls is multiplied by height. This approach is more suited to buildings, such as, hospitals, factories, and other buildings with large variation in heights. This approach also depends on the conditions and items of maintenance.

(iii) **Cost per Unit of Accommodation:** The unit on which this method is based is the number of persons using the premises. This is useful as an overall measure of maintenance costs for buildings of similar age, use and construction specifications, such as, **schools, hostels-per room, hostels-per** student, hospitals-per bed, etc. Cost per unit helps in estimating maintenance cost of similar building.

(iv) **Cost per Element (item-wise):** This method is based on the cost of each element (item wise) involved and related to an overall classification with reference to total floor area. Each element is broken down into the number of items to be considered for overall accuracy of maintenance estimate. Typical sub-elements or items can be, such as, floors, walls, ceilings, joinery (door and windows), roof terrace, structural components, etc. This approach provides more accurate estimate for maintenance budget purposes.

2.6.1 Estimating for a Future Budget .

Estimate of cost is necessary for allocation of money to carry out maintenance and adaptation work in the future. This requires an assessment of cost estimate for appropriate allocation of money in advance. Such an estimate can be carried out based on the trade or department, activities or the project as a whole. Information on nature and cost is collected from previous years and future cost is estimated by considering rate of inflation and other changes in conditions, contents and specifications. Two important components of any estimate are **labour** cost and **material** cost. Size of labour force required can be determined by using appropriate **labour output** under the specified conditions. It may be noted that labour output is likely to be smaller for piece meal jobs. Material costs can also be calculated by calculating quantities of materials allowing for material wastages, inflation and price trends of materials likely to be used for the maintenance. The amount of budget required can be based on the following costs:

(i) All wage rates of labour force including basic wages plus bonus or other statutory contributions and fringe benefits;

(ii) Estimated material costs considering price trends;

(iii) Maintenance establishment and overhead charges;

(iv) Any element of profit required;

(v) All-in costs to cover supervisory and other non-productive staff of maintenance wing.

The object of this budget estimation is two fold-first to allocate budget and second, to use it for comparative statement for tenders from outside contractors.

2.6.2 Estimating for Current Budget

This is required for assessing the cost incurred on a specific item of job from the total budget allocation. Sometimes this is required to compare with the predicted cost used in the preparation of the budget and previous feedback information obtained from the experts. This cost estimation may also be required for apportioning the budget for carrying out a specific job within certain allocated budget for the different trades, sections or departments. An estimate of how much money should be spent on any of these specific tasks could be prepared and this would enable a certain control to be exercised over the current budget money allocation for unspecified maintenance works.

The actual method of calculation will depend on the nature of work and can range from spot item pricing to a unit rate pricing system based on actual quantities of labour and materials required. A similar procedure may be adopted for requesting extra finance over and above the current budget allocation except that the reasons and costs would be very detailed to convince management and justify such extra expenses.

2.7 SPECIFICATIONS FOR ADAPTATION AND MAINTENANCE WORKS

A specification may be defined as a **written document** setting out in detail the exact nature and contents of the work, the minimum **acceptable standard of workmanship** together with the **materials** to be used. This document needs to be written and read in conjunction with any drawings, schedules and bills of quantities prepared for the proposed work. A specification should not supersede the particulars or information given in other documents but should clarify the information where it is not clearly given, or else it can be subject to misinterpretation. Drawings usually show the general arrangement and details of the proposed works and the specification should describe the exact requirements of these details and arrangements. All the important dimensions and annotation should be included on the drawings and, therefore, it is not usually necessary to include this in the specification. The general dimensions, such as, room sizes and the sizes of individual members, such as, joists/beams and reference should always be quoted. On small works where drawings or details are not provided the specification must fully describe the works and include all relevant information and dimensions. A specification which is detailed, but at the same time accurate will help to prevent mistakes and misunderstanding during work execution.

A specification can be prepared by the designer, quantity surveyor, maintenance manager, building maintenance contractor or the client, depending on the size and nature of maintenance job. Specifications are prepared before undertaking any maintenance task to avoid disputes and ensure suitable quality standards for the job.

Generalizations in specification, such as, "as required" and "as necessary" should be avoided, since such generalizations lead to different interpretations and indicate lack of clarity on the part of person preparing the specifications. It is also unfair to expect an estimator to give an accurate price on the basis of a vague statement and this, in turn, could lead to claims for extra payment when the accounts are settled after execution of the work. When the person preparing the specifications is not sure of proper description of the exact nature of the work or what will actually be required on the job site, a provisional sum should be included which can then be adjusted by agreement between the parties concerned when settling the final accounts.

It follows, therefore, that as far as possible all descriptions and clauses contained in a specification should be **definite, clear** and **comprehensive**.

2.7.1 Preparation and Contents of a Specification

The actual preparation and format of a specification will be governed by the following three factors:

- Who is to carry out the preparation of specifications ?
- What is its purpose ?
- Who will use it ?

If the preparation is to be carried out by an owner or occupier for the purpose of internal instruction or to obtain an estimate it is most likely to be non-technical in its contents and may be grouped in areas of activity such as repairs to a roof, repairs of toilets, etc. The descriptions would probably be all embracing that is, not separated into materials and labour components, or classified by trades and without any indication of quantity and quality of materials and workmanship. As an internal document, it is usually satisfactory since the maintenance manager or worker would be familiar with this form of presentation and could analyse the document in a suitable manner for estimating, costing, ordering and job instruction purposes. If this type of specification is used as a basis for obtaining an estimate from an outside maintenance contractor, it could lead to ambiguities and misunderstandings of both the **content** and **intent** of the descriptions. This in turn could lead to an estimate which is unrealistic and ultimately to extra claims to contractors. If estimates are to be obtained from several contractors, the limitations of this type of specifications can result in wide variations in cost estimates. It is, therefore, suggested that specifications should be prepared by a person having necessary technical knowledge and expertise to set out the requirements, in clear and unambiguous terms, for the sake of quality.

Specifications commences with clauses covering:

- Preliminaries regarding conditions
- Materials type and quality
- Workmanship and methodologies
- General clauses about scope and requirements.

Preliminaries: These clauses would include information regarding site of the works to be executed, any drawings to be used in conjunction with the specification, general contract conditions and responsibilities for insurance, protection of existing works, hoardings and general storage facilities.

Materials: Clauses giving general description of the materials to be used, stating the **standard quality** required by reference to a particular code, or agreement certificate. To save unnecessary repetition, materials and components which are to be used for several activities could be fully described at this juncture, instead of repeating the description every time the items appears in the specification.

Workmanship: The requirements of workmanship should be precise, setting out in detail the manner in which the work is to be executed in terms of **quality and method**. Allowances or restrictions for such items as inclement weather and the use of alternative methods should also be stated. Reference to **relevant standard code** of practice should also be included, so as to clarify the minimum **acceptable standards**.

General Clauses: In view of new works, specifications are prepared in the same order and groupings as for taking up a bill of quantities. But specification for small or isolated maintenance work is often written to cover an area of activity to enable contractor to see the extent of work at glance for accurate estimation of cost. Specifications for maintenance and adaptation works are written to cover the exact conditions, circumstances and requirements of the work involved. Specifications for the new jobs and works can be developed by extracting typical clauses from manufacturer's literature, text books and similar sources. Main objectives of a specification is to bring out clarity and details for execution of work, so that all persons from investigator to the operater on site, understand the requirements and how to accomplish them. Maintenance managers and engineers should practice writing specifications for maintenance tasks for satisfactory execution.

Specifications form very important consideration for accurate estimation of **cost**, achievement of **quality** of maintenance job and appropriate relations amongst building users, maintenance staff, and contractors, if any. Thus, specifications need to be prepared accurately and considering ground realities incorporating various components viz. **preliminaries, materials, workmanship** and **general clauses**.

2.8 HEALTH AND SAFETY REQUIREMENTS IN MAINTENANCE

There are many legislations (Acts of Parliament) which set out the legal obligations of management as to their responsibilities with regard to the health and safety of their work force and work places. These aspects are covered in the enabling Act entitled "The Health and Safety at Work"-Act 1974. It applies to all persons at work including employers, employees, and self employed persons, with the exception of domestic servants in private households.

This Act also ensures that the health and safety legislation protects not only people at work but also the general public who may be affected by any work or maintenance activities.

This act relates to health and safety and sets out the duties of the employer, which are to:

- provide and maintain **plant in a safe condition** without any risk to health;
- arrange for the safe handling, **storage** and **transportation** of goods;
- as far as practicable make any **place of work safe** and maintain its safe condition without risks to health;
- ensure that the means of **access and egress** from a work place are maintained in a safe condition without risks to health;
- provide and maintain a **safe and healthy environment** in which to work;
- provide **instruction**, training and supervision of persons so that their health and safety is safeguarded;
- **set standards** of maintenance to safeguard the well-being of all employees.

This Act/legislation requires all employers to prepare and revise a **written policy** statement in respect of health and safety of employees and its organisation at the work site. The policy statement must also include the arrangements for ensuring safety under the stated policy implementation. These policy statements may include maintenance and provision of:

- Maintenance of plant, equipment and safety in works;
- Safe arrangements for the use, handling, storage and transport of articles, materials and substances;
- Information, instruction, training and supervision to enable all employees to contribute positively to their own and others' safety and health at work to avoid hazards;
- Provision and maintenance of safety equipment, safety warning system, fire fighting and protective clothing and ensuring that all employees are informed of their obligations with regard to care and use;
- A safe and healthy place of work and safe access and egress for employees and members of the public;
- Adequate welfare facilities, such as, canteen, water coolers, recreation places etc.

Written policy statements of any organisation would also include the policy objectives on:

- Safety of organisation;
- Employees' responsibilities;
- Safe operating procedures of equipment, plant and tools;
- Accident handling procedures;
- Fire fighting procedures;

- Inspection and checks;
- Training in safety;
- Means of communication

Safety representatives are appointed legally as independent nominees of a recognised trade union in any organisation, as defined in the Employment Protection (Consolidation) **Act 1978**. Health and Safety Act require consultation between employer and employee to ensure the health and safety of workers. It is required to appoint **safety representative** to check on the effectiveness of the arrangements made. Act requirements also include maintenance work and staff.

A safety representative has powers to **investigate hazards, complaints, carry out inspection, make representations** to employers, **receive information** from inspectors and **attend safety committee meetings**. Safety representatives and Safety Committee Regulations Act 1977 specifies safety representative's duties and rights along with safety committees set out regulations.

The management must keep themselves well informed of the changes as they occur for successful maintenance policy planning and execution. Complete understanding of **Health and Safety Act** at works is necessary for maintenance manager for successful policy formulation and execution of maintenance works. This Act provides for wide range of powers and includes maintenance works also. Observance and understanding of health and safety requirements under the Act plays a very critical role in successful maintenance management in any organisation.

2.9 QUALITY MAINTENANCE

In the present age of competition and globalisation every organisation is struggling for its survival and growth. Quality and productivity play vital role in survival and growth of any organisation. Quality and productivity form critical components of production system in any organisation. For enhancing quality and productivity we need to provide and maintain excellent environment around production system. Apart from keeping production equipment and plant in excellent operative conditions, we need to provide and maintain the total building environment in excellent condition. Quality of building maintenance directly influences the quality of production or service. It is, therefore, most important to practice and achieve quality in maintenance jobs. A small neglignece in quality of maintenance of tools, plant and building may lead to a very high loss in quality and productivity. The loss of quality and productivity may result in the down fall of any organisation. Thus, quality in maintenance is vital to any organisation.

Quality of any product or service may be defined as **fitness for use of customers'** requirements, satisfaction and delightment. Here production staff becomes customer while maintenance staff becomes supplier and quality in maintenance is specified by the production staff (internal customers). Maintenance staff has to provide and maintain excellent work environment by achieving quality in maintenance works. Thus quality in maintenance works has a direct role in achieving high standards of quality and productivity in the production system of any organisation.

2.10 LIFE EXPECTANCY OF BUILDING

Expected useful life of any building is determined by the life of its individual elements. Shortest life of any element infact determines the useful life of the building. Life of building elements depends on the quality of basic construction and adequacy of its repair and maintenance. If the quality of basic construction is poor, actual life of building get reduced tremendously causing loss in terms of building cost. It may be understood that normal repair and maintenance can not rectify defects in the basic construction. Highly specialized methods and materials for rectifying defects in basic construction may cost many times more than the cost of quality in basic construction. **It may be understood that normal repair and maintenance are no substitute for good quality basic construction.** It is, therefore, most important for building contractors and owners to ensure **quality in basic construction itself as it is much cheaper than rectifying defects by repair and mainte-nance later.** The quality of basic construction and repair and maintenance plays a vital role in deciding the expected life of any building.

Weathering forces (rain, heat, cold, wind and chemical gases) cause deterioration in building elements and, hence, reduce its life of useful occupancy. Repair and maintenance enhance the life expectancy of buildings by improving weathering resistance of the building elements. Thus, the life expectancy of any building depends on the following basic factors:

(i)   quality of basic construction;

(ii) type of occupancy;

(iii) quality of repair and maintenance;

(iv) weather and environmental conditions, and

(v)   natural disasters.

Table 2.1 of life expectancy is given with the assumption that the basic construction is of good quality and appropriate cycles of maintenance depend upon the weather conditions, quality of basic construction and types of occupancy. Life expectancy of building depends on its elements and approximate life expectancy of elements is indicated in the **Table 2.1.**

2.11 CORRECTIVE AND PREVENTIVE MAINTENANCE

One approach to repair and maintenance is to set right deteriorated portions of any building element after weathering action has already occurred. Another approach to building maintenance is to prevent or retard the weathering action by weather protective coating or treatment. There are many chemicals and paints which can be used for developing surface coat or film which will obstruct the contact of weathering forces with the building elements without adversely affecting the normal functioning of such elements. Most common weathering forces are :

● Polluted air, gases and fumes;

● Rain water or seepage water causing dampness;

● Heat and frost (high and low temperature) causing cracks;

- Wind causing surface attrition and stresses; and
- Earthquake, floods and fire causing severe stresses and cracks.

TABLE 2.1: AGE UNDER AVERAGE CONSTRUCTION UNDER NORMAL CONDITIONS

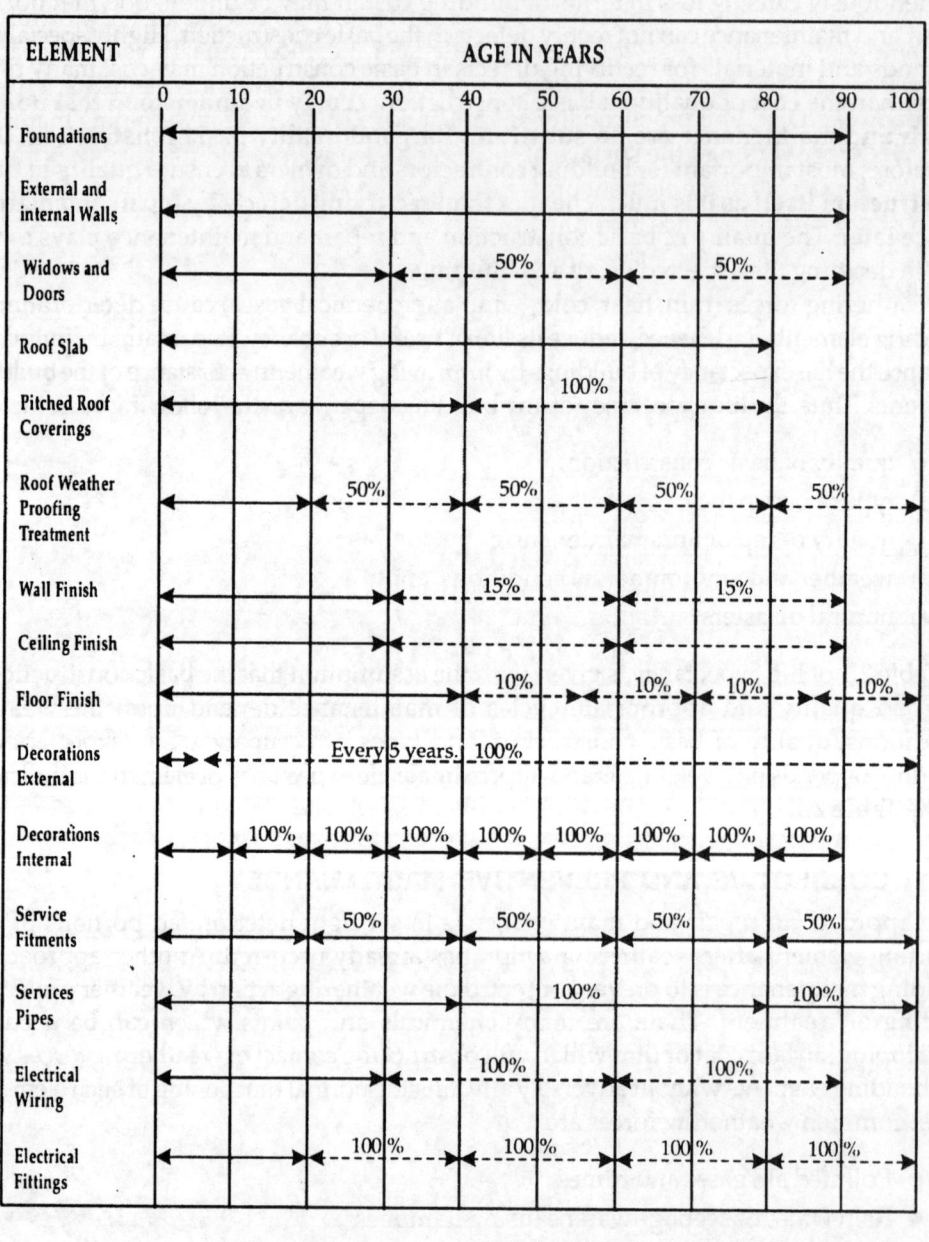

ELEMENT	AGE IN YEARS
	0 10 20 30 40 50 60 70 80 90 100
Foundations	◄──────────────────────────────► (to 90)
External and internal Walls	◄──────────────────────────────► (to 90)
Widows and Doors	◄──────────────►◄ 50% ◄───────►◄ 50% ───────►
Roof Slab	◄──────────────────────────► (to 80)
Pitched Roof Coverings	◄──────────────►◄ 100% ───────────────►
Roof Weather Proofing Treatment	◄──────►◄ 50% ───►◄ 50% ───────►◄ 50% ───►◄ 50% ───►
Wall Finish	◄──────────────►◄ 15% ──────────►◄ 15% ───►
Ceiling Finish	◄──────────────►◄ ───────────────────────►
Floor Finish	◄──────────────────►◄ 10% ──►◄ 10% ►◄ 10% ──►◄ 10% ►
Decorations External	◄►◄ ──── Every 5 years. 100% ─────────────►
Decorations Internal	◄─►◄100%►◄100%►◄100%►◄100%►◄100%►◄100%►◄100%►◄100%►
Service Fitments	◄──────►◄ 50% ───►◄ 50% ───►◄ 50% ───►◄ 50% ──────►
Services Pipes	◄────────────►◄ 100% ──────────►◄ 100%►
Electrical Wiring	◄──────────────►◄ 100% ──────►◄ 100% ──────►
Electrical Fittings	◄──────►◄100%►◄ 100%►◄ 100%►◄ 100%►

◄──────► Assesed life   ◄------► Projected life with percentage of total element to be renewed within the period

Preventive or corrective maintenance will nullify or retard and reduce the effect of these weathering forces. Most of these preventive or corrective coatings are used on the surface of the building elements to create barrier to weathering forces. Life of the element will depend on the effectiveness of these preventive treatments. Most critical weather force is rainwater and dampness which affect steel reinforcement, concrete and bricks. Preventive or corrective maintenance creates barrier to flow and passage of moisture to eliminate contact of moisture with relevant material. Different materials for surface coating may be considered based on weather conditions, type of building elements, type of occupancy and use, and practicability of application. These protective coating materials form water proof film at the surface of the building element or fill the surface pores, to keep the inside core of the element protected from the weathering action. Thus, protective and corrective maintenance forms an important part of maintenance management.

2.12 MAINTENANCE MANUAL

Maintenance manual is prepared at the time of construction of a building or design and manufacture of any tool and plant. This manual illustrates clearly various building components, services, and equipment giving their locations and drawings. The manual also specifies the material of construction of all these elements and method of their repair and maintenance. The manual describes maintenance cycles for various building elements. Maintenance manual also provides hidden details of building elements, such as, location of embedded pipelines, electric conduits, structural reinforcement. This manual becomes quite useful for carrying out adaptation work and other modifications. Maintenance manual becomes an important document for facilitating maintenance throughout the life of the building.

Maintenance manual must contain drawings as constructed, details of hidden portions, access to ducts, maintenance instructions, bill of quantities for maintenance items, and any of the specific information useful for maintenance. Path and locations of rain water pipes, sanitary and drainage lines, water supply lines, electric conduit line and telephone lines must also be very clearly indicated. Most of the problems that arise in the maintenance are related to these service lines and hence appropriate instructions to show access to these is essential, for effective maintenance management.

2.13 SUMMARY

Maintenance management is considered as an **act of dealing, planning, controlling, directing and guiding** maintenance team for **keeping the production process in good working condition,** so as to accomplish desired goals without interruptions. Maintenance management involves basic functions of:

- Planning and designing of maintenance system;
- Coordinating, communicating and controlling of maintenance jobs;
- Directing, guiding and motivating maintenance team

A suitable organisation is established for effective maintenance management based on the **size of organisation, anticipated maintenance functions, work load, type of jobs, budget** and **policy** of direct or contract labour. Direct labour force has advantages of **flexibility, intimate knowledge** of jobs and **better control**. Major disadvantage of a direct labour force may be lower productivity.

Effective maintenance depends on the development of information and communication system which provides feedback at various stages of construction and maintenance. Information on **the build drawings** and **specifications** used must be available in the form of maintenance manual. Choice of direct or contract maintenance work force depends on various factors such as **size, nature of job** and **available funds**, etc.

Property inspections play a **vital role** in maintenance management. These inspections may be carried out for the purpose of building survey, defect rectification, planning **preventive maintenance** or controlling. Control inspections are carried out to check **quality, specifications** and **progress** of maintenance jobs.

Budget estimates are essential for approval of any maintenance or adaptation job. Budget preparation depends on nature of work, scope, method, constraints, labour cost, special requirements and standards. Generally budget is prepared on the basis of cost per unit based on past experience of similar jobs. Budget estimates are based on projected future requirements, functions, technologies and prices.

Specification for maintenance and adaptation jobs are prepared giving accurate details of **materials, methods, standards of workmanship, drawings, scope of work, dimensions** and **constraints,** to avoid disputes and bad work. Specifications comprise of general and specific clauses.

Health and safety at work are all very important considerations in maintenance management. Health and safety at work - Act 1974 ensures protection to employers, employees and the public in general at the work site in respect of **handling, storage, transportation, access to work** place and work environment. This Act also provides for **policy statements, responsibilities** of employers, employees and safety representatives.

Life expectancy and serviceability of buildings are highly influenced by maintenance management and environmental forces. Life of building depends on original construction quality and type of occupancy. Preventive maintenance facilitates enhancement of serviceable life of buildings.

Building **maintenance manual** should be prepared to provide details of building components, services, and equipment with locations and drawings as constructed. Layouts of all hidden service lines must be shown very clearly for future repair and maintenance.

QUESTIONS

2.1 **Describe** maintenance management cycle.

2.2 Describe importance of maintenance management.

2.3 List basic functions of maintenance management.

2.4 Describe briefly:

(a) Planning and designing of maintenance system

(b) Coordinating, communicating and controlling of maintenance jobs

(c) Directing, guiding and motivating maintenance team

2.5   Explain main functions of maintenance team

2.6   Describe briefly type of maintenance work force and advantages of each.

2.7   Explain briefly need for communication and information system in maintenance.

2.8   Explain briefly importance of specifications in building maintenance.

2.9   List type and purpose of property inspections.

2.10   Differentiate between :

(a) Planned inspection and inspection to rectify

(b) Building inspection and control inspection

2.11   Explain the purpose of maintenance budget estimates.

2.12   List the factors considered in preparation of maintenance budget estimates.

2.13   Describe briefly important clauses included in maintenance specifications.

2.14   Describe important provisions of "the health and safety - Act 1974" in relation to maintenance job.

2.15   Explain how quality of building maintenance affects production in any organisation.

2.16   Explain how maintenance of building elements affects serviceable life of a building.

2.17   Differentiate between preventive and corrective maintenance

2.18   Describe importance of preventive maintenance.

2.19   Explain the need for maintenance manual.

Chapter 3

AGENCIES CAUSING DETERIORATION

AGENCIES CAUSING DETERIORATION

LEARNING OBJECTIVES

After studying this chapter, the learner understands the agencies causing deterioration in building elements and will be able to:

- **Define** deterioration/decay;
- **Explain** mechanism of deterioration;
- **Classify** the factors causing deterioration;
- **Explain** the process of deterioration due to human factors;
- **Explain** the process of deterioration due to chemical factors;
- **Explain** the process of deterioration due to adverse environmental conditions;
- **Explain** the miscellaneous factors causing deterioration in building elements;
- **Explain** the effect of various agencies causing deterioration on building materials.

3.1 INTRODUCTION

All the materials deteriorate with age either due to weathering action or other factors such as poor selection of materials, the human ignorance, vandalism and biological agencies. For successful design of buildings and maintenance policies, it is essential to understand the agencies causing deterioration. This will facilitate quality construction at the design stage and provide appropriate maintenance techniques, for prolonged and efficient service life of the building.

3.2 MECHANISM OF DETERIORATION

Deterioration means that the condition of a structure or a building or its components has degenerated or has become unusable. Deterioration of building or its components, if allowed to occur may result in complete decomposition where replacement becomes the only solution. Deterioration is nothing but gradual disintegration on account of any destructive action from aggressive waters and soils, exposure conditions to weathering agents and relative movements of components. The rate of deterioration of building varies with resistance of the materials used. Permeability is one of the critical characteristics influencing the durability. There are several ways in which mechanism of deterioration sets up. One is the external environment to which structure is exposed and second is internal causes within the material. Thus, there can be external and/or internal cause of deterioration. The process of deterioration can also be classified as under:

(a) Mechanical	Wear and tear	
	Fatigue, impact, or over loading	
(b) Physical	Thermal change	
	Volume change	
	Cracking	
	Deformation of shape	
	Freezing and thawing	
(c) Chemical	Reaction of aggressive substances in contact	
	Reaction of harmful chemicals present within material	
	Electro-chemical process like corrosion	
(d) Biological	Bacteriological growth	

The process of deterioration is very complex and may set in due to one or combination of above mechanisms. The various factors responsible for initiation of the process of deterioration are:

- Human aspects
- Chemical factors
- Furring
- Environmental aspects
- Fire hazards

- Faulty design
- Faulty construction
- Faulty materials
- Faulty system of maintenance
- Inappropriate cleaning
- Misuse of building

These factors are discussed and explained in detail in the subsequent paragraphs.

3.2.1 Human Aspects

(i) **Maintenance Staff Lacks Maintenance Culture:** The effects of deterioration can be minimized or slowed down by taking the corrective actions, at appropriate time, by the persons responsible for maintenance of structure. Deterioration occurs due to lack of appropriate maintenance culture on the part of the maintenance staff.

However, knowing what, where and how to carry out corrective measures does not ensure that they are, in fact, undertaken. This requires fixing of responsibility of executing the repairs and maintenance. Failure to provide the necessary capital, either due to inappropriate budgeting or inadequate allocation of financial resources hamper maintenance and thus lead to deterioration. Delay in attending the maintenance job can also lead to more severe problem of maintenance.

The deterioration may also be enhanced due to following human factors:

- failure to carry out routine maintenance, well in time,
- lack of knowledge about factors causing deterioration,
- poor planning, budgeting and allocation of inadequate monetary resources to enable maintenance activities to be undertaken,
- lack of sharing of responsibilities, and accountability, towards maintenance
- poor security leading to misuse,
- lack of awareness of maintenance needs among the users,
- using casual approach to repairs,
- failure to establish acceptable standards of maintenance,
- having a **negative attitude of waiting** until emergency measures become necessary.

(ii) **Occupants Misuse Buildings:** Lack of security, lack of awareness among occupants consequences of deliberate vandalism; and failure to repair the areas damaged by vandalism are some of the causes which become sources of further deterioration in structures. Blatant abuse of building, its fittings, furnishing and finishes may lead to deterioration. Lack of proper cleaning causing formation of injurious materials (alkaline and acidic) which may attack the building components in contact is an example of misuse.

3.2.2 Chemical Factors

Interaction of certain materials with surrounding environment is one of the main factors of deterioration. The critical chemical factor in the context of deterioration is that of corrosion. The corrosion is the result of instability of metals with its surrounding which tends to reach more stable state by combining with surrounding elements, such as, air, water, soil, sulphate and carbon dioxide.

Corrosion occurs in several forms :

(i) Local corrosion (called pitting of surface) limited to pitted area.

(ii) General corrosion where whole surface is attacked.

(iii) Electrolytic corrosion is due to contact between a metal and nonmetal. Moisture needs to be present to initiate this form of attack, where the potential difference between the metals sets up a galvanic action.

To overcome the problem of corrosion (rusting) it is necessary to know its causes and protective measures.

3.2.3 Furring

Furring is the depositing of mineral scale in vessels and pipes in which lime and magnesium bearing water is heated or conveyed. It depends on amount of temporary hardness in water. In the absence of evaporation, the fur which is precipitated at temperatures above 60 °C, adheres to the walls of vessels/pipes. This type of deterioration affects building services, such as, water supply. It is undesirable because:

(i) it makes pipe/vessels poor conductor of heat.

(ii) it diminishes the bore diameter and changes the flow patterns in pipes.

To overcome the problem, the designer should use treated water or can use oversized diameter pipes in the primary pipelines.

3.2.4 Environmental Aspects

Environmental factors refer to exposure of building components with atmospheric agencies, such as, air, rain, moisutre, gases, radiation and surrounding soil. Environment generally results in weathering, which is the action of atmospheric climate on the exposed materials and components. Weathering is the **process of decomposition** caused by components of weather, such as, **radiation, rain, snow, hail, wind, gases** and **contaminants** of air. Biological agencies, ground water and salts also affect the durability of materials, components and parts of the building. The various environmental factors causing deterioration are :

(i) **Solar Radiation:** Solar radiation is received at earth's surface both by short (direct) and long wave radiations. Most building materials absorb the radiation, depending upon the nature and colour of the surface. The absorption by some materials can lead to degradation, in particular, of plastics, paints and bitumen based materials.

(ii) **Temperature Effects:** Temperature changes cause dimensional changes in the materials, particularly when the coefficient of expansion is high. The resistance to these changes causes stresses which can exceed the strength of some materials and results in rupture and failure.

(iii) **Moisture:** Moisture in solid, liquid or vapour form can be regarded as principal agent causing deterioration. It is always present in atmosphere and when surface temperature falls, condensation can occur. It can cause excessive damage even under covered areas. Water frozen in the pores of materials can cause spalling of the surface, general cracking or disintegration. Proper preventive measures should be taken to reduce the effect of moisture and its deterioration process to avoid failures of building.

(iv) **Biological Agencies:** Many construction materials like concrete, bricks, and timber get affected by biological agencies, such as, algae, moss, etc. They attack generally wet timber. Timber with moisture content less than 20% is not generally attacked by biological agencies. Wet rot generally occurs in damp conditions. Once germination has occurred, it penetrates the cracks and make the timber lose its strength. The conditions which favour germination are darkness, stagnancy, moisture content above 20 per cent and temperature above 20 °C. Once germination occurs, it spreads very fast. As timber loses its cellulose, it loses its strength and shape resulting in cracks, shrinkage and loose fittings.

(v) **Gaseous Constituents and Pollutants of Air:** Sulphur dioxide is the most aggressive gaseous pollutant and can promote corrosion of some metals and cause some stones to blister and to spall. Carbon dioxide also forms a weak acid capable of slowly eroding lime stone. The extent of carbonation in concrete can have a marked influence on the corrosion rate of reinforcement. The ever increasing pollution is the major source causing deterioration of structures.

(vi) **Solid Contaminants:** The dirt from the atmosphere causes adverse effect on the buildings. It may increase the corrosion rate of metals and deterioration of some stone surfaces. The dirt also contain some soluble salts. It absorbs water from the atmosphere and accelerates corrosion process by maintaining moisture contact. Dirt deposited on building also causes an adverse effect on its appearance.

(vii) **Ground Salts and Waters:** Salts present in the ground, can rise with water and by capillarity in porous materials. On subsequent evaporation of solvent, salts remain deposited on the surface which can damage the buildings. Usually it causes efflorescence and defacing of building surfaces and finishes. More seriously, if magnesium sulphate is present, disintegration of rendering and masonry surfaces can occur. Acidic ground water can cause concrete to disintegrate and loss of strength.

3.2.5 Fire Hazards

Fire, if not controlled, can be hazardous to both the building and the occupants. Building components and elements should be tested to establish their degree of fire resistance, which is usually measured in hours or parts of hours.

All materials and components can be adversely affected by fire. These can be directly burnt wholly or partially, or by losing some of its properties, particularly structural strength and its aesthetic appeal.

At the design stage the designer must consider what materials or components are suitable for a particular situation in terms of resistance to fire. Designer should also consider the final consequences of out break of fire. The heat of fire, along with water used can lead to swelling, distortion, spalling and cracking of nearby materials and components.

3.2.6 Faulty Design

Faulty design leads to faster deterioration of structures. Unsuitable materials due to lack of knowledge of their characteristics and use of inadequate size of structural members will result in failure of structures. Lack of adequate attention to maintenance needs of future at the time of design will result in faster deterioration of structures.

3.2.7 Faulty Construction

The factors which contribute to deterioration due to faulty constructions are:

- Lack of supervision during construction period,
- Failure to monitor the work adequately,
- Failure to understand and follow exactly the specifications/drawings,
- Failure to replace the defective work,
- Lack of skilled labour,
- Over-emphasis on need for quantity rather than quality output.

The above factors, if not attended during construction, will lead to faster deterioration of structure during its service life.

3.2.8 Faulty Materials

Following factors result in poor quality of construction which, in turn, lead to faster deterioration:

- Wrong selection of material and specifications;
- Use of substandard material;
- Inadequate inspection of materials;
- Provision of inadequate facilities for storage at site;
- Inconsistent mixing of materials at site;
- Use of in approprite materials in relation to use;
- Use of stale/expired materals;

3.2.9 Faulty System

Inadequate knowledge on the part of the designer, unsatisfactory design details, inability of the builder to follow specifications/drawings, inadequate testing of the system before commissioning, failure to follow maintenance instructions and inability of the owner to operate the system as instructed can cause faster deterioration of building structures.

3.2.10 Inappropriate Cleaning

Maintenance starts with cleaning activity by considering some questions and issues. How easy is it to clean? How fast will it become dirty? It must be taken into consideration to reduce the effect of agencies causing deterioration. Improper/inadequate cleaning may be due to:

- Failure to carry out routine cleaning operations;
- Use of incorrect cleaning materials/techniques;
- Inadequate supervision to carry out effective cleaning;
- Insufficient time or incorrect equipment used for cleaning operations;
- Failure to employ specialists for cleaning special fittings and equipment.

3.2.11 Misuse of Buildings

Lack of security, lack of awareness among occupants of the consequences of deliberate vandalism; and failure to repair the areas damaged by vandalism.These are some of the causes which become sources of deterioration in structures. Balatant abuse and misuse of building, its fittings, furnishing and finishes may result in deterioration.

3.3 EFFECT OF DETERIORATION OF MATERIALS

Knowledge of various causes and sources of decay and deterioration is necessary to plan prevention of faster deterioration of structures. It is also necessary to understand the effects of various agencies causing deterioration of building materials to take proper protection against these agencies. The choice of material is governed by its:

- (i) Ability to withstand the effect of climate;
- (ii) Ability to fulfil the designed functions;
- (iii) Reaction with surrounding material;
- (iv) Ease of maintenance and/or replacement;
- (v) Overall economic acceptability.

The materials which have acceptable physical, chemical and economic advantages may be selected. To carry out analysis, one must know how the various materials deteriorate. The effect of various agencies of deterioration on major materials is briefly discussed in subsequent paragraphs. The designer's task is to find an effective solution to the above factors.

3.3.1 Bricks and Clay Products

Clay products which can be used on external surfaces include roofing tiles, coping, terra-cotta tiles and bricks. Generally all bricks and clay products have good durability. Most common effect of weathering on these products is change in appearance. Usually these materials give a dry look.

The most common form of crystallization of soluble salts in the context of clay products and brickwork is of efflorescence. **Efflorescence is white surface deposit of salts in the** form of loose powder. Generally surface deposit is not harmful, but it causes disfiguring. It occurs when dry weather follows wet weather. However, if the soluble salts crystallize inside the body of brickwork instead of on the surface, the expansion can cause spalling of the external layer. Sometimes brick walls may also deteriorate due to mortar composition.

3.3.2 Timber and Timber Products

Timber is a term used for wood of sufficient size for commercial purposes. Trees which produce suitable timber belong to one of two classes viz., conifers and deciduous.

Conifers: These yield softwood and include pines, spruces and larches having more or less needle like leaves.

Deciduous: These are broad leaf trees and yield hardwoods such as oak, Sal, Seesham and Teak.

Knots, shakes and too many edges are the main defects of natural timber and can be eliminated during the process of conversion into scantling sizes.

Timber for all purposes is a long lasting material and will not normally deteriorate unless attacked by fungi, insects or fire. Timber decays as a result of destructive action of fungi growth. The most well known being called dry rot, which requires moisture content of about 20% and spreads very rapidly. Wet rot requires moisture content of 30% and it spreads fast comparatively on wet surfaces.

Insect infestation can also weaken and destroy timber in building. The damages by insects are done much before these become apparent by the exit holes left behind by emerging beetles. It is the wood boring grubs which do the damage by eating timber fibers and extracting the cellulose.

It is important that the right grade or quality is specified, since most timber products are produced for either external and exposed conditions or for internal use. The use in external situations could lead very quickly to an advanced state of deterioration.

3.3.3 Concrete

The resistance of concrete to deterioration is dependent on the quality of ingredients used, the mix design and concreting operations viz. mixing, compacting, curing, etc. Concrete needs only small quantity of water to promote the necessary hydration of cement but it will be unworkable. As the water content increases, so does the workability and permeability. Protection of reinforcement is provided largely by cover of concrete over the reinforcement. The effectiveness of concrete cover largely depends on environment and its impermeability qualities. The corrosion of reinforcement could be accompanied by expansion of metal which in turn could result in spalling or cracking.

Chemical attack on concrete results in breakdown of the mix which is usually due to sulphates of calcium and magnesium. It is important to use right type of cement with correct amount of w/c retio on concrete cover. Thus, in terms of weathering, the major requirement is to produce a concrete which has sufficient resistance to water and moisture penetration. If the concrete is allowed to dry too quickly, the shrinkage cracks could lead to deterioration at an early stage.

Concrete also expands and contracts with changes in temperature. The amount of movement is governed by mix design. Expansion of concrete also affects the appearance due to unsightly cracks the appearance.

Other factors which affect its appearance are staining and fungus growth at the surface.

Staining: Running water in contact with the surface carries dirt or impurities. This results in series of streaks due to water drying out before it reaches its discharge end. Staining is very often a result of poor drainage and detailing. Projections such as sills and copings will give a certain degree of protection from rain and discolouration/staining beneath the surface.

Surface Growths: These largely consist of algae, lichens and mosses which are usually green or brown in colour. They can be killed by using toxic washes.

Alkali-aggregate Reaction: Some forms of silica found in aggregates can in the presence of water react with alkalis produced from cement hydration and may rusult in expansion and subsequent damage to concrete. The reaction is known as alkali/aggregate reaction(AAR).

Freezing and Thawing: Water freezing within the pores of concrete can cause disintegration due to increase in volume of ice. Susceptibility to such attack is greatest with poor quality concrete used in severely exposed conditions.

Sub-soil Salt Attack: Water soluble sulphates in soil can result in concrete expansion, spalling and disintegration. The extent of damage will depend upon the amount and types of sulphate present, the ground water conditions and the quality of concrete.

Corrosion: Steel reinforcement in concrete is inhibited from rusting by high alkalinity of surrounding concrete. Carbon-dioxide is always present in atmosphere and reduces this alkalinity by carbonating the alkalis, and thus increases the vulnerability of the steel to corrosion. In good quality dense concrete, penetration by carbon dioxide is extremely slow, and rusting is prevented. When concrete is permeable, or cover is inadequate, corrosion of reinforcement can occur. It can also occur if calcium chloride is used as an admixture. Rusting of reinforcement leads to spalling and cracking and sets the process of deterioration.

3.3.4 Paints

Paint is one of the most vulnerable building material and it needs regular maintenance. Defects in paint usually arise from one of following three causes:

(i) Incorrect selection: All the paints should be selected in relation to their exposure condition and backing material.

(ii) Application to damp surfaces: Dampness breaks down the adhesion of the paint with the surface of the component causing flaking and cracking from the surface.

(iii) Poor workmanship: It is one of the main causes of paint deterioration and defects. Poor workmanship can be attributed to incorrect, inadequate or non-existent surface preparation. Over thinning of paint, improper brush selection, poor brushing techniques and failure to apply the specified number of coats may result in deterioration of paint.

The deterioration of paint may be visual or a breakdown of the material itself. A loss of colour or loss of gloss is the first visible sign of weathering. If weathering is allowed to continue, paint film will eventually become brittle, resulting in loss of bond and cracking of the paint film. It is possible to prepare the base surface by simple rubbing before applying new coat of paint. In the absence of surface preparation, deterioration of paint spreads to lower layers and re-painting of whole surface becomes necessary which is costly and time consuming.

If the specified paint fulfils the primary function of protection, the critical areas are the contact faces between the backing material and the paint. If this is good, the paint will deteriorate by surface erosion which is often called chalking and not by loss of adhesion, cracking or flaking. The main source of deterioration of paint is presence of moisture. This can be minimized by selection of good quality paint and surface preparation. It is very necessary to maintain the painted surfaces of various components to increase the overall life of the building as a whole.

3.3.5 Asphalt and Bitumen

Asphalt is a natural or manufactured mixture of bitumen with substantial proportion of inert mineral matter. When heated, asphalt becomes plastic and can be moulded by hand into any shape. Bitumen is a complex mixture of hydrocarbons having both water proofing and adhesive properties. Bituminous based materials have a very long natural life, but this can be affected by sunlight, acids, and by impact.

When asphalt and bitumen are exposed to light and heat, oxidation occurs. This results in slow hardening of material, giving rise to cracking through thermal movements. Bitumen being a thermoplastic material, will soften on heating with tendency to flow. In horizontal situations, an insulating water proof film should be incorporated between asphalt and supporting structure to allow for differential movements. If insulating layer is omitted, the continuous expansion and contraction could result in cracking of asphalt, allowing moisture to penetrate. Vertical application should be adequately keyed to surface at a close interval to prevent flow of asphalt as it soften with rise in temperature. To lessen the effect of sunlight and solar heat, a reflective covering should be specified usually in the form of a suitable light coloured stone aggregate.

When set, mastic asphalt is brittle and can therefore be easily damaged by impact loading. It should be protected by a **screed** or similar coating if there is risk of impact damage. Under permanent loading, suitable protection should be provided to support the load.

Asphalt and bitumen are not affected by biological agencies, or by pollution, but contact with oil can be damaging. Moisture has no direct adverse effect but trapped moisture-vapour pressure can cause blistering in asphalt and bitumen layer.

3.3.6 Mastics and Sealants

Many modern buildings and structures use large prefabricated cladding elements. The normal joint sealers such as cement and mortars are unacceptable, since they can not accommodate the movement associated with larger structural elements. Mastics and sealants can be effectively used for this purpose. Mastics are materials which are applied in plastic stage and form a surface skin (film) over the base element and the film remain pliable for a number of years. Mastics can accept only small degree of movement.

Weathering of mastics and sealants depends upon the type and formulation of materials, but most of them harden/or craze with age, and thus lose some of their movement accommodation properties and allow the moisture to penetrate.

Incorrectly specified, or poorly applied sealants may deteriorate rapidly. Correct specifications and detailing in terms of gap width, minimum depth, adhesive properties, material and surface preparation are very important for their long term durability.

3.3.7 Metals

Steel, aluminum, copper, lead and zinc are the metals most commonly used in buildings. The most injurious agencies affecting performance of metals are those which cause corrosion such as **moisture**, and **gaseous, solid** or **liquid pollutants**. Under completely dry conditions, corrosion does not take place. In most situations in buildings, moisture is present and corrosion is a potential risk. It is wise to assume that moisture will be present at some time or the other. Moisture acts as an electrolyte and a galvanic cell is formed which leads to the loss of metal forming the **anode** of that cell. The most common example of this type of situation arises when two different metals are in contact with one another in the presence of water. However, galvanic cells can be created when some metals are in contact with other building materials in **moist condition,** such as bricks or plaster. When an alloy has elements of different electrolytic characteristics, corrosion action starts e.g. brass has many elements with different electrolytic characteristics. Glavanic action may also occur in single element metals when a difference in **oxygen concentration** occurs at the surface. For example, in pitted steel where the base of the pit has less access to oxygen than the metal surrounding the pit. Galvanic cells may also be created, and subsequent corrosion occurs when particles of a metal or other substance are transported and deposited on other metals. Corrosion is a complex **electrochemical reaction** which can be affected by the presence of dissolved atmospheric **gaseous pollutants, dirt,** manufactured admixtures (in particular, calcium chloride) and by temperature. At certain temperatures, bimetallic corrosion reactions may be reversed. The sensitivities to corrosion of different metals commonly used in building is described in subsequent paragraphs.

(a) Corrosion

    (i) **Aluminium:** Aluminium is used mainly for cladding, flashings and window frames. Aluminium in contact with copper and its alloys get readily attacked, resluting in severe damage caused by the chemical corrosion. This can happen even at a distance. Examples are known of aluminium flashings and window frames suffering severe pitting corrosion through rainwater drainage in copper-covered roofs by depositing small particles of copper on the aluminium. Similar attack can occur from lead also. It is essential to use the correct aluminium alloy for a specific purpose. Many prefabricated aluminium dwellings built in 1950 used a high strength aluminium-copper-magnesium alloy. The presence of copper in aluminium together with damp conditions caused by heavy and repeated condensation, will lead to severe corrosion.

Wood preserved with copper-containing preservatives can also attack aluminium when it comes in contact with it. Unprotected aluminium should not be embedded in cement mortars or in concrete. Direct exposure to sea water will also cause corrosion. Aluminium is not attacked by zinc, galvanized steel or stainless steel. Aluminium can be anodised, by treating it electrochemically to thicken the natural oxide film which may be formed during normal atmospheric exposure. This thickened film confers upon the metal an improved appearance and resistance to corrosion. The anodised layer is readily disfigured by splashes of cement or lime and it needs to be well protected at site, during any construction or repair work.

(ii) **Copper:** The main uses of copper are for roofing, cladding and plumbing. Copper is very resistant to corrosion. Copper when used in mixed metal system in buildings, it forms the non-corroding cathode. It can be attacked by flue gases containing sulphur dioxide if in close proximity to chimneys. This can be readily prevented by good chimney design or by the use of a copper - silicon alloy. Copper is corroded by ammonia and some mineral acids but the likelihood of such a combination or circumstances is not very common in buildings. However, it is good to keep copper away from direct contact with latex cements used for fixing some types of flooring. Waters with high carbon-dioxide content may dissolve copper but the rate of loss of copper is small.

(iii) **Lead:** Lead is used mainly for roofing (sometimes as a covering to steel sheets) for flashings, DPC, and in main water supply. It is highly resistant to corrosion through the formation of a dense protective film of basic lead carbonate or sulphate. It is not, however, wholly immune from attack. Organic acids released from damp oak and western red cedar can cause corrosion. Direct contact with these materials should be avoided. Lead can also be attacked by free alkali present in cement sand mortars. Severe corrosion has occurred within 10 years in case of DPC constructed with such a mortar. It is advisable to use protected lead sheet in such circumstances.

(iv) **Steel:** Steel is used in building for structural purposes, either as reinforcement in concrete, or for window frames, cladding, rainwater fittings, and plumbing systems. Ordinary mild steel, without substantial alloying metals, rusts easily when exposed to the atmosphere. Mild steel is seldom directly exposed but has a protective coating of paint, bitumen or zinc, or is enclosed by other materials, such as concrete. Protective coatings reduce the rate of corrosion. Steel protected by zinc galvanising has good resistance to corrosion. If steel is alloyed with small amounts of copper, about 0.25 per cent, corrosion rate in air is roughly halved. Alloying improves weathered appearance with tenacious coating. A greater resistance to corrosion can be imparted when steel is alloyed with at least 10 per cent of chromium, together with one or more other alloying elements. Steels so alloyed are known as stainless steels.

(v) **Zinc:** Zinc is used in building for roofing, cladding, flashings and sometimes, for rainwater fittings. Additionally, much of the total zinc usage is in the form of protective coatings to steel. These may be applied by several techniques, of which hot-dipped galvanizing is the most common. Both hot-dipping and metal-spraying can achieve thick coatings at reasonable cost. The corrosion rate in unpolluted environments is slow but increases markedly when zinc is exposed to sea environment or to sulphur gases and compounds. Rapid corrosion of galvanised wall ties occurs when these are embedded in black ash mortar, a source of sulphur compounds. Zinc is slightly attacked by wet concrete but progressive attack is inhibited by the formation of calcium zincate which is insoluble in the prevailing alkaline conditions. This resistance to attack has enabled galvanised steel to be used as reinforcement in concrete when enhanced protection is necessary where difficulties are expected in providing the full depth of concrete cover. It should be noted, however, that bond strength can be reduced by hydrogen bubbles liberated in the initial reaction between zinc and alkalis in the concrete. To prevent this reduction, it is desirable to use galvanised reinforcement which has been `passivated' by a chromium based treatment.

Galvanised steel is also used for cold-water cisterns. Soft water, containing dissolved oxygen and carbon dioxide attacks zinc. Such circumstances need protection by bituminous paint coatings. Severe pitting corrosion occurs due to copper and brass debris from plumbing installation activities inside unprotected galvanised cisterns. This should always be prevented. Zinc alloyed with copper can provide a range of brasses and one such, known as alpha-beta brass, is commonly used for fittings in water services. If the water is acidic or alkaline with high chloride content, zinc can be removed from the brass (dezincification), leaving behind a spongy copper residue. This will lead to penetration of body of the fittings causing water seepage, or blockage by zinc corrosion compounds. If such water cannot be treated to improve their temporary hardness, the use of hot-pressed alpha-beta brass fittings should be avoided.

(b) Effects of Fatigue and Creep on Metals

While corrosion remains the principal cause of deterioration of metals, some problems of fatigue and creep are worth describing. Creep is a permanent deformation caused by continuous loading beyond elastic range. Fatigue is a phenomenon caused by large number of cycles of loading and unloading causing reversal of stresses. Creep in metals is caused due to sustained loading and is a slow process and depends on temperature at the time of loading and range of loading. Phenomenon of creep does not pose a very serious problem. Failures of lead flashings have occurred, however, partly as a result of its low resistance to creep, its high thermal movement and low resistance to fatigue. This tendency can be checked by good design, with appropriate limitations on the size of any one piece of lead sheet used or by the use of lead containing a small proportion of copper.

3.3.8 Glass

Glass is a very durable material but brittle under impact and is seldom affected by any of the agencies of deterioration mentioned. Surface etching can occur if sheets are closely stacked under damp conditions. If alkali from paint remover splashes on to glass and is not removed it causes scratches. Thermal stresses in framing materials may cause cracking of glass panes. Proper allowance must be made in design and construction of glass products to accommodate differential movements.

3.3.9 Mortars and Renderings

Most of present-day mortars consist of cement (or lime) and sand—with or without an air-entraining agent. Mix proportions depend upon the types of masonry or block used and their exposure conditions. The effect of shrinkage caused by drying and by carbonation are generally insignificant. Mortars can be affected by frost but their resistance can be increased by the use of an air-entraining agent.

Serious disintegration of mortar can occur when soluble sulphates present in bricks react with tricalcium aluminate present in portland cement mortars. A considerable increase in volume of mortar occurs due to chemical reaction of cement compounds (C3A) with soluble sulphates which results in disintegration. Similar sulphoaluminate reaction can occur when mortar is exposed to condensed water vapour containing sulphates derived from flue gases. This is a particular hazard when slow-combustion fuel appliances are used with unlined chimneys.

Typical external renderings are generally similar in composition to mortars and have similar properties. Sulphate attack can occur through salts derived from wet brickwork. Good quality rendering does not attract frost action. Drying shrinkage can be more of a problem in rich cement/sand mortars which can craze and crack, particularly if applied in warm and dry weather.

3.3.10 Plastics

A wide range of plastics are used in buildings. Poly Vinyl Chloride (PVC) has the widest application, in either a plasticised or unplasticised form. Plasticised PVC is extensively used as floor covering, false ceiling under pitched roofs, membrane covering in flat roofs and plastic membranes for water proofing. Rigid unplasticised PVC is used principally for domestic soil and vent systems, rainwater disposal, drainage, wall cladding, translucent or opaque corrugated roof sheeting, ducting and skirtings. Rigid PVC has a more restricted application for window frames but mostly in combination with metal or timber. Expanded PVC is also available for thermal insulation. Glass Reinforced Plastics (GRP) are also used for structural purposes.

Polyester resins reinforced with glass fiber are used principally as cladding sheets. These are also used for cold-water cisterns and floats, cold-water tubs, pipes, bath tubs, basin and sink. In sheet form, it is used as damp-proof membrane and for covering concrete and hardcore surfaces. Acrylic resins are used for sinks, drains and bath, tubs, corrugated sheeting and for roof light coverings. Special plastics (acrylonitrile butadiene syrene copolymers) are used for large drainage chambers, plumbing, drainage fittings and wall ties. Polycarbonates are used for glazing. Phenol formaldehyde resins are used to impregnate paper and fabric to provide wall and roof sheets. Foamed plastic provide a cellular material used for thermal insulation. Epoxy resins are used for in situ flooring and for concrete repair. Polystyrene, polyurethane and urea-formaldehyde in expanded form are used for thermal insulation. This is not a complete list, as many other plastics have found some use in building construction, for example, nylon, cellulose acetate and polyacetate in taps and miscellaneous fittings.

Short-wave solar radiation degrades plastics by causing embrittlement and change in surface appearance. The risk is greatest in coastal and rural areas. The effects can be reduced or increased by additives incorporated into the plastic. The addition of fire retardants improves resistance to degradation. On the evidence so far available, the durability of plastic seems to be good. Moisture in general has little effect but can reduce bond strength between glass fibre and polyester resin. The extent of any weakening depends greatly upon the control exercised during manufacture. Plastics are not harmed, in general, by contact with other building materials, though cracking of polyethylene in cold-water cisterns is known to have been caused by the use of oil-based jointing compounds.

PVC and polycarbonates have high thermal expansion. Unless properly allowed for, the movement of PVC gutters and down pipes can cause joint failure and leakage. Varieties of polyethylene have greater thermal movement. Plastics creep under continued loads and special precautions are needed when stresses are high.

The long-term behaviour of many of the plastics is uncertain. More work is needed to enable better prediction of fatigue and creep under long-term loading. However, external performance has been generally good by nullifying short wave solar radiation by proper shielding from direct sunlight.

3.3.11 Natural Stones

Natural stones are classified as belonging to one of the three main groups - igneous, sedimentary or metamorphic. Igneous rocks are formed by solidification and modification of a molten magma. Common examples are granite, dolerite, bassalt and pumice. Granite is highly resistant to all the agencies of deterioration.

Sedimentary rocks may be formed from particles produced from older rocks by the normal processes of weathering, accumulation of skeletons (usually marine organisms) and by chemical deposition. Limestones and sandstones are the principal examples of sedimentary rocks and provide most of the building stones. The atmospheric pollution causes deterioration of limestones and sandstones. Sulphuric gases dissolved in rainwater react with calcium carbonate to form calcium sulphate. When this crystalises under dry conditions, it generates a stress which causes disintegration. If wetting and drying are frequent due to rainfall, the surface of the stone gets slowly eroded as the calcium sulphate gets continually removed. Under sheltered conditions, the calcium sulphate, which is only sparingly soluble, may build up to form a hard skin which causes more unsightly damage. When the skin eventually blisters and breaks off, it pulls away limestones with it. Dissolved sulphuric gases in rain water can also weaken the bonding of calcium carbonate in calcareous sandstones and severely weaken the stone. Magnesium carbonate is also attacked by sulphuric gases leading to similar deterioration of magnesium limestones and sandstones. Ground salts, and those in sea environment, can also cause damage. The former can cause expansive damage when penetrating into limestones and calcareous sandstones. Attack from atmospheric gases is usually manifested by a general powdering pattern of the external surfaces. Frost may also attack some limestones.

A phenomenon known as contour scaling can cause damage to sandstones. Calcium sulphate can be deposited within the pores of sandstone, even though the latter may not contain calcium carbonate. The calcium sulphate is derived from limestones attacked by sulphuric gases which when washed penetrate into sandstone surface causing internal stresses in the surface crust. These internal stresses arise from difference in thermal expansion and/or moisture movement between the sandstone and the crust block with **calcium sulphate**. The crust breaks away, usually at a depth of between 5 mm and 20 mm, and follows the contours of the surface.

Deterioration through pollution and frost becomes more sever if sedimentary stones are laid with the natural bed parallel to the vertical face of the wall. The greater durability can be achieved when the natural bedding plane lies parallel to the horizontal courses in a wall.

Metamorphic rocks are derived from sedimentary rocks due to the action of great heat and pressure arising from movements of the earth's crust. The structure of the rock then changes radically. The only metamorphic rocks of significance for building purposes are **marble** and **slate**. Marble is used for cladding and is very durable. It is not immune from attack by sulphuric gases. Slates are used for roofing, cladding and DPC. Roofing slates are exposed to the most severe conditions and can be affected by sulphuric gases.

One major cause of damage in all stones can arise from corrosion of embedded fixtures. In the past, the rusting of iron and steel cramps and dowels has caused extensive damage, particularly to limestones and sandstones.

3.4 SUMMARY

All materials deteriorate with age due to action of various forces. Deterioration is the gradual disintegration on account of any destructive action from aggressive waters and soils, exposure to weathering agents and relative movements. The cause of deterioration can be external or internal. The process of deterioration can be classified as mechanical, physical, chemical or biological and the actual process may set in due to one or combination of these mechanisms.

Various factors responsible for initiation of the process of deterioration are **human**, **chemical**, **environmental**, alongwith causes like **faulty design**, **faulty construction**, **faulty materials**, **faulty drainage** and **misuse of buildings**. Preventive measures need to be taken during design and construction stages to prevent rapid deterioration of buildings. Regular maintenance reduces deterioration of buildings. Solar radiation, temperature variations, moisture movement, biological agencies, gaseous and air pollutants, solid contaminants, ground salts and water are some of the environmental factors which contribute towards deterioration of buildings.

To prevent deterioration, it is necessary to understand the effects of various agencies on building materials. Bricks and clay products are generally very durable. Common effect of weathering on these products make changes in appearance. **Efflorescence**, which is white surface deposit of salts in the form of loose powder, occurs when dry weather follows wet weather.

Timber deteriorates under the attack of fungi, insects or fire. Proper moisture content must be maintained in timber to prevent spread of dry or wet rot. The resistance of concrete to disintegration depends on quality of **ingredients** used, **mix design** and concreting **operations**. Change in temperature and seepage of water are the major causes of deterioration in concrete. Factors like **staining, surface growth, alkali-aggregate reaction, freezing and thawing**, subsoil attack can affect concrete **expansion,** spalling and **disintegration**. Permeable concrete leads to **corrosion of reinforcement**. Bituminous based materials are affected by sunlight, acids and impact.

The performance of common metals is affected by corrosion which may be caused by **moisture**, and **gaseous, solid** or **liquid pollutants**. Durability of plastics is generally good. Radiation, thermal changes and high stresses may lead to deterioration of plastics.

The main cause of deterioration in materials is due to atmospheric pollution. The problem of deterioration requires timely action in reducing its rate and effect on buildings and structures.

QUESTIONS

3.1 Describe the term "deterioration" and the process of deterioration in buildings.

3.2 Describe the classification of the process of deterioration in buildings.

3.3 List down various factors responsible for initiation of the process of deterioration.

3.4 Explain briefly the effect of lack of maintenance culture in buildings.

3.5 Explain the following factors on deterioration of building elements:

(i) Chemical factors
(ii) Furring

3.6 Explain the process of deterioration in building elements caused by **environmental factors**.

3.7 Explain the importance of adequate attention to design and construction for slowing the deterioration process.

3.8 Explain how does the following contribute to the process of deterioration?

(i) **Faulty drainage**
(ii) Inappropriate **cleaning**
(iii) Misuse of building

3.9 Explain the effect of various deterioration agencies on the following:

(i) Bricks and clay products
(ii) Timber and timber products
(iii) Concrete
(iv) Paints
(v) Natural stones

3.10 Explain how does the following agencies affect Asphalt and Bitumen used in buildings:

(i) Sunlight and Solar Heat
(ii) Impact loading
(iii) Moisture and biological agencies

3.11 Describe the process of deterioration and its effect on building in following building materials:

(i) Mastics and Sealants
(ii) Metals
(iii) Mortar and Renderings
(iv) Plastics

Chapter 4

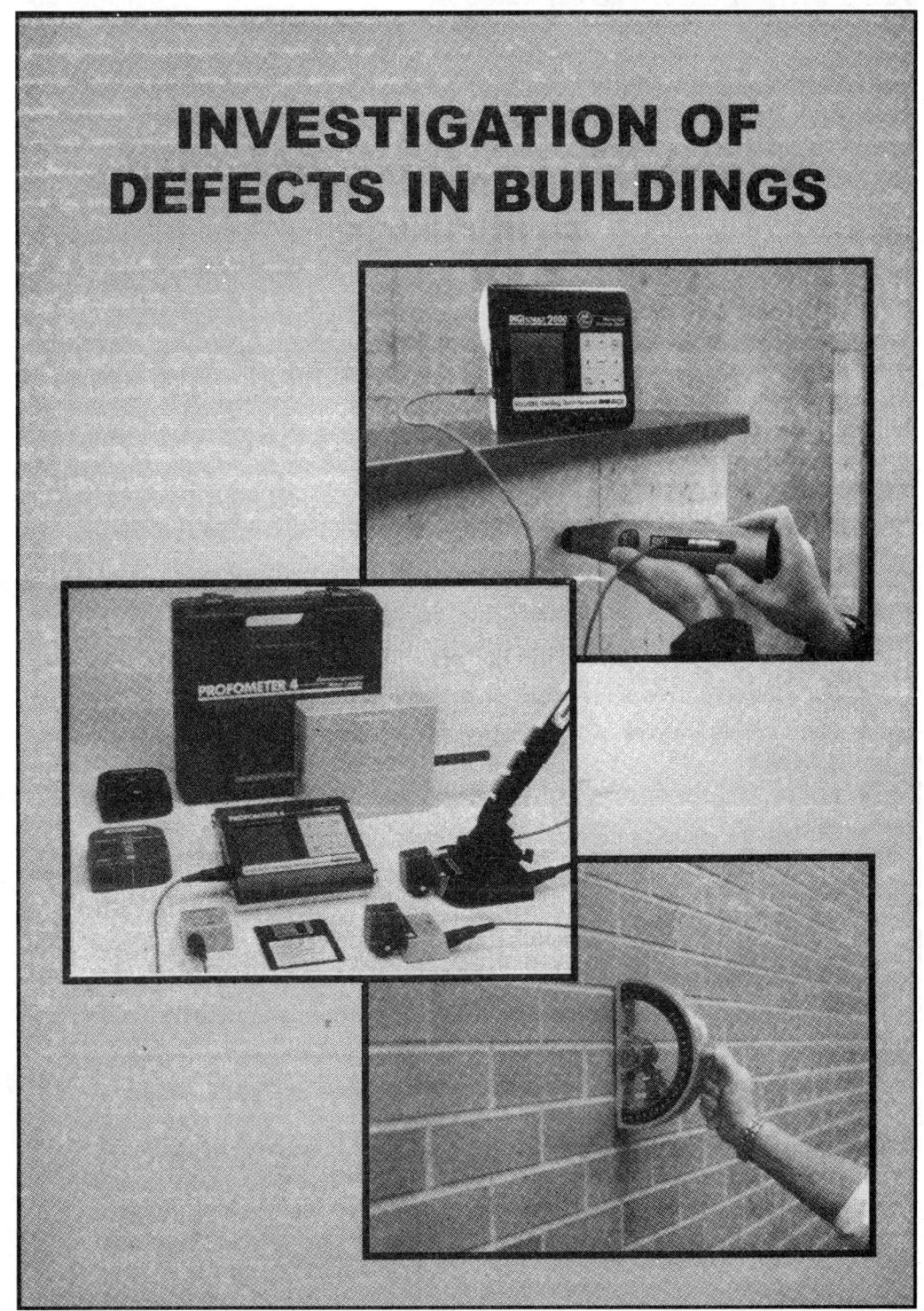

INVESTIGATION OF DEFECTS IN BUILDINGS

INVESTIGATION OF DEFECTS IN BUILDINGS

LEARNING OBJECTIVES

The learner understands investigations of defects in buildings and will be able to:

- **Describe** systematic procedure of investigation of building defects;
- **Prepare** a checklist for identification of building defects;
- **State** the objectives of investigation of building defects;
- **List** various sources of information for building defects investigation and Analysis;
- **List** sequence of detailed steps for diagnosing building defects/problems;
- **Explain** the visual symptoms to be observed during inspections;
- **Specify** the relevant tests to be carried out for correct diagnosis of building defects.
- **Describe** various tests on materials to identify defects;
- **Explain** four most commonly used **NDT** for building defect investigation;
- **Explain** the procedure to identify defects from information and data collected.

4.1 INTRODUCTION

Building structures are created at a considerable expense of time and money. Concrete and masonry structures are susceptible to cracking, while steel gets corroded. Steel structures also manifest loss of serviceability by excessive deformation and buckling. Other than these, many times deterioration occurs due to lack of adequate care and coordination in enforcing the primary objectives of, at the time of construction:

 (i) Safety;
 (ii) Durability of structure, and
 (ii) Serviceability of structure.

Maintenance of building structure is required to ensure proper functioning of its elements as required by the users and these elements should not fail at any inconvenient moment. The maintenance department in the organisation must keep or restore the building to an acceptable standard as per functional needs, rules, regulations and acts. As discussed earlier, all materials deteriorate either by ageing, or by the action of other injurious forces of destruction, such as, poor selection of materials, the occupational, and human factors. After understanding the ways in which deterioration occurs, it is necessary to investigate the exact cause of deterioration, so as to take appropriate remedial measures. In this chapter, we will discuss the **essential requirements** and **scope of investigations** that need to be carried out in assessing the deterioration in buildings, and plan repairs, systematically. Generally, lack of awareness leads to neglect of timely remedial measures and consequently the same problem manifests repeatedly. Hence, timely investigation will facilitate effective remedial measures to prevent further deterioration.

4.2 SYSTEMATIC APPROACH OF INVESTIGATION

The damaged structure develops visual signs of cracks of different pattern and sizes resulting in peeling of plaster, spalling of concrete and other surface finishes.

Primary task in investigation is to determine whether the damage is structural or non-structural. Structural repairs are undertaken to restore the structural stability to sustain the actual stresses under the service conditions. Non-structural repairs are undertaken for the long term durability and aesthetic appearance, but these do not increase the load bearing capacity of the structure in question. A non-structural repair, if not carried out timely, can lead to structural distress in times to come. However, before taking up repairs it is necessary to systematically carry out investigations to identify and assess the exact cause of deterioration.

The various steps involved in the investigation of causes of deterioration, before undertaking repair work, are:

 (i) Preliminary investigations
 (ii) Physical inspections
 (iii) Material tests
 (iv) Non-destructive tests
 (v) Detailed diagnosis of defects

   (vi) Study of available documents

  (vii) Estimation of actual loads and environmental effects

 (viii) Checking errors in design

  (ix) Retrospective Analysis

   (x) Strengthening requirements

  (xi) Relevant approach to repair.

4.3 SCOPE AND OBJECTIVES OF INVESTIGATIONS

The primary aim of investigation is to determine the extent of damage or distress, classify the damage as structural or non-structural and assess its root causes. Unless root causes of distress are established, remedial measure may not be meaningful. The factors to be ascertained are:

    (i) Carbonation of affected concrete

   (ii) Depth of carbonation and cover

  (iii) Chloride levels of concrete

  (iv) Permeability of affected concrete

   (v) Degree of corrosion

  (vi) Load carrying capacity of affected structure

 (vii) Spread of defects (localized or spread over the entire area)

(viii) Appearance and type of cracks

  (ix) Difference in design loads and actual service loads

   (x) Type of defects in timber work

  (xi) Extent and source of dampness in building

 (xii) Causes of leakage/overflow in water supply and drainage pipe work

(xiii) Causes of moisture and leakage in air conditioning service.

Before starting detailed investigation, it is necessary to decide the scope and objectives of investigation. The visual examination and diagnostic tests are to be carried out, depending on the scope of investigation. Diagnostic tests for buildings are similar to 'clinical tests' to diagnose the disease in human body, to plan and administer the appropriate medication to cure the patient, as best as possible, within the prevailing constraints.

4.3.1 Scope

The scope of investigation of a defect is dependent primarily on the amount of money and effort that can be spent on it. The scope is further related to:

- Nature of the defect
- The accuracy with which causes of defect needs to be identified
- The main reason for the defect investigation (such as remedial work required).

The scope of investigation becomes wider with uncommon defects and requirements of greater accuracy.

4.3.2 Purposes of Investigations

Before taking up investigations for assessing the causes of deterioration of building structures/services, it is necessary to understand reasons for carrying out such investigations. The purposes for investigations are:

(a) **Legal:** Commission of inquiries are normally appointed to go into the reason of such deterioration and failures. The emphasis is on 'who' **rather than on "what"** went wrong. But more often the aim of these investigations is to find out the "Culprit". If deaths have occurred due to deterioration/failure, these investigations become part of **criminal procedures** also.

(b) **Insurance Surveys:** Many times important works under construction are insured against **"Contractor's All-Risk (CAR) Policies"**, Centering, form-work, and works under water are insured by the contractors against failure risks. Even Architects and Consultants take **"Professional Indemnity Insurance"** to protect themselves against claims from the clients by insuring their designs. Here again the investigations concentrate more on Assessment of damages including damages in terms of money.

(c) **Structural Failures:** These investigations are primarily done out of scientific curiosity to find out **what went wrong with regards to design and construction** and to pin point the cause of deterioration.

Investigations to find causes of deterioration or failure of building structure/services are very important for planning maintenance and repair of buildings and other service elements.

4.4 PRELIMINARY CONSIDERATIONS

Building deterioration means certain building elements have become incapable of serving the expected or intended use due to some damaging forces. The main consideration of investigation is to find the source and the root cause of the damaging force. For the purpose of investigation, it is important to keep the **collapsed/deteriorated building site undisturbed**. The first visit to site is of great importance for collecting the photographic data. Nothing should be taken for granted and one should not go by external appearance of concrete surfaces. The state of mind at this juncture should be purely for recording things as they are, and **not to jump to conclusions.**

4.4.1 Building up Data

Having understood preliminary considerations and purposes of investigation, now go to the most important phase of data collection. There are two types of data. First, the factual data collected from site and records, and the second, generated data based either on the factual data or from other situational considerations.

All factual data at site must be collected meticulously and, as far as possible, under investigator's own direction. When one realises the inadequacy of such data, it is too late because by the time one realises the inadequacy of such data, it may have become too late as everything might have been removed and disposed off by then.

Even such a simple looking thing as photographs needs attention. A photographer may be interested in the scenic beauty of a bridge, a press-reporter will emphasize the utter shambles in which a structure has gone down to emphasize tragedy, but a maintenance engineer will look out for the 'Cone of Failure', he may like to note down certain dimensions (e.g. spacing of cracks) which will be first painted on concrete element before photographing. In fact all pieces and photographs are properly numbered and a list made.

Next in order, comes the interviews of concerned people instead of highly placed executives alone. Aim should be to get the maximum factual information and not putting any one in the witness box. People have their own perception of the situation and it is surprising to note that there could be many aspects/versions of the same happening. The best policy is to listen more and speak less. All these interviews must be recorded on the spot as one often forgets details much more than what one will like to concede.

Correspondence, Design Calculations and Drawings, Material Testing Registers can be brought home and gone through quietly. A last word of precaution is that all data collected at site must be appraised and authenticated carefully at home.

4.4.2 Sources of Information

The information required in diagnostic work should aim at providing the investigator with data concerning the actual materials used and the conditions to which the materials and the building elements have been exposed or subjected before, during and after construction.

Various sources of information are:

(i) **Drawings and specifications** provide the details of materials used and methods of construction. In practice, they cannot be completely relied upon to give all the revisions during construction and thus provide only basic clues. Actual details of construction done/undertaken will have to be determined by observation, inspection, other site notes and study of "as constructed" drawings.

(ii) **Consultant/Architect's instruction** provides clue to **variation** in materials and **specifications** changed during the course of construction and contract execution.

(iii) **Site notes, Minutes and Reports** provide the sort of difficulties encountered, the quality of workmanship achieved, the extent to which precautions were taken to protect materials stored, and weather conditions experienced during construction.

(iv) **Maintenance Manual and other Records** provide accurate details of alterations and additions carried out since the completion of building, and history of any defect.

(v) **Interviews** with the users and those connected with the design, construction and maintenance can provide number of aspects associated with building defects such as its history and indications of conditions of exposure

4.4.3 Aspects to be Investigated

From the varied sources of information, data must be collected as correctly as possible. The important aspects which need attention are:

- Correct/incorrect materials specified;
- Correct materials specified, but inappropriately detailed;
- Correct materials specified and appropriately detailed, but incorrectly used and handled on site;
- Inappropriate design loading; and
- Exposure conditions.

4.5 DETAILED STEPS FOR DIAGNOSIS OF DEFECTS

Diagnosis plays key role in assessment of distress or damage. Assessment of distress or damage is necessarily an inductive and interactive procedure. When not applied properly, it is liable to serious error. To minimize the likelihood of error, distress/damage assessment must be carried out carefully in a scientific manner. Systematic steps of diagnosis of defects are given in Fig. 4.1.

Diagnosis of the structure can be done on the basis of available information, physical inspection, study of documents, and interviews with the knowledgeable people. To confirm this diagnosis, a careful and detailed diagnosis ought to be made including retrospective analysis, so as to check whether the observed symptoms indeed follow from the postulated causes. It may be mentioned that though the symptoms of distress in buildings are common, the causes responsible for such distress can be varied. The symptoms of distress in buildings may be common, but diagnosing the exact causes of distress, out of a large number of possibilities, is truly a demanding task.

4.6 PHYSICAL MEASUREMENTS

This is carried out in order to discover all possible evidence of defects and damages. The inspection also provides information on ageing, deterioration, and the standard and quality of original construction work. Deflections, deformations and relative displacements are observed and measured during inspection. Details of crack patterns (width, locations, orientation and progressive propagation) are recorded. A list of all symptoms of defects and damages and their probable causes are carefully noted and analyzed. Signs of crushing, spalling, splitting and other modes of deterioration of the structure are to be noted together with the evidence of construction deficiency in a systematic and scientific manner. The possible causes and symptoms of deterioration in concrete structures are given in Table 4.1.

It is further emphasized that to determine/investigate the causes of distress in a structure the engineer should have comprehensive knowledge of the various causes that can lead to distress. These can be broadly categorised, as under:

(a) Damage/distress arising due to inadequacy of the foundation design,

(b) Damage/distress arising due to construction faults,

(c) Damage/distress arising due to inadequacy of design,

(d) Damage/Distress arising due to natural calamity and vagaries of nature, and

(e) Damage/distress arising after occupancy of building.

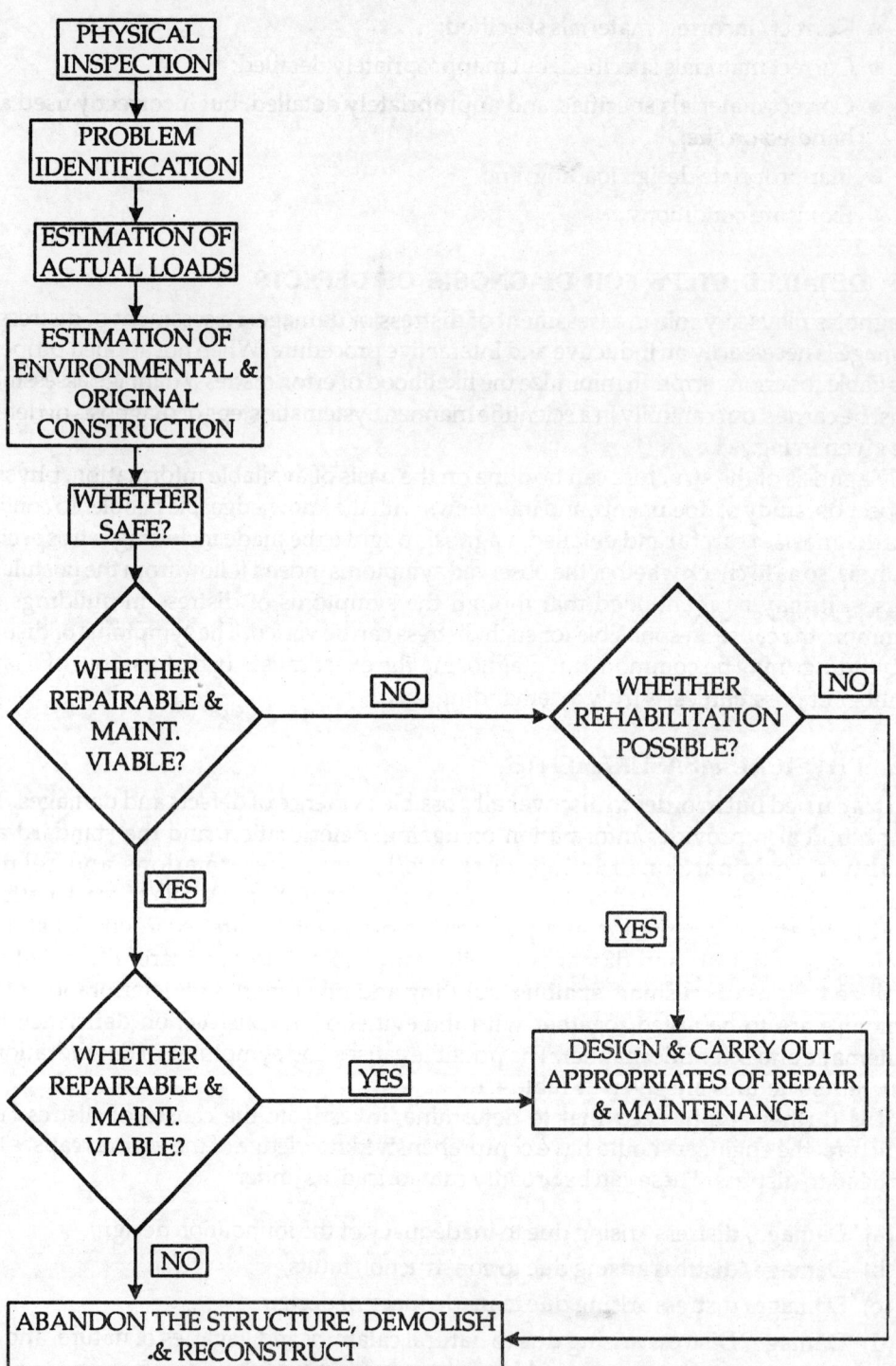

Fig. 4.1: Systematic Approach to Defect Diagnosis

TABLE 4.1: POSSIBLE CAUSES AND SYMPTOMS OF DETERIORATION IN CONCRETE STRUCTURES

Cause of Deterioration		Signs/Symptoms
Design errors	Incorrect joint, spacing of bars, restraints, in-correct load calculations, excessively slender designs	Cracking and buckling
Poor Concrete Mix Design	Poor aggregate grading, too high or low cement content, In-correct water/cement ratio.	Cracking, increase in permeability, surface shrinkage, increased carbonation, porous concrete, spalling, corrosion.
Poor workmanship	Excessive addition of water (incorrect W/C ratio), Poor compaction, inadequate cover to steel, Poor shuttering, Omitted or insufficient curing.	Pores, voids, cracks blemishes, exposed Reinforcement, low strength, Surface shrinkage, premature carbonation and corrosion of reinforcement.
Excessive mechanical stresses	Static or dynamic overloading that is, collision, explosion and abrasion.	Disintegration, cracking, shrinking, worn out surface.
Thermal stresses	Temperature changes, freeze/thaw cycles, fire damage	Cracking, reinforcement corrosion, surface shrinkage.
Chemical effects	Aggressive gases (CO_2, SO_2), corrosive soils or water, acids, salts	Disintegration, spalling, surface sanding, carbonation, reinforcement corrosion.
Biological effects	Plants, Micro-organisms	Cracking, expansion effects, flaking.

4.6.1 Visual Examination for Problem Identification

It is usually the visual appearance of a defect in the building that first alerts the owner to a problem within the building structure. There are a number of visible factors which can provide a useful guide as to whether corrosion is occurring, or there is probability of corrosion. Unfortunately, on bridge decks which are normally covered with asphalt wearing and running courses these phenomena cannot be observed without removing the surfacing. There is also evidence that the presence of surfacing impedes the development of spalling. The main phenomena resulting from corrosion of reinforcement can be observed at the concrete surface are: rust **staining, scaling, cracking, spalling, delamination and leaching** (water leakage).

Rust staining is obviously a direct indication of corrosion but unless present on a large scale, it can be misleading. For example, localized rust staining of high colour intensity can be caused by foreign bodies (e.g. nails, nuts, bolts) which are often found near the surface of concrete. Iron pyrites present in the aggregate, or resulting from low levels of reinforcement corrosion can also indicate rust stains. The intensity of rust staining cannot be used as a measure of the intensity of rebar corrosion.

Scaling is breakdown of the surface mortar accompanied with the loosening of surface aggregate. Scaling is usually the result of frost damage which increases the rate of transport of oxygen, water and salt through the concrete to the reinforcement thus enhancing the possibility of corrosion. Scaling is not caused by corrosion, but will hasten the phenomena of corrosion.

Spalling is breaking off of concrete pieces from the reinforcement to the surface and is caused by the forces produced in the concrete, as a result of the bar expansion due to corrosion. Obviously once spalling has occurred, the rate of further corrosion can proceed more rapidly due to increased formation of differential corrosion cells.

The influence of crack width is uncertain, but it is reasonable to assume that cracks perpendicular to the reinforcement will speed the corrosion of intercepted bars by facilitating the ingress of **air, water** and **salt** through the concrete. **Cracks which run parallel with the reinforcement are caused by corrosion** and the crack length is a good guide to the extent of corrosion along the bar. This type of cracking is important since it reduces the concrete cover by spalling, and is often found in conjunction with the reduction in resistance to delamination of the concrete. Construction joints, unless satisfactorily sealed, can act like cracks in facilitating the transfer of corrosion activators to the reinforcement. Consequently the concrete adjacent to joints should be examined carefully for any signs of corrosion. Leaching of surface water through cracks and joints in concrete is quit common. There is increased probability of corrosion near the crack or joint due to increasing of moisture and salt levels and decreasing of alkalinity of concrete.

Careful examination of the concrete surface can indicate of rebar corrosion (spalling, rust staining and longitudinal cracking) and factors likely to increase the probability of corrosion (scaling, leaching and cracking).

A particular damage/distress may fall under more than one category mentioned above. Each category contains a number of causes for damage/distress. A thorough understanding of various **root causes** of damage/distress is a **pre-requisite** for a meaningful investigation into the cause of damage in any particular building structure.

The investigation work should be planned ahead and the observations of the investigation must be recorded in a systematic manner. The visual observations should be recorded on the following aspects:

   (i) **Position, direction** and **extent** of cracks in R.C.C. slabs, floors, beams, columns and walls;
  (ii) Position of honey-combing in R.C.C. Work;
 (iii) **Excessive** and also **relative deflections** in R.C.C. components and **tilts** in walls or columns;
 (iv) **Cracks** and **sinking** in floors;
  (v) **Position of leakage** and **extent** of damage;
 (vi) Abnormal **loads, pattern** of loading and **deviations** from design loads;
(vii) Type and quality of masonry,
(viii) R.C.C. work having excessive voids and hollows;
 (ix) **Inspection** of other **adjoining existing buildings** in order to ascertain whether they have also undergone similar damage which might be caused by natural forces.

4.6.2 Investigation Kits For Diagnosis

A kit is required in all investigations which would include, equipment for recording (mainly of dimensions, levels, temperature and relative humidity),making observations (generally of hidden parts), and limited opening up of a construction element. A specialised kit is required where more accurate measurements and extensive observations and laboratory testing are required. Investigation kit includes :

- Basic kit; and
- Specialised kit.

Basic kit

Basic kit includes following:

   (i) Sketch pad, notebook, clipboard, scale, pencil and eraser for recording and sketching
  (ii) Rulers and tapes for measuring overall dimensions and profile of defects should be available. A calliper gauge is most useful for measuring widths of joints and cracks, outer dimensions of frame pipes, and depths of hidden parts. Micrometer is used to find thickness of the thin material elements.
 (iii) Adhesive material for temporary fixing of rulers and labels.
  (iv) Cameras can also be used for photographs to reveal important clues missed during visual inspection. A video camera can cover much of the affected portion.
   (v) Torch or other light sources can be used to survey dark areas.
  (vi) Mirrors can be very useful for visual inspection of otherwise inaccessible exposed parts such as underside of parapet coping projections or voids in floor and roof spaces. Wide angled mirror or mirror with appropriate source of light help greatly in investigations.
 (vii) A magnifying glass should be used for identification of surface finishes, cracks and nature of holes.
(viii) Binoculars are required for studying surface defects and details that are inaccessible without the use of ladders.
  (ix) Screwdriver, small hammer, plier and other tools are useful for scraping surface finishes, removal of debris for exposing hidden or covered surfaces for investigating a limited area.
   (x) Small spirit level and angle indicator are also useful for assessing surface levels.
  (xi) **Moisture Meter** indicates the dampness in the structure. It also measures the average depth of dampness. This meter provides usually accurate results.
 (xii) Thermometer is used for measuring surface temperature.

Specialised kit

Specialised kit consists of **electronic** and **non-destructive test equipment** and is required for more accurate analysis. The specialised equipments are discussed under non-destructive tests.

4.7 MATERIAL TESTS

Following are the mandatory material tests for an analytical study of damaged or distressed building portion:

(a) Testing of materials is necessary for correct **strength** evaluation of samples of concrete and steel used in the building.

(b) **Bearing capacity** of the soil at the depth to which the foundations have been taken must be tested.

(c) The bricks used in the masonry must be **tested** in the laboratory for **strength** and **efflorescence**.

(d) The structural design of the building components and R.C.C. components may then be **re-evaluated** for the **present load** carrying capacity.

(e) Type of pipe work, its capacity and **water tightness** must be checked throughout.

The results of the material tests and analytical study are put in the form of tabulations for effective analysis system. Tabulations make it convenient to compare the actual results with the desired results.

4.7.1 Load Testing of Structure

Many times it becomes necessary to examine the safety of the structure by subjecting it to full load tests. In case of cracked structure, such tests are carried on a cracked structure and also on uncracked member, so that the effect of crack can be established without reference to effect of design calculation.

The test should be done as per IS 456-1978 which states that "the structure should be subjected to load equal to full dead load of the structure plus 1.25 times the imposed load". The test load can be applied by using materials such as bricks, stones, concrete blocks, sand bags, water and steel plates. The test loads on the structures/buildings shall be arranged in one or more patterns so that the maximum bending moments, shear forces and thrusts due to the imposed loads are produced at the critical sections. During testing, supports strong enough to take the whole load should be placed in position leaving a gap under loaded members so that in the event of a failure, **damage** to other members of the structure is minimized.

Simple dial gauges, deflectometers, and clinometers can be used. The results of tests are checked with the actual assumed data to plan proper repairs and rehabilitation.

4.8 NON DESTRUCTIVE TESTS

Non-destructive techniques, which are less time consuming and relatively inexpensive can be used to assess the *in situ* strength, quality, location and the extent of cracks, voids and honeycombs. NDT facilitates confirmation of location of suspected distress or deterioration, assessment of partial strength and durability, and determination of extent of corrosion in the structure. Generally the results of non-destructive tests are most useful when supplemented by a limited number of destructive tests. These tests are more useful in cases where data of structure (like drawings, concrete grades) is not available. These techniques are generally used in investigation for repair works in structural elements for the following purposes:

(i) Determination of **extent of corrosion** of reinforcement,

(ii) Determination of **chloride ions,**

(iii) Determination of **permeability,**

(iv) Determination of depth of **concrete cover, bar diameter** and **spacing,**

(v) Determination of **concrete strength, extent of cracks, voids** and **honeycombing,**

(vi) Determination of **defects in piles,**

(vii) Determination of *defects* in metals and **welded joints.**

4.8.1 Determination of Extent of Corrosion of Reinforcement

Corrosion is the main cause of deterioration in reinforced concrete structures. It mostly occurs in marine environments, building facades, car parks, bridge decks and tunnels. Deterioration of structure starts with the setting in of corrosion. It is very important to diagnose the cause and extent of corrosion occurred in structural elements, for its safe maintenance.

The **corrosion of reinforcement** in concrete is affected by the following:

1. High pH encourages a **protective oxide film** on the steel surface.

2. Concrete provides a **physical barrier** to the penetration of moisture containing dissolved salts, carbon dioxide and oxygen.

3. Concrete provides a **chemical barrier** to the ingress of moisture containing acidic substances such as carbon dioxide and sulphur dioxide.

4. The passage of **electrolytic corrosion** currents is limited by the low electrical conductivity of concrete.

A good quality concrete will have **low permeability, high pH** and a reserve of alkalinity. It follows that low cement concrete has high permeability leading to faster corrosion. The corrosion of reinforcement leads to cracking, scaling and spalling. This type of damage can only be observed following a significant level of reinforcement corrosion. Corrosion of reinforcement is undesirable as it results in cracking and spalling of concrete leading to severe structural failures. Corrosion of reinforcement may not always result in scaling, cracking and spalling, particularly where the corrosion is localized and has just started.

The repairs of structures where spalling, cracking and scaling have occurred are more difficult and expensive as compared to repairs when performed at the onset of rebar corrosion. For these reasons, proper diagnosis must be done to predict and detect the extent of corrosion of reinforcement on site. These techniques are predominantly non-destructive (**ND**) and must be used in combination with other data obtained by appropriate observations and interpretations of the root causes of the problem.

There are many techniques for examining reinforcement for corrosion. These can be classified into **direct** and **indirect, non-destructive, semi-destructive** and **destructive.** These can measure the rate and extent of corrosion.

The important techniques of **electric potential** measurement is explained for determination of corrosion.

Electric Potential Measurements for Corrosion

The purpose of potential measurements is to find the areas with risk of corrosion. Pathfinder equipment or ordinary multimeter and a reference electrode measures the potentials at various points.

Half cell potential method measures the **electrode potential** of steel reinforcing bars in concrete environment by comparison with the known **reference electrode** in open. The electrode potential value of steel in concrete is an indicator of corrosion activity. The electrode potential value indicates corrosion occurrence and extent, however, it gives no indication of the rate at which corrosion is occurring. The reference electrode generally used in measurement on reinforced concrete is the saturated copper/copper sulphate electrode. However, the silver/silver chloride can also be used at times. The values of electrode potential of reinforcement (E), as representing corroding and non-corroding conditions, are as shown in table 4.2:

TABLE 4.2: DETERMINATION OF REINFORCEMENT CORROSION BY HALF-CELL POTENTIAL MEASUREMENT

E (millivolts) as Measured by		Probability of Corrosion
Silver/Silver Chloride	*Cu/Cu SO$_4$*	
Zero	Zero	No corrosion
Between 0 and 500 mV	Between 0 and 200 mV	Corrosion is in very early stage. No corrosion on bars
Between 500 and 700 mV	Between 200 and 300 mV	First signs of corrosion and will be visible on bars.
More than 700 mV	More than 300 mV	Corrosion is in a notorious stage, with evident corrosion signs on the bars and in some cases on the concrete too.

An electrical connection is made to the reinforcement at a convenient position enabling electrode potentials to be measured at any desired location. The surface of the concrete is normally divided into a grid system of suitable dimensions. In case of reinforcing bars are embedded in the concrete, the current discharging and current receiving areas are set up at the point. Current receiving areas indicate the rust. The voltage difference between two areas can be measured by sensitive voltmeter. Generally, for making electric connections, select strips. Most convenient area is where the cover is minimum. For measurements on large areas, a grid size of **500 mm × 500 mm,** may be chosen.

The potential difference between the reinforcement and the half-cell is measured using a high impedance volt-meter.

This is the only non-destructive test which examines a specified portion of affected reinforcement. The results can be plotted to form potential contour diagram to indicate the corroted sections of the structure. This method detects the likelihood of present state of corrosion of steel but can not indicate the rate of corrosion. Now-a-days sophisticated commercial equipment such as potential wheel, path finder, etc. are available with higher speed and accuracy.

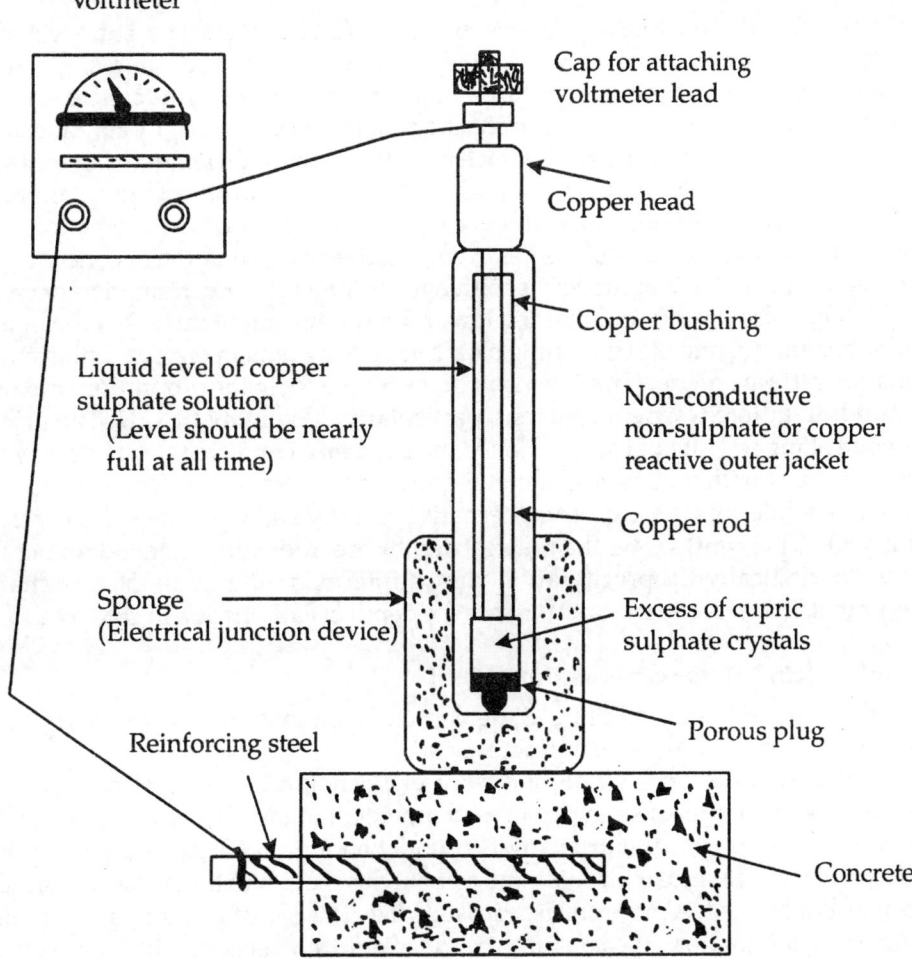

Voltmeter

Cap for attaching
voltmeter lead

Copper head

Copper bushing

Liquid level of copper
sulphate solution
(Level should be nearly
full at all time)

Non-conductive
non-sulphate or copper
reactive outer jacket

Copper rod

Sponge
(Electrical junction device)

Excess of cupric
sulphate crystals

Porous plug

Reinforcing steel

Concrete

Fig. 4.2: Half Cell Potential Meter

Sometimes, it may not be feasible to give connection to reinforcement due to concrete cover. In such cases, surface potential measuring technique may be suitable. The method uses two reference electrodes and no electrical connection is required for reinforcement. A potential difference of more than **30 mV indicates** that steel remains in passive condition of corrosion and if it is more than 100 mV it indicates active corrosion conditions. Commercially available corrosion analysis meters and resistivity meters can be used for this purpose. Resistivity measurements predict whether significant corrosion is probable. Following values give an indication regarding probability of corrosion with resistivity:

Resistivity > **12,000 ohm-cm**, corrosion is unlikely

Resistivity > **5,000 to 12000 ohm-cm**, corrosion is possible

Resistivity < **5,000 ohm-cm**, corrosion is almost certain

4.8.2 Carbonation

When exposed to air, concrete will release a portion of its free water content. Most of this free water is lost from the surface layers of the concrete. The progress and extent of this evaporation depends on the age, exposure conditions and density of the concrete. During drying process the pore water in the concrete evaporates and is replaced by air. Air contains carbon dioxide and other acidic gases which can react with the alkaline constituent of the concrete. This process is known as carbonation. The normal **protection against corrosion** provided by the **concrete** is lost as a **result of carbonation**. Corrosion of steel reinforcement will occur if moisture and oxygen comes in contact due to carbonation of concrete.

The depth of carbonation increases with age, but the rate of carbonation decreases with age. It has also been found that the **lower the water cement ratio** the slower is the rate of carbonation. Concrete containing blast furnace slag cement or pozzolanic cements carbonation rates are higher. This is a result of the difference in the proportion of alkaline compounds in different types of cement, particularly calcium hydroxide. The primary factor controlling the rate of carbonation is the **exposure conditions,** particularly with respect to contact with moisture. Concrete saturated with water is effectively free from carbonation, while concrete exposed alternative to a dry and wet atmosphere (relative humidity 50-75 per cent) shows the greatest depths of carbonation. Concrete exposed to moist air periodically by precipitation shows intermediate depths of carbonation. Phenolphthalein test is carried out to assess the depth of carbonation in concrete.

Phenolphthalein Test for Carbonation Depth

On site the best method of measuring the depth of carbonation is by exposing a fresh concrete surface created by a clean fracture using a hammer and chisel. Then spraying this surface with a 0.2 per cent pH indicator of phenolphthalein in alcohol solution. Magenta colour areas represent uncarbonated concrete and the colourless areas represent carbonated concrete. The colour change occurs at about pH 10. It is important that a clean fracture is made since if un-hydrated cement dust gets onto the surface, misleading results may be obtained. A more precise measure of depths of carbonation can be made by microscopic examination, for calcium carbonate in thin sections cut from concrete specimens. This, of course, would have to be done under laboratory conditions and is a time consuming operation.

The above discussion has considered the main aspects of carbonation and how it is measured. This will now be related to the corrosion of the reinforcement. Uncarbonated concrete has a pH in the range of 12.5-13.2 and provides a good protection against corrosion even when moisture and oxygen are freely available. If sufficient chloride ions are present in the concrete adjacent to the reinforcement, the protection to steel is lost. As the concrete becomes progressively carbonated with age, a stage will be reached where the zone of concrete containing the reinforcing bars will start getting affected. The pH will then gradually fall. There is evidence to show that **protection is lost when the pH falls below 11.0** and corrosion will then result if moisture and oxygen are present. If chloride is also present in the reinforcement zone then **protection is lost even at a higher pH value**. Consequently in order to predict the probability of corrosion it is necessary to know the **state of carbonation in vicinity of reinforcement** and the level of corrosion, it is necessary to measure chloride composition and carbonation along with those of depth of concrete cover (cover meter). It should be remembered that phenolphthalein distinguishes concrete having pH above and below 10.0, *i.e.* it isolates the fully carbonated concrete, but partially carbonated concrete appears in uncarbonated region. The prospect of corrosion is increased substantially if chloride is also present in partially carbonated concrete around the reinforcement.

4.8.3 Chloride Ions in Concrete

Chlorides are found in both old and new concrete. In old concrete the source may have been calcium chloride used as accelerator and/or marine sand aggregate, or marine water spray. In new concrete the chloride develops from the use of de-icing salts and in special cases as a result of marine exposure or the use of marine aggregate. The relevant standards have now severely limited the use of accelerators and marine aggregate to avoid future problems, which may originate predominantly from exposure to salt-laden environments. Chlorides penetrate the hardened concrete, during service, as a result of repeated applications of de-icing salt or exposure to marine spray. When chlorides reach rebar level corrosion occurs with a progressive accumulation of rust around rebars causing **increase in volume resulting in tension cracks** in the concrete thus permitting further ingress of chloride. This causes further acceleration of the corrosion process and subsequent damage to concrete. Deterioration is sometimes caused by faulty bridge deck drainage that allows contamination of the structural members by de-icing salts.

The total chloride content of the concrete mix arising from all sources should not exceed 0.35 per cent by weight of the cement in the mix for 95 per cent of the test results and with no result exceeding 0.50 percent for reinforced concrete made by using cement complying with the IS269 specification. For prestressed concrete, steam cured concrete and concrete made using Super-sulphated cement, the chloride limit is restricted to **0.06 per cent** by weight of the cement. If these limits are approached or exceeded the risk of reinforcement corrosion is increased. Generally, it can be said that the higher the chloride ion content the greater the probability of corrosion. An assessment of the hazards arising from the presence of chloride therefore requires a method for measuring the chloride content of hardened concrete. There are two suitable methods:

(i) An acid extraction followed by the Volhard volumetric titration.

(ii) An acid extraction followed by the use of 'Quantab' chloride titrator test strips.

Test for Chloride Content of Concrete

Quantab method measures the total chloride content of the concrete. This includes the chloride combined in cement hydration products which are only slightly soluble and hence do not normally result in significant amounts of corrosion. The portable equipment have currently become available for conducting rapid test on site for the measurement of chloride content in concrete. A rotary percussion drill is used to collect a pulverized sample of concrete and the chlorides are extracted by a special acid. The amount of acid soluble chloride is determined directly by a chloride sensitive electrode connected to an electrometer. If different samples are obtained from concrete *e.g.* from 0.30 mm depth, 50 to 100 mm depths, it can be established whether there is diffusion of chlorides into the concrete from the environment. The electrode measures the potential of the solution and from the correlation graph chloride contents are known. The chloride content of concrete should be determined at different depths.

4.8.4 Determination of Permeability

The permeability of concrete has been recognised as a major factor responsible for the durability of concrete structures. Many types of commercial permeability testers are available to determine this property at site.

The Figg method is used to determine the air permeability of concrete and the apparatus can also be modified to determine water permeability. The equipment is simple to use and can thus be used on site. An advantage of the system is that it measures permeability inside the concrete rather than on the surface. The apparatus is illustrated in figure 4.3.

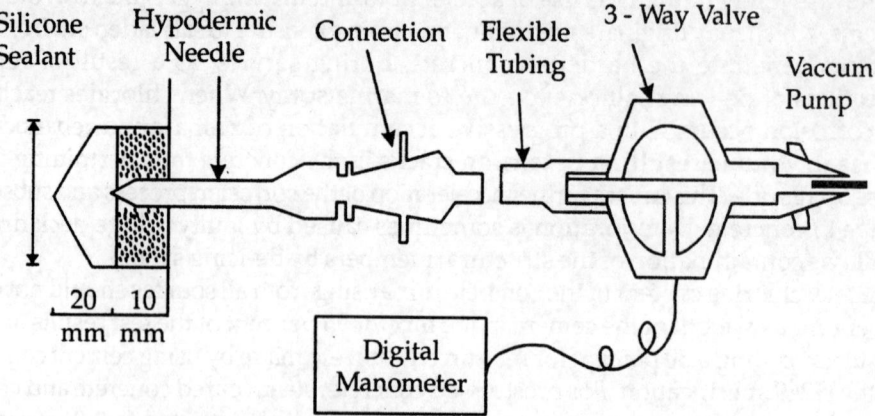

Fig 4.3: Figg Apparatus for Measuring Air Permeability

Figg Method for Determination of Permeability

A 5.5 mm diameter hole is drilled in the sample to a depth of 30mm. This is then `blown-out' using a small compressor and air pipe to remove any loose material. Next a polystyrene plug is inserted in order to form a cavity approximately 20mm deep. A catalyzed silicone rubber (*e.g.* Elastosil 105) is then put into the hole to produce an in situ rubber plug; this provides an airtight seal to the cavity. Once cured a hypodermic needle is inserted through the rubber and into the cavity, a cannula is then passed through the needle and into the cavity. The needle is then removed. Alternatively, a hypodermic can be used which is attached to a fine plastic tube. In this case the hypodermic is left in place. The cannula or fine plastic tube is attached to a hand operated vacuum pump and a digital pocket manometer via a three way vacuum tight tap. A vacuum of-55 kpa is then created using the vacuum pump and a stop clock started, the time taken for the pressure to increase from 55 to 50 kpa is noted. This is repeated five times per hole, on three individual holes, per sample. In order to check the apparatus for leaks the test can be carried out using a rubber plug as a control, the time taken for the desired pressure increase is recorded.

The longer the time taken for the vacuum to drop, the better the quality of the concrete (as also the lower the permeability).

Low permeability is beneficial as this reduces the rate at which aggressive ions permeate through the concrete to the steel.

4.8.5 Depth of Concrete Cover, Reinforcement Details and Spacing

These measurements are very important in investigation, particularly the depth of cover is of great significance for interpreting chloride contents and carbonation depth in the concrete. The depth of cover in relation to factors, such as type of exposure, concrete permeability and the existence of waterproofing layers, becomes important for predicting corrosion. The depth of cover for concrete placed *in situ* is rarely uniform and this leads to problems of the formation of differential corrosion cells, particularly if the nominal depth of cover tends towards the lower acceptable limit.

Concrete Cover Measurement

The cover meter method is based on the principle that the presence of steel reinforcement affects the electromagnetic field. Several equipment known as perfometer, pachometer, Fe-depth meter are available for this purpose. The covermeter consists of a unit containing power source, amplifier, meter and a search unit containing electromagnet. The procedure for measuring depths of concrete cover is easily employed on site as the test is non-destructive and requires knowledge of the rebar diameter only. Sometimes it can be difficult to distinguish between a small diameter bar at low cover depth and a large diameter bar at high cover depth. The covermeter is also frequently used simply as a means of detecting reinforcement position for the purpose of core cutting or drilling. The detection of location and depth of the rebar are measured by the degree of distortion of a magnetic field caused by the reinforcement. The covermeter is generally effective for depths of cover up to 80mm. If covermeter measurements are carried out on a grid system over the concrete surface, equio-depth of cover contours can be prepared which clearly illustrate the variability in-depth of cover and regions of less than satisfactory values.

Cover depth, w/c ratio and cement content requirements for different conditions of exposure are given in Table 4.3. Most external concrete above ground will be subjected to severe environmental conditions, such as exposure to driving rain, alternate wetting and drying and freezing, depending on the climatic zone.

TABLE 4.3: NOMINAL COVER TO MEET DURABILITY REQUIREMENT

Exposure Condition	Nominal Concrete Cover in mm Not Less than
Mild	20
Moderate	30
Severe	45
Very severe	50
Extreme	75

Notes:

1. For main reinforcement up to 12 mm diameter bar for mild exposure the nominal cover may be reduced by 5 mm.
2. Unless specified otherwise, actual concrete cover should not deviate from the required nominal cover by +10 mm.
3. For exposure condition 'severe' and 'very severe', reduction of 5 mm may be made, where concrete is M35 and above.

TABLE 4.4: ENVIRONMENTAL EXPOSURE CONDITIONS

Sr.No.	Environment	Exposure Conditions **
1.	Mild	Concrete surfaces protected against weather or aggressive conditions, except those situated in coastal area.
2.	Moderate	Concrete surfaces sheltered from severe rain or freezing whilst wet
		Concrete exposed to condensation and rain
		Concrete continuously under water
		Concrete in contact or buried under non-aggressive soil/ground water
		Concrete surfaces sheltered from saturated salt or air in coastal area
3.	Severe	Concrete surfaces exposed to severe rain, alternate wetting and drying or occasional freezing whilst wet or severe condensation
		Concrete completely immersed in sea water
		Concrete exposed to coastal environment
4.	Very severe	Concrete surfaces exposed to sea water spray, corrosive fumes or severe freezing conditions whilst wet
5.	Extreme	Surface of members in tidal zone
		Members in direct contact with liquid/solid aggressive chemicals.

**Cover depth may be reduced for higher grades of concrete and increased for higher w/c ratios than specified.

Thus, for example, the minimum grade of concrete required for a severe environment is **M30** with **45mm** cover, maximum free w/c ratio of 0.45 and minimum cement content of 320 Kg/m³. Evaluate concrete cover depths, with reference to the specified condition to detect inadequacy of cover causing corrosion of rebars.

4.8.6 Determination of Concrete Strength, Extent of Cracks, Voids and Honey Combing

The surface characteristics play an important role in measuring the strength and other properties of concrete structures. The surface hardness, rebound of an elastic spring mass or penetration by a probing mechanism are measured by these tests. Rebound hammer, pull out tests, Windsor probe and ultrasonic concrete tests are widely used to find the quality of insitu concrete and detection of cracks, honey combing and voids. The accuracy of prediction of concrete strength/quality from such tests, is of the order of ±25% only.

(a) **Rebound Hammer Test:** This method is based on the principle that the rebound of an elastic mass depends on the hardness of the surface against which the mass impinges. Surface hardness is considered to be related to the compressive strength of the concrete. The rebound number is read from a graduated scale on the rebound hammer and corresponding strength is read from the related standard curves.

(b) **Windsor Probe Method:** It is basically hardness tester. It provides an excellent means of determining the relative strength of concrete in the same structure or in different structures. The equipment may be used to assess the uniformity of in situ concrete, to delineate the zones of poor quality or deteriorated concrete. In this method, the penetration of probe in the concrete is measured from a power activated gun or driver. The exposed length of probe is related to compressive strength of concrete.

(c) **Pull out Test:** This test measures the force required to pull out a specifically shaped steel rod whose enlarged end has been embedded in concrete. The force is measured with the help of a dynamometer. The pull out strength is calculated from the ratio of the pull out force to the idealized area of the frustum of the cone. Concrete strength is determined with calibrated correlation between pullout strength and compressive strength.

(d) **Pull off Testing:** This method can be used for estimation of compressive strength of concrete. A circular steel probe is bonded to the concrete. Tensile force is applied using a portable mechanical system until concrete fails. Compressive strength is estimated using calibration charts.

(e) **Core Test:** Among many methods to find *in situ* strength correctly, core testing has been recognised by most of the countries. This is the most direct way of measuring actual strength of concrete in structures. The core taken by a special core cutter is tested in the laboratory by using compressive test machine. However, this is a partial destructive test, because holes are made in the structure.

(f) **Ultrasonic Pulse Velocity Test:** In this method, the ultrasonic pulse is produced by the instrument which travels through a known distance (L) in the concrete and is received at the other end. An electronic timing circuit enables the measurement of transit time (T) of the pulse. The velocity is given by $V = L/T$ in Km/sec. The underlying principle of assessing the quality of concrete is that comparatively higher velocities are obtained, when the quality of concrete in terms of density, homogeneity and uniformity, is good.

In case of poor quality, lower velocities are obtained. If there is a crack, void or flaw inside the concrete which comes in the path of transmission of pulse, the pulse strength is attenuated and it passes around the discontinuity, thereby making the path longer. Velocity is mainly related to the density and the modulus of elasticity. The modulus of elasticity is also related to the strength of concrete. The zones of poor quality in the structure can be identified for repair works by using ultrasonic concrete testing. .

(g) **Ultrasonic Pulse Echo Test:** This provides estimation of compressive strength, uniformity and quality of concrete, location of reinforcing bars, voids, delamination and concrete thickness. It works on the principle that the original direction, amplitude and frequency of stress waves introduced into the concrete get modified by the **presence of cracks, objects** and **sections** which have **different acoustic impedence**. It can be used in situations where surface is accessible. It allows one to investigate inside body of the concrete. Presently, it is not a common test method.

(h) **Resonant Frequency Testing:** This method is used to determine various fundamental modes of vibration for calculating modulus in the laboratory and in field to detect voids, cracks, delaminations, etc. This method allows one to investigate the inside of a structure. It operates in sonic range and does not have resolution of ultrasonic. This method is still at a developing stage.

(i) **Acoustic Emission Method:** This method can be used for continuous monitoring of structure during service life to detect impending failure and for monitoring performance of structure during testing. It works on the principle that during crack growth the rapid release of energy produces acoustic waves (sound waves) that can be detected by sensors in contact with or attached to the surface. Thus, this method can be used to monitor structure's response to applied loads, and for locating source of possible failure. However, the method is still at laboratory stage.

(j) **Acoustic Impact Method:** This method is used to detect debonds, delaminations, voids and hairline cracks. The frequency and dampening characteristics of sound resulting from the impact on the surface give an indication of presence of defects. This is a simple equipment and requires low level of expertise, whereas the electronic system requires training.

(k) **Maturity Method:** This is a technique to estimate insitu strength which accounts for the effects of temperature and time on strength development. The measurement is based on the internal temperature of body of concrete. This method is particularly valuable during adverse weather conditions when the knowledge of strength development in concrete is important.

(l) **Radioactive Method:** This method is used for identification of defects in concrete, hidden from eye. This method is based on the principle that when x-ray or gamma radiation passes through concrete, more radiation is absorbed by the denser part of concrete. The two main techniques in this category are radiography and radiometry. In both cases gamma radiation is preferred because its source is easier to use at site. This method can be used to locate voids and poor compaction in concrete. A limitation of gama radiography is that only concrete within 50 mm of the surface is examined effectively.

(m) **Radar Method:** This method is used to detect substratum voids, delaminations and embedment measurements of thickness of concrete pavements. Electromagnetic impulse signals are used for void detection. It can be used when only one surface is available. However, this method requires high degree of expertise.

4.8.7 Determination of Defects and Irregularities in Piles

The main component of the pile foundation is the correct construction of piles according to design. Structures, having piles of questionable integrity, defects and irregularities in form of voids, separation of section, necking, honeycombing, and cracks, require careful estimation of load carrying capacity of the foundation. Non-destructive integrity of piles is useful for both cast-in-situ and pre cast concrete piles. The method involves recording of pile head acceleration caused by blow of a hand held hammer on the pile head. The wave generated by blow travels down the length in axial direction with the speed of sound. Any irregularity in the concrete pile make the wave to reflect back. If there is no variation, the wave will reflect back from the toe only. Monitoring and analysis of these reflections form the basis of this test. Thus, the major defects such as cracks, necking, soil inclusions and changes in cross-sections can be identified by this method.

4.8.8 Determination of Defects in Metals and Welded Joints

The various properties such as hardness, dimensional tolerance, discontinuities in materials and welded joints are to be tested in metals. Following techniques are used for assessing the defect in metals:

(a) Ultrasonic examination is useful in detecting the defects in structural materials and in the welded joints.

(b) Radiographic examination is useful in controlling the quality of welded joints, specially the butt joints between plates and the structural sections.

(c) Magnetic particle examination is useful for detecting defects open to surface but difficult to be seen with an unaided eye. To a limited degree, it can also be used to detect defects which are close to the surface, but not open to the surface.

4.9 STUDY OF AVAILABLE DOCUMENT

After carrying out various possible and required tests, the next step is to study document concerning the in-service history of building. It will be helpful in explaining unexpected cracks and damages. Abnormal events such as earth tremors, flooding, demolition, alterations, etc. give clues of the crack nucleation and crack propagation.

The available document on the analysis, design and construction details of the structure under consideration can provide useful information for the diagnosis of structural inadequacy. The design documents usually provide information on simplifying assumptions concerning the structural behaviour, methods of analysis, design loads, concrete strengths, and type and quantity of reinforcing steel. Construction information, which is likely to be useful, include **mix details, concrete slump tests,** cube tests, the age at loading and compressive strength of bricks.

4.10 DIAGNOSIS OF PROBLEM

From the results of various tests and study of documents, the cause(s) of defect in the structure can be established. After knowing the root causes of the problem, it will be necessary to know the actual loads and various environmental conditions to which the structure is exposed during its present and future service life. This will facilitate proper planning for repair and maintenance of the structure.

4.11 ACTUAL LOADS AND ENVIRONMENTAL EFFECTS

After the various defects have been identified, it becomes necessary to assess actual loads and environmental conditions which may have caused the defects observed during tests. Generally, loads acting on the structure under distress, partial failure or total loss of serviceability are much different from the loads assumed in the original design calculations. As a result, unexpected cracking and damage may occur. These forces and load combinations ought to have been considered at the design stage. In cases of overloads structural rectification is simplified and adequacy can be demonstrated for restoration and corrective maintenance work at a minimum cost.

Specifications and codes are generally silent on environmental effects on the structures. Vibration effects on structures located by the side of railways tracks and on busy traffic roads are required to be investigated. Extreme temperatures, extreme aggressive atmosphere in coastal areas or highly industrialised areas can create a range of serviceability problems. Foundations are an important part of the structure that can be in an environment far different from those assumed in the design. Thus, it is necessary to estimate the actual loads and environment conditions and predict their adverse effects on the structure before undertaking any repair and maintenance work.

4.12 ORIGINAL DESIGN AND CONSTRUCTION PRACTICES

After carrying out various tests to find defects and assessment of actual loads and environmental conditions, sometimes it may be necessary to check design details and errors occurred in original construction of the structure. Many a times they are the main cause for the failure of steel and concrete structures. Following are some of the main causes for cracking in metal or concrete structures due to errors in design and construction errors and must be thoroughly investigated to evaluate the root causes of deterioration/ defect:

(i) Sudden changes in cross-section responsible for stress concentration crack nucleation.

(ii) Inadequate structural connections in steel framed buildings and Poor design of moment resisting connections in gable-frames and portals may result in movement causing large unsightly deflections.

(iii) Imperfect geometry and deflections arising from poor quality fabrication and construction, causing over stressing and brittle fracture.

(iv) Omitted, damaged or removed bracings in the columns and beam-columns results giving warning of impending failure.

(v) **Soil-structure interaction causing rotation and or differential settlement** results in redistribution of moments and forces and consequent failure of-connections or members.

(vi) Poor **detailing of reinforcement** near corners resulting in **shear cracking** and/or **crushing** of concrete (*e.g.* water tank).

(vii) **Large span/depth** ratios in lintels over large openings showing **excessive deflections** and cracking.

(viii) **Poor design of connections** in pre-cast units. Pre-cast lintels over doors and windows showing cracks due to poor design and execution.

(ix) Little importance is given to the **form-work design**, especially ballies used in strutting and shoring.

(x) **Wrong sequence** of removal of form work may also result in cracking.

(xi) Concrete in cluster of reinforcement inside the joints may not be **compacted** fully while casting and may result in **cracking and honey combing**.

(xii) Improper **placement of reinforcing bars** in correct position resulting in serious problems in construction especially in slender columns.

(xiii) **Localized settlement** of subgrade, over vibrations and setting shrinkage are defects at the construction stage.

(xiv) Premature removal of shuttering and form work causing **over stressing and cracking** of the structure.

Tentative diagnosis is made in view of the above and later confirmed, before corrective repair measures are taken. This confirmation should be obtained by retrospective analysis and substantiated by material testing or structural load tests.

4.13 RETROSPECTIVE ANALYSIS

The purpose of the retrospective analysis is to confirm the initial diagnosis and to know whether the structure will be safe or not. The actual data concerning loads, structural geometry, material property, and climatic conditions are used to arrive at the actual load carrying capacity instead of the normal design values. It may be noted, however, that this analysis is also a **structural idealization** and must be treated as such. The analysis may also be extended to evaluate the **seriousness of any defect or damage**, thereby facilitating the remedial corrective measures.

Considering the **limitations of validity** of the above procedure for analysis, appropriate restoration measures may be taken. However, full understanding of structural behaviour coupled with an **appreciation of specifications**, codes and regulations, is also necessary. The retrospective analysis helps to decide whether structure is safe or not and whether **rehabilitation or repair** is viable. Decision should be taken as to whether to repair/rehabilitate the structure, or change the occupation, or to abandon it, on the basis of economic considerations.

4.14 CONFIRMATION OF DIAGNOSIS

Once the diagnosis is confirmed, the remedial measures for repair can be planned accordingly. The building structure should be repaired and strengthened, so as to keep it in serviceable condition. It is necessary to follow a proper approach for carrying out repairs or remedial work. This includes:

  (i) Proper **surface preparation,**
 (ii) Proper selection of **repair materials,**
(iii) Actual **repair practices,**
 (iv) Record of maintenance work and **maintenance reports.**

4.15 SUMMARY

Identification and diagnosis of defects and their root causes, play a critical role in deciding about the required repair and maintenance. Correct diagnosis of building defects facilitates selection of **correct materials and methods** for repair. Lack of awareness leads to neglect of timely repair.

Systematic investigation involves steps of **preliminary investigation, diagnosis of defects, inspections, material tests, NDT, study of documents, actual loads** and **environment,** errors in **design,** strengthening needs, and repair approach.

Objectives of investigation is primarily to determine the **extent of damage** or distress and **classify** problem into structural, or non structural. Factors assessed include: **carbonation** in concrete, **depth of cover, chloride** level, **permeability** in concrete, **corrosion, load carrying capacity,** affected areas, **cracks,** defects in timber, **dampness,** leakage in drainage pipes, etc. The purpose of investigations may be **legal, insurance requirement** or **structural safety.**

Preliminary investigations are carried out visually and data collected from various sources, such as drawings, specifications, architect's instructions, maintenance manual and records, interviews and site notes. During visual inspections attention is given to correctness of material detailing, design, loading and other data of exposure conditions.

Detailed investigation and diagnosis includes assessment of **distress nature, its extent,** physical tests including NDT, exact conditions of exposure and loading. Various damages are measured and specified in detail with reference to root causes. Records are prepared to decide the correct materials and methods of repair. Kits may be used for investigation.

Conditions of material must be evaluated by physical, chemical and other testing techniques. Some times overall structure may also be tested by full scale loading and observing conditions of various elements and their materials.

In most of the investigations of existing structures, different type of non destructive tests **(NDT)** are carried out. A limited number of destructive tests are used to confirm NDT results and diagnosis for correct remedial repairs. Corrosion of reinforcement can be determined by using electric potential measuring instruments such as copper sulphate half cell. Carbonation in concrete can also be measured by phenolphthalein. The portable equipment can be used to assess chloride contents in concrete. Figg's apparatus may be used to determine the permeability of concrete.

ND tests mainly comprise of **ultrasonic pulse** velocity test, concrete **reboundhammer test**, pull out test, **core tests**. NDT can be carried out, to approximately assess the strength of damaged concrete elements.

Original document of building must be studied to evaluate construction specifications and other material details. This document also clarifies the loads and other conditions considered for the design. Various tests, inspections, and study of documents must facilitate systematic investigation and accurate diagnosis for deciding on correct approach to repair and maintenance.

QUESTIONS

4.1　Describe importance of systematic investigation for effective maintenance.

4.2　List 11 steps in sequence involved in systematic building repair.

4.3　List 10 important factors to be ascertained for effective remedial meaures of building defects.

4.4　Describe different purposes of investigations for building defects.

4.5　List sources of information for building defect investigation.

4.6　Show flow chart of defect diagnosis.

4.7　Describe importance of physical measurement for defect diagnosis.

4.8　Describe sign/symptoms for various defects in concrete elements.

4.9　Explain the importance of visual examination for diagnosis of building defects.

4.10　Describe brief details of investigation kit for building defects.

4.11　Describe importance of NDT in building material assessment.

4.12　Explain determination of corrosion in reinforcement.

4.13　Explain briefly working of copper sulphate half cell for assessment of corrosion in steel.

4.14　Explain briefly phenolphthlein test for carbonation depth in concrete.

4.15　Explain briefly determination of chloride-ions in concrete.

4.16　Explain briefly determination of concrete permeability with Figg. Apparatus.

4.17　Explain determination of depth of concrete cover in RCC elements.

4.18　List methods for determination of strength of concrete in existing RCC elements.

4.19　Explain the importance of study of building documents and actual loads in building maintenance.

4.20　List down various causes of cracking in concrete elements.

Chapter 5

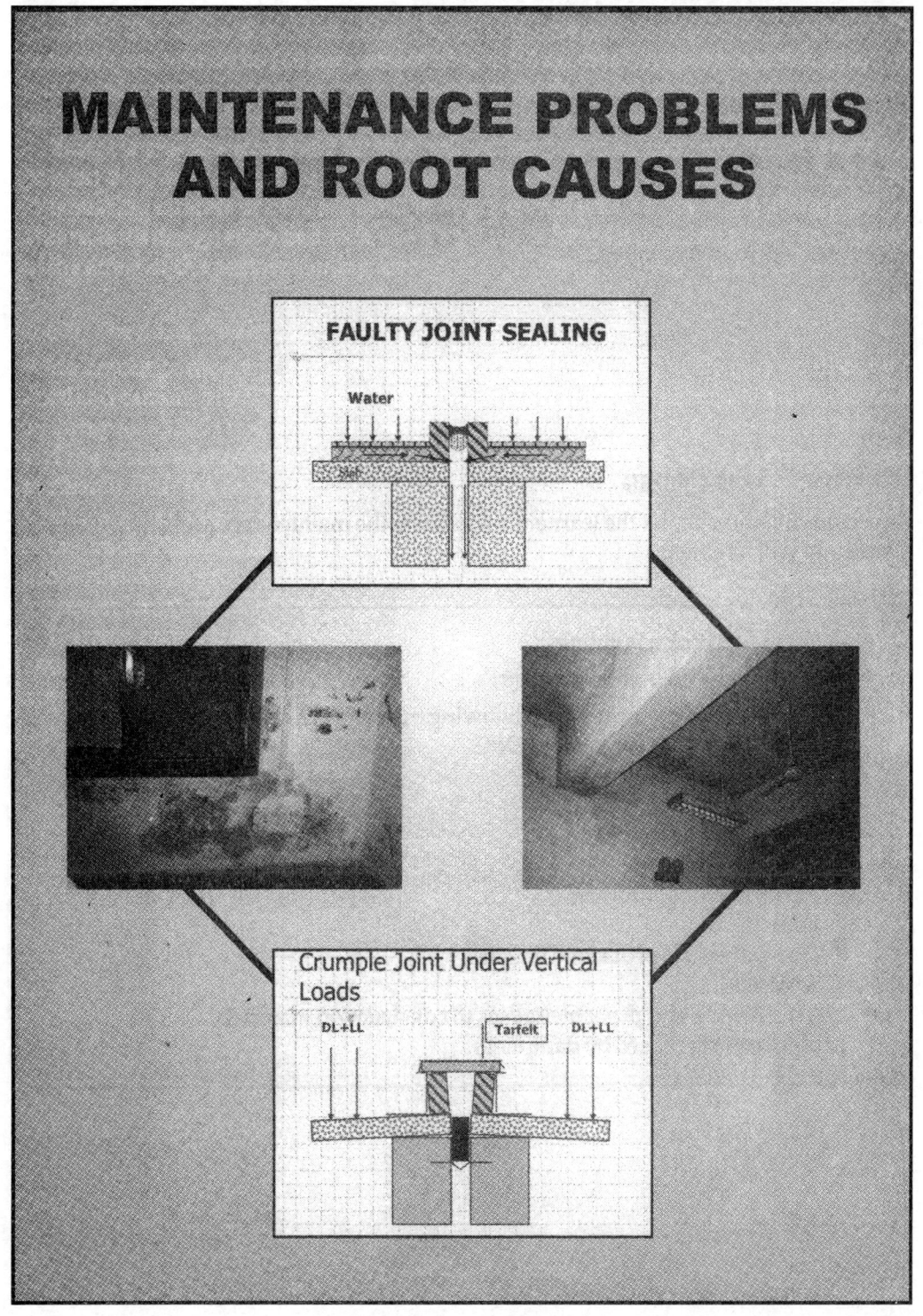

MAINTENANCE PROBLEMS AND ROOT CAUSES

FAULTY JOINT SEALING

Water

Crumple Joint Under Vertical Loads

DL+LL   Tarfelt   DL+LL

5

MAINTENANCE PROBLEMS AND ROOT CAUSES

LEARNING OBJECTIVES

After studying this chapter the learner understands the maintenance problems, their root causes and will be able to:

- **Describe** defects in buildings;
- **Explain** causes of building defects;
- **List** defects and their causes in following building elements:
  - Foundations, basements and DPC;
  - Walls;
  - Chimney stacks and shafts;
  - Columns and beams;
  - Roof and terraces;
  - Floors and floor finishes;
  - Joinery;
  - Decorative and protective finishes;
  - Services.
- **List** the defects and probable causes in construction materials;
- **Explain** defects caused by dampness.

5.1 INTRODUCTION

The extent of defects and failures in buildings are yardsticks to evaluate the performance of buildings and construction practices. There is a very thin dividing line between **defects and failures**. Defects represent lack of appropriate functional service. Defects in reference to buildings can be termed as anything that hampers the functional performance of buildings. Where as failure is defined as behaviour not in agreement with the expected conditions of structural safety and stability and non compliance with the desired usage and occupancy of the completed building structure.

It is well known fact that most of do not building defects are avoidable. These defects do not occur, because of lack of knowledge but occur due to **negligence** and **non-application** of **correct knowledge** during construction and maintenance. Building defects are results of **incorrect construction practices** using new materials.

Much has been published about building defects, their causes and cures but unfortunately the information is scattered over a vide range of research papers, business publications and books. However this literature has not been easily available to students and civil engineering professionals at various levels. Even practicing engineers have been devoid of this because of lack of culture of on the job training. Most of the published literature for remedial work often assumes that the causes of the failure and defects are already known but it is not so in most of the cases. If a wrong diagnosis is made without analysing main causes, **the treatment may be unsuccessful, leading to the waste of funds, materials, labour and time.**

There is well known saying that **prevention is better than cure**. The designers and builders can reduce the frequency of occurrence and extent of defects by using correct practices. The defects can be further avoided or minimised if the field engineers are careful in design implementation and adopting **sound construction practices**.

The defects can be categorised based on the two approaches viz., elemental and characteristic performance.

Elemental Approach	Characteristic Performance Approach
• Components • Materials	• Structural • Hygrothermal and • Comfort and Environmental performance.

Elemental defects represent specific problem in certain component of the building or any construction material. **Structural** Technology aspects include the design and construction practices in buildings. Defects in this area are caused by deficiency in the structural components, its materials and effect of external environment. Lack of supervision by competent persons during construction also contributes to these defects.

Hygrothermal deficiency of the building includes both lack of weather proofing and appropriate thermal behaviour of the building. Now a days building users are laying more emphasis on the provision of comforts. Buildings, which perform badly in structural and hygrothermal functions, are not comfortable. Such buildings are assessed as defective.

These defects become severe when we consider the **comfort along with environmental performance**. In the total performance of the building the role of health and safety aspects must be considered. The environmental performance is also linked to the building services provided within it. There can be wide range of deficiency in these services also.

The complete avoidance of defects is almost impossible but can only be minimised. A compilation of report on various defects along with its causes provide a valuable source of information for actual building structures for repair and maintenance. Main causes of defects in building structures are:

- Lack of **careful design**, *e.g.* choice of wrong materials and inappropriate provision of sections and details.
- Lack of **sound construction practices**, *e.g.* non adherence to standard construction specifications; poor workmanship; inadequate inspections and supervision.
- Lack of suitable **maintenance practices**.

Having known the factors which contributes to the defects, it is also necessary to investigate and identify the root causes. The analysis of defects in building requires sound and extensive knowledge of building materials and construction practices. This knowledge can be gathered from a combination of text books, research publications, latest trade literature and field experiences.

Various defects encountered in buildings have been identified along with their probable causes. These defects have been classified according to elemental approach in tabular form for ease of understanding and better assimilation for repair under the following groups:

- Foundation, Basement and Damp proof course (DPC)
- Walls
- Wall Finishes
- Chimney Stack and Shafts
- Columns and Beams
- Roof and Roof Terraces
- Floors and Floor Finishes
- Joinery Work
- Decorative and Protective Finishes
- Services
- Materials
- Dampness in Various Building Elements

These defects are relevant and common for identification in older buildings for maintenance and repair. Building defects are required to be diagnosed prior to suggesting effective remedial measures. Defect analysis provided in this chapter will facilitate the engineering fraternity in guiding them for the suitable remedial measures in maintaining buildings to perform most satisfactorily. Defects and causes listed in different building elements are most generalised and may not be existing independently. For one particular defect, there may be one or more than one causes. Since the defects and causes are generalised, different tables indicate probable causes only. Maintenance engineer must analyse the defects critically by observations and detailed local testing. Most of the defects in buildings are caused due to lack of proper supervision during construction and later during occupation.

5.2 CAUSES OF DEFECTS

Most of defects are caused by dampness. Dampness is one of the most difficult faults to trace and diagnose. There are innumerable ways in which water can be present in a building component and several of these causes may occur at the same time and even be responsible for the same damp patch. The identification of one source must not be considered enough for discontinuing further investigation into the other sources of water. In many instances where the detailed analysis has been neglectd before repairing, the wet patch has quickly reappeared even after repair work.

Most buildings are assembled from different components comprising of variety of materials. Many of these materials are water absorbent. Main causes of defects in building elements are related to:

- Water or dampness
- Physical movement due to forces
- Effect of environmental factors

5.2.1 Water

Different sources of water in building elements causing problems of dampness can be attributed to:

(a) Construction water
(b) Intruding water
(c) Condensation
(d) Occupational

(a) **Construction Water:** Considerable quantity of water is used in the construction of a building. Some of this water is consumed in the setting of materials such as Portland cement and gypsum plasters. Some of water gets dried out by evaporation by the time the building gets occupied, while much water still remains in the structure. Areas adjacent to relatively large masses of wet building materials such as concrete may be susceptible to moisture absorption particularly on the underside of concrete floors. **Construction water** may also accumulate under impervious finishes such as paint.

Each occupant exhales
½ litre of water as vapour
every 9 hrs.

Burning 1 litre of parafin
produces 1 litre of water
as vapour

Fig. 5.1 (a): Source of Water Vapour

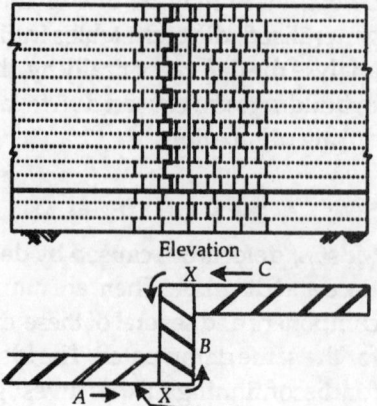

Elevation

This end of wall A has expansion
at DPC level causing rotation of
teturn wall B, resulting in cracks
at X in wall C

Fig. 5.1 (b): Cracking due to expansion
of brickwork

(b) Intruding Water: Water enters many buildings as a result of **rainfall**, either directly through **defects in the roof and walls** or indirectly by **absorption**. It may then be transferred for some distance via **cracks and voids** in the structure, or along hollow sections. An investigation should always be done for locating such paths because dampness may occur at points away from the source of water.

The water tightness of the joints is a key factor in preventing rain penetration. Though rain falls by gravity it can be blown by wind in all directions against a building, this is particularly true in the case of large tall buildings in high rainfall areas. Driving rain finds easy entry through the weaknesses in any element, though it may be difficult for the investigator to locate such points of entry.

Dampness may enter a building from the ground in case of inadequate and improperly constructed damp proof courses and membranes. A number of defects are made in DPC which lead to severe defects in joinery and walls. These defects indicate the first symptom that water has entered the building.

(c) Condensation: Water vapour is normally present in the atmospheric air in varying quantities. Some of water vapour gets converted into liquid water when air comes in contact with relatively cold surfaces. Such condensed water may appear on any of the internal surfaces of a building or it may form within the building element or material which is known as interstitial condensation. This form of condensation can be very troublesome partly because of the difficulty in locating and recognizing it and partly because of its adverse effect on the insulation. It may also give rise to other defects causing damage to the construction materials.

The amount of water, which condenses in some buildings, is very considerable leading to a state of almost permanent dampness. This provides favourable conditions for mould growth. Some of the materials have the property of absorbing moisture from the air. This absorption results in transferring of chlorides present in soot specially deposited within a flue.

(d) Occupational: Apart from above three type of waters the water vapour produced by the occupants also aggravates the problem of condensation (one adult produces half a litre of water in 9 hours simply from breathing). Water may also be present in a building from leaking pipes, tanks and cisterns. The use of excessive amounts of cleaning water can also contribute especially when it seeps underneath floor coverings. The spillage of liquids in industrial premises takes place at many locations. This spillage may also be contaminated with chemicals which are harmful to the construction materials.

5.2.2 Movement

Damages are caused in building structures due to movement of moisture which occurs due to following factors:

- Externally applied forces (*e.g* wind, dead and live loads);
- Vibrations (*e.g* earthquake, machines);
- Temperature changes (*e.g* expansion, contraction);
- Physical and moisture changes (*e.g.* shrinkage, creep, swelling)

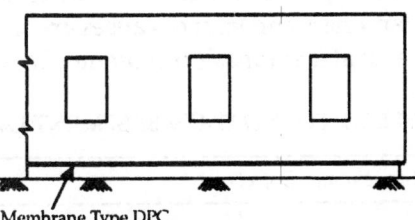

Membrane Type DPC

(b) Oversailing of Brick Masonry at DPCL Level Due to Expansion

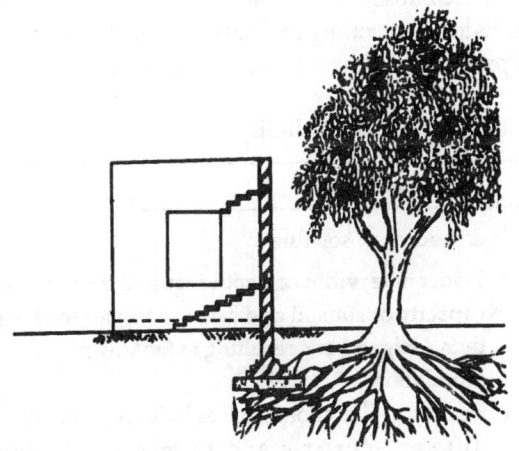

(a) Trees Growing Close to a Building on Shrinkable Soil May Cause Cracks in the Walls

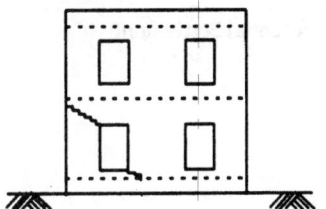

(c) Cracks at the Corner of a Building Due to Foundation Settlement

Fig. 5.2

Movement in building elements may occur due to any of above or combination of these factors. The severity of damage depends on nature and intensity of factors and resistance of building elements.

5.2.3 Environmental Factors

Environmental factors affecting building components and materials include:

- Solar radiation;
- Temperature;
- Humidity
- Biological agencies;
- Air pollutants (solid, liquid or gaseous);
- Ground salts.

Effect of these factors will cause deterioration and damage depending on severity and combination of these factors.

5.3 INVESTIGATION OF DAMPNESS

There are many points, which may have to be considered when investigating dampness. These points include the quality of construction, the exposure conditions, protection features on the facade of the building, the internal location of the dampness, and co-relation of dampness with rainfall. It is always useful to examine the building while rain is actually falling, but it must be remembered that the water may not be visible internally for some hours after the onset of the rain.

It can be said that most of the buildings suffer from the dampness and its associated defects. The cure of dampness can be said half the battle won in maintenance. The defects of dampness and its root causes are further listed separately in Table 5.1 to 5.12, so as to lay a greater emphasis on commonly faced problem of dampness in buildings.

TABLE 5.1: FOUNDATION BASEMENTS AND DAMP PROOF COURSE

Element and Defects	Causes
Foundations Excessive settlement Or Deferential Settlement	• Over loading of soil due to ▸ Inadequate width or depth required for bearing capacity of subsoil and causing higher loading than designed one resulting in Sinking of foundation. ▸ Structural alterations such as building used for a different purpose and formation of large openings in supported walls causing concentration of loading on adjacent foundations. ▸ Constructing additional storey without strengthening of existing foundations.

TABLE 5.1: FOUNDATION BASEMENTS AND DAMP PROOF COURSE

Element and Defects	Causes
Cracks and fracture in foundation	◆ Inadequate bearing capacity
	◆ Local variations in soil under foundations.
	◆ Adjoining construction activities.
	◆ Overloading and quick sand phenomenon.
	◆ Differential settlement in foundation.
	◆ Shrinkage of clay sub-soils due to.
	▸ Withdrawal of ground water by prolonged drought
	▸ Growing trees nearby
	▸ Site drainage
	▸ Tunneling
	▸ Long time settlement of peat or made up ground if present under foundation.
	▸ Down hill creep of clay soils if present
	▸ Foundation of extension bays, screen walls laid with shallower foundations than those of the main buildings.
	▸ Dynamic loading on foundations due to e a r t h quake.
	▸ Vibrations from : Heavy traffic, Machinery, Pilling operations and Tunneling operations etc.
	▸ Freezing of water in upper layers
	▸ Non provision of expansion joints below DPC level
	▸ Vegetation-growth and thrust exerted by tree roots.
Disintegration of foundation	◆ Chemical reaction such as sulphate attack on cement concrete.
	◆ Freezing and thawing
Basements · Inward bulging of walls	◆ Excessive lateral pressure due to combined effect of active earth pressure and water pressure from out side.
	◆ Increased pressure on basement walls due to
	(a) Plugging of drains
	(b) Excessive rain
	(c) Inadequate design
	(d) Submerged soil pressure
	(e) Drainage quality of the sub-soil.

Element and Defects	Causes
Cracks in raft slab	◆ Inadequate design
	◆ Construction joints not treated
	◆ Non provision of expansion joints
Vertical cracks in raft walls	◆ Drying shrinkage
	◆ Excessive thermal stresses due to temperature variations
Vertical or inclined Cracks in wall	◆ Deferential settlement of foundations
Horizontal and diagonal cracks in wall	◆ Lateral pressure and excessive shear force
Efflorescence	◆ Poor quality materials and continuous dampness or alternative wetting and drying.
Damp Proof Course	◆ Poor quality materials in DPC
Rising dampness	◆ Poor construction practices
	◆ Non provision of damp proof membrane
	◆ Failure of damp proofing systems
	◆ DPC not wide enough to cover full wall width
	◆ Damage caused in DPC while providing service pipes, etc.
	◆ Bridging of DPC due to
	(a) Inadequate plinth level i.e. less than 15 cm from DPC.
	(b) Continuous external rendering over foundation and DPC
	(c) Improper level of floor (floor level above DPC)
	(d) Droppings and improper ties in cavity walls
	◆ Non provision of vertical DPC in split level floors.
Cracks	◆ Excessive loading
	◆ Differential loading
	◆ Shrinkage
	◆ Physical damage by impact
	◆ Improper construction and expansion joints
	◆ Improper design of DPC.

TABLE 5.2: WALLS

Element and Defects	Causes
Load Bearing Walls	
Diagonal cracks along horizontal and vertical joints in masonry	◆ Differential Settlement of foundation ◆ Vegetation in walls ◆ Loose Soil under foundations.
Horizontal and vertical cracks	◆ Thermal movement due to omission of expansion joints in long walls ◆ Foundation laid on shrinkable clay ◆ Moisture movement due to expansion of freshly used bricks in Masonry ◆ Sulphate attack on joint mortar ◆ Corrosion of metal ties if used in wall masonry for strengthening.
Splitting	◆ Excessive horizontal movement because of earthquake, Explosion.

Outward bulging of walls	◆ Displacement of wall through one or more of the following reasons
	  ▸ High slenderness ratio i.e. thickness of wall insufficient in relation to height
	  ▸ Inadequate lateral support i.e. cross walls or floors not built into the walls to restrain movement
	  ▸ Overloading of structure
	  ▸ Vibrations from heavy traffic or machine.

Movement of wall at DPC level	◆ Irreversible expansion due to absorption of moisture in new clay bricks and lack of mechanical key/bond at DPC.
Spreading (outward displacement of wall at top)	◆ Spreading of beam bottom or rafter feet/ring beam, exerting horizontal thrust on supporting walls due to non existence of bearing plaster/Kraft paper
Efflorescence	◆ Presence of soluble salts in construction material ◆ Alternate wetting and drying
Non Load Bearing walls-fill panels Crack at head and sides	◆ Drying shrinkage in light weight concrete block/panel ◆ Inadequate support at base ◆ Deflection of upper floor joints

Element and Defects	Causes
	◆ New portion build on old floor
	◆ Differential movement between partition and surrounding load bearing structure
Brick panels Outward bowing	◆ Differential movement between concrete frame and brick panels
Concrete panels	◆ Inadequate projections to keep rain water clear off surface
Uneven weathering and dirt streaks	◆ Uneven surface finish
Rust stains	◆ Inadequate cover to reinforcement or poor quality porous concrete
Surface disintegration	◆ Sulphate attack
	◆ Weathering
	◆ Physical damage
	◆ Inner bar rusting if reinforced.
Glass panels	◆ Failing of holding material viz. puty, sealant, clips
Cracks and Fixing defects	◆ Natural calamity
	◆ Erroneous fixing and alignment.
	◆ Improper edges and non-rectangular corners
	◆ Physical impact.
	◆ Non-matching shape of glass and the frame.
Parapet walls	◆ Moisture penetration.
Cracks	◆ Sulphate attack
	◆ Deflection and tilt due to wind/earth quake force.
	◆ Rusting of reinforcement.
Oversailing of parapet	◆ Over all movement due to
with or without cracks	▸ Expansion of bricks due to moisture
	▸ Frost attack of the bricks/ mortar
	▸ Sulphate attack of mortar
	▸ Thermal movement of roof
	◆ Rusting of reinforcement, if any
	◆ Improper shape during construction
Boundary walls Collapse	◆ Lateral stability not considered in design for wind

Element and Defects	Causes
Masonry disintegration	◆ Inadequate foundation
	◆ Poor bonding of mortar.
	◆ Height/ thickness ratio high.
	◆ Natural calamity viz. floods, earth quake
	◆ Scouring of foundation soil.
	◆ Unsuitable bricks and mortar to exposure conditions
	◆ Sulphate attack
	◆ Rising dampness.
Wall leaning	◆ Inadequate foundation and lateral support
	◆ Scouring of foundation soil
	◆ Uneven swelling of foundation soil.
	◆ Differential settlement of foundation
Cracks	◆ Moisture movement in walls
	◆ Vegetation growth in foundation or wall.
Splitting of coping	◆ Expansion of metal work due to corrosion in RCC lifting of top course coping and iron fencing.
	◆ Penetration of moisture

TABLE 5.3: WALLS FINISHES

Element and Defects	Causes
Plaster	
Crazing of surface	◆ Drying shrinkage.
	◆ Shrinkage of substrate light weight concrete
Large cracks	◆ Differential movement of structural members
	◆ Cracking of substrate due to settlement of foundations
	◆ External vibrations and physical impact.
Continuous cracks in	◆ Non provision of groove in plaster
plaster at junction of walls	◆ Rotation of junction and deflection of roof slab/ slab/beam and wall beam
	◆ Differential movement of structural elements.

Element and Defects	Causes
Cracks at the junction of different substrate elements (e.g. lintels, Columns and embedded pipes)	◆ Different thermal movement of backing materials subjected to high temperature variations (e.g. concrete, metal and timber).
	◆ Different shrinkage due to material properties and time of construction of different elements.
Pitting or blowing in lime plaster	◆ Inadequate slaking and maturing of lime mortar
Flaking	◆ Different thermal movement of backing materials subjected to high temperature variations
	◆ Differential shrinkage due to material properties and time of Construction.
Flaking of finishing coat	◆ Excessive shrinkage of final coat.
	◆ Delay in final coat
Unsoundness and de-bonding	◆ Plaster material not strong enough to resist movement of substrate
	◆ Excessive absorption of water from plaster by porous substrate
	◆ Inadequate key to substrate due to smooth surfaces.
	◆ Differential thermal movement of plaster and substrate.
	◆ Impurities in mortar
	◆ Differential temperatures of substrate and finishes.
Powdering of plaster	◆ Improper setting and hardening of plaster.
	◆ Inadequate curing.
Bulging of ceiling plaster	◆ Excessive deflection/vibration in substrate
	◆ De-bonding due to moisture penetration at the interface of substrate.
Efflorescence	◆ Alternate wetting and drying
	◆ Presence of soluble salts in plaster mix.
Pointing	
Hollowness and erosion of mortar	◆ Improper joints during construction
	◆ Poor workmanship
	◆ Inadequate curing
	◆ Joints not filled with mortar
	◆ Weak/poor mortar mix

Element and Defects	Causes
	◆ Severe climate-wind driven rainstorm, floods
	◆ Debonding of earlier pointing from the substrate
	◆ Freezing and thawing during construction
	◆ Long term ageing of mortar.
Rendering	
Debonding and cracks in rendering	◆ Inadequate bond to wall face
	◆ Weak/poor mortar
	◆ Differential movement of substrate
	◆ Transfer of cracks from substrate
	◆ Shrinkage of rendering mortar
	◆ Contamination resulting in high shrinkage.
Horizontal and Vertical cracks in rendering	◆ Sulphate attack in the joint mortar
	◆ Expansion of mortar in the joints
Efflorescence	◆ Sulphate attack
	◆ Alkali aggregate reaction in background material
Cladding	◆ Inadequate bond/adhesion with base
Displacement and cracking of cladding	◆ Differential movement of substrate
	◆ Shortening of individual beams and columns of RC frame due to shrinkage, creep and loading
	◆ Loss of adhesion due to bending of base structure.

TABLE 5.4: CHIMNEY STACK AND SHAFTS

Element and Defects	Causes
Chimney stacks Cracking, splitting and bending	◆ Condensation of the water vapour in the flue gases resulting in water migrating in to the walls of the stack causing mortar expansion.
Damp penetration	◆ Exposed stacks not weather proof.
	◆ Flues open to roof space/flue opening not covered
	◆ DPC in chimney absent or incorrectly placed.
	◆ Roof flashing incorrectly detailed..
Staining of chimney breast	◆ Flues not vented and condensation leaking from flue in to brick work.
	◆ Room ventilation inadequate.
	◆ High differential temperature inside and outside.

Element and Defects	Causes
	◆ Masonry and flue liner not joined properly.
	◆ Flue terminal unsuitable.
Service Shafts	◆ Design fault-Lack of drainage.
In sanitary condition	◆ Misuse of access
	◆ Lack of suitable access.
	◆ Birds and insect breeding.
	◆ Human factor-throwing of litter.
	◆ In-sanitation and damp condition.

TABLE 5.5: COLUMNS AND BEAMS

Element and Defects	Causes
Columns	
Cracks	◆ Shrinkage
	◆ Differential temperature
	◆ Corrosion
	◆ Excessive loading
	◆ Eccentric loading.
Spalling	◆ Corrosion-Carbonation or Chloride attack
	◆ Physical damage
Crushing of column	◆ Excessive loading beyond compressive strength of concrete
Buckling	◆ Inadequate design
	◆ Excessive slenderness ratio
	◆ Excessive bending/eccentric loading
	◆ Joint rotations
Defective lift joints and Joint failures	◆ Poor workmanship
	◆ Inadequate design
	◆ Excessive beam reaction/shear force
	◆ Bearing failure
	◆ Lateral physical impact
Beams	◆ Shrinkage

Cracks	◆ Shrinkage
	◆ Inadequate design.
Spalling Crushing of bearing	◆ Thermal movements.
Excessive deflection, Lateral bulging	◆ Corrosion of reinforcement.
	◆ Inadequate bearing surface.
	◆ Design or construction failure
	◆ Excessive span/effective depth ratio.
	◆ Excessive loading.
	◆ Loading pattern and load positions.
	◆ Defective workmanship
	◆ Excessive lateral loading unaccounted in design .
	◆ Poor form work
	◆ Excessive vibrations

TABLE 5.6: ROOF AND ROOF TERRACES

Element and Defects	Causes
Flat Roof	
Cracking	◆ Shrinkage in concrete
	◆ Adverse end/support conditions
	◆ Incorrect placement of reinforcement
	◆ Inadequate design
	◆ Improper mix design
	◆ Inadequate compaction and curing
	◆ Delay in providing weather proofing course
	◆ Differential expansion of hidden conduit pipes
Spalling	◆ Freezing and thawing
	◆ Corrosion of reinforcement
	◆ Sulphate attack
Deflection	◆ Over loading.
	◆ Improper design-high span/depth ratio
	◆ Improper construction practices i.e. Poor form work or Excessive cover provided
Crushing of bearing	◆ Inadequate bearing area

Element and Defects	Causes
	◆ Inadequate reinforcement at bearing
	◆ Simply supported bearing against continuous support conditions
Shearing failure	◆ Inadequate design details for shear force
Corner lifting	◆ Inadequate design
	◆ Harsh initial drying
	◆ Excessive bending and deflection
Ponding	◆ Inadequate slope and drainage
	◆ Inadequate design of rain water pipe
	◆ Blockage of the drainage system
Pitched Roof	
Sagging/and deformation	◆ Inadequate design of frame and its joints
	◆ Deterioration in strength of structural members
	◆ Failure and excessive bending of purlins
	◆ Overloading
	◆ Load of the roof causing the spread of the outer walls
	◆ Horizontal movement at feet of rafters
	◆ Bearing wall plates not properly secured to the wall
	◆ Removal of support purlins and internal struts
Sagging of purlins and rafters	◆ Under size members
	◆ Over loading and point loads on roof
Slipping roof tiles and slates	◆ Failure of fixing
	◆ Loosening of fixing due to frost action
	◆ Damage and corrosion of frames and fixtures
	◆ Pollution attack on slates
	◆ Too steep slopes
Joint failures	◆ Failure of welding, riveting, nailing, bolts, etc. due to inappropriate design and poor workmanship
Mould growth, corrosion and rot in roof spaces	◆ Inadequate provision for ventilation
	◆ Leakage and dampness
	◆ Passage of water vapour and gases into the roof space from inside
	◆ Condensation in roof space

Element and Defects	*Causes*

Ceiling Plaster

Cracks
- ◆ Shrinkage
- ◆ Deflection of structural substrate panels
- ◆ Physical impact

Loss of adhesion
- ◆ Weak undercoat
- ◆ Too smooth substrate of roof slab
- ◆ Absence of mechanical key

Cementitious Water Proofing systems

Improper bond
- ◆ Poor surface preparation
- ◆ Improper primer and bonding material

Cracks
- ◆ Transmission of substrate cracks
- ◆ Poor crack bridging capacity
- ◆ Excessive deformation and stresses in substrate

Deterioration
- ◆ Effect of ultra violet(UV) rays.

Liquid Membranes

Blistering
- ◆ Entrapment of air and /or moisture
- ◆ Poor breathing capacity
- ◆ Substrate moist

Deterioration
- ◆ Ageing/effect of UV rays

Factory produced membranes

Folding and Ridging/Rippling
- ◆ Incorrect layout
- ◆ Incorrect use of bonding media(adhesive)
- ◆ Improper storage of membranes

Blistering
- ◆ Expansion of entrapped air or moisture by solar heat

Embrittlement of the membrane
- ◆ Ultra violet radiation
- ◆ Atmospheric oxidation of the bitumen
- ◆ Excessive differential movement between the membrane and the substrate

Loss of adhesion
- ◆ Improper application (during cold/wet or windy day)
- ◆ Damp substrate

Leakage
- ◆ Pressure on blister or folded surface resulting in cracks and Openings.

Element and Defects	Causes
	◆ Inadequate provision to isolate felt from building movement.
	◆ Deterioration/mechanical damage of membrane.
Polymer Roofing	
Leakage	◆ Inadequate bonded joints
	◆ Puncture of polymer sheet
Splitting	◆ Mechanical damage by dropping of stones, nails and tools on Polymers
Ferrocement	
Cracks	◆ Shrinkage (Rich cement mortar)
	◆ Corrosion due to improper cover
	◆ Inadequate curing
	◆ Non provision of bonding agent
Day joints	◆ Poor surface preparation
De-bonding	◆ Inappropriate bonding
Brick bat coba	
Cracks	◆ Improper detailing of construction/ day joints
	◆ Shrinkage of mortar
	◆ Inadequate curing
Cracks along the periphery	◆ Non provision of fillet/Gola.
Lime terracing	
Cracks	◆ Non provision of fillet/Gola at junctions
	◆ Inadequate maturing of lime Dhar
	◆ Inadequate provision of supplementary materials like fenugreak Powder, Gur water and fibers in lime terrace surface
Blisters	◆ Inadequate slacking of lime
Mudphuska	
Opening of surface tile joints	◆ Shrinkage and thermal movements
	◆ Abrasion
	◆ Human activities
Ponding	◆ Improper drainage slope of tile surface
	◆ Improper maturing and compaction of mud
	◆ Inadequate use of supplementary materials like natural fibers and crude oil in base mud mortar

Element and Defects	Causes
Lifting of tiles	◆ Ingress of water below tiles
	◆ De-bonding of tiles with base mud mortar
	◆ Evaporation of locked moisture
	◆ Tile joints not sealed with proper mortar
Deterioration of tiles	◆ Chemical action due to environmental pollution

TABLE 5.7: IPS/MOSAIC/STONE FLOORING

Element and Defects	Causes
Cracks and opening of joints	◆ Thermal movement(Expansion/Contraction)
	◆ Shrinkage
	◆ Mechanical impact
	◆ Poor compaction of base mortar
	◆ Human activities
Breaking of floor separators	◆ Alternative wetting and drying
Slab on grade	
Bulging Cracks	◆ Swelling of soils
	◆ Incorrect specification of mix
	◆ Inadequate thickness of screed
	◆ Deformation of base and sub-base
	◆ Inadequate curing
	◆ Slab subjected to harsh initial drying
	◆ Non provision of contraction and isolation joints
	◆ Large size floor panels
Joint's edge breaking	◆ Deterioration of floor separators /strips
Sinking of floor/ Collapsing	◆ Design failure
	◆ Improper compaction of sub-base
	◆ Settlement of subsoil
Wetting of floor	◆ Leakage in buried pipes
	◆ Improper mix used
	◆ Inadequate curing
Lifting/curling	◆ Sulphate attack
	◆ Floor laid in too large bays
	◆ Poor workmanship

Element and Defects	Causes
	◆ Defective/deteriorated lean sub-base concrete
	◆ Poor bonding with sub-base
Upper Floors	
Deflection in slab	◆ Inadequate design
	◆ Large span/depth ratio
Springy floors	◆ Incorrect joint length.
	◆ Perimeter support omitted.
Unaccepted sound transmission	◆ Insufficient insulating media, inadequate isolation/through floor resilience against impact and sound
Granolithic finish	
Cracking and lifting	◆ Poor bond with base concrete especially when old concrete is not cleaned, hacked and wetted before laying floor finish
	◆ Shrinkage in finish material
Dusting	◆ Unsuitable materials and impurities in the mix such as dust and clay
	◆ Poor workmanship
	◆ Excessive bleeding and rapid drying
	◆ Inadequate curing
Terrazzo	
Cracking and lifting	◆ Poor bond with base
	◆ Omission of movement joints
	◆ Panels laid in large sizes
De-bonding of top layer	◆ Inadequate substrate preparation and poor mechanical key
Spalling at joints	◆ Abrasion
	◆ Improper level at joints
	◆ Corner lifting
	◆ Concrete disintegration
	◆ Due to weathering
Disintegration of floor separators	◆ Abrasion
	◆ Poor quality material
	◆ Ageing

Element and Defects	*Causes*

Linoleum

Irregular surface
- Inadequate preparation of base slab
- Poor workmanship in laying

Flexible PVC

Lifting of edges and
Brown stains on surface
- Alkaline moistures from screed attacking adhesive surface bubbles at edges
- Entrapped air or moisture bubbles in primer or bonding coat
- Tar from adhesive diffusing through micro holes and cracks in PVC

De-bonding
- Poor surface preparation
- Improper use of bonding material

Scratching and pitting
- Movement of sharp metallic parts
- Pointed loads on the surface

Scuff marks on surface
- Black rubber in foot wear, caster wheels/ tyres

Rubber

Surface deterioration
- Contact with oil fats and greases
- Mechanical deterioration
- Weathering effect

Cement concrete/ Terrazzo/ Stone/ Ceramic tiles

Arching and lifting
- Drying shrinkage of base screed
- Expansion of tiles due to moisture absorption
- Differential thermal movement of concrete base and tiles

De bonding
- Poor surface preparation
- Improper application of bonding material

Surface deterioration
- Attack of solvent based cleaners and chemicals.
- Physical wear and tear.

Wood tiles

Staining and lifting of tiles
- Rising dampness from floors causing expansion of tiles
- Omission of expansion strip around perimeter
- Use of inadequately seasoned tiles
- Unsuitable moisture content and shrinkage

Element and Defects	Causes
Industrial Concrete floors	
Blistering	◆ Finishing started before the base concrete has stiffened
	◆ Floating and trowelling far too soon
	◆ The surface was sealed too early
	◆ Entrapped air and water continue to rise during finishing
	◆ Formation of voids between the surface mortar and the base concrete due to water and air
	◆ Improper use of surface hardeners
Crazing	◆ Shrinkage of the cement paste on the surface caused due to too high or too low w/c ratio in the mix
	◆ Excessive bleeding and rapid drying
Curling and wavy surface	◆ Top concrete cools while the lower layer remains warm
	◆ Rapid shrinkage at top than at bottom due to different rate of drying
	◆ The concrete beneath the surface not allowed to harden before finishing
	◆ Rapid drying due to Sun, Wind or low humidity
	◆ Cold sub-grade or use of retarder delaying setting below the surface
	◆ Bulging of cracks in base concrete as the power trowel moves across the slab
	◆ Non uniform concrete surge ahead of the screed
	◆ Addition of extra water at the concreting site to increase slump
	◆ Change of slump between lots
Dusting	◆ Excessive presence of free lime in concrete matrix
	◆ Chemical action
	◆ Concrete surface dries without hardening
Disintegration	◆ Abrasion
	◆ Impact damage
	◆ Freezing and thawing
Plastic shrinkage cracking	◆ Rapid evaporation of water from the surface due to

Element and Defects	Causes
	▶ Low humidity
	▶ High wind velocity
	▶ High concrete temperature
	▶ Moderate to high air temperature
Uneven colour	◆ Variation of Mix, Slump, Finishing and Curing
	◆ Excessive bleeding producing lighter surface
	◆ Low water cement ratio producing dark surface
	◆ Change in cement brand
	◆ Early troweling on wet surface
	◆ A troweled surface or a broomed surface
	◆ Plastic sheets in contact leave a mottled appearance
Scaling	◆ Air entrained concrete is not used.
	◆ Poor quality of finish due to high w/c ratio
	◆ Too soon and too much troweling
	◆ Curing not done properly
Uncontrolled Shrinkage cracking	◆ Shrinkage during drying and cooling causes uncontrolled cracks especially in large panels
Indentation	◆ Point loads applied to the finish for an appreciable period
Slipperiness	◆ Excessive presence or unsuitable polish
Uneven wear	◆ Excessive traffic such as in aisle
	◆ Poor workmanship
Popouts	◆ Internal pressure from absorptive particles near surface
	◆ Excessive presence of soluble salts
Settlement	◆ Inadequate soil compaction under the substrate

TABLE 5.8: JOINERY WORK

Element and Defects	Causes
Frames	
Surface damage/deterioration	◆ Inadequate protection to timber
	◆ Use of inappropriate grade of timber
	◆ Moisture ingress into timber

Elements and Defects	Causes
Distortion	◆ Varying moisture content causing expansion and contraction
	◆ Opening up of partly formed joints/poor workmanship
	◆ Extreme temperature and humidity
	◆ Inadequate lintel design(lintel load transferred on frame)
	◆ Improper fixing with adjoining masonry
	◆ Corrosion of hold fast/anchors
	◆ Alternate wetting and drying
Cracking of glass panels	◆ Corrosion of steel screws, clips and frame
	◆ Excessive deflection of lintel causing distortion of frame
Loosening/Deterioration	◆ Excessive pressure on joint fixtures-Jointing failures
of individual members	◆ Poor workmanship
	◆ Excessive moisture content in wooden members
Cracking and softening	◆ Dry and wet rot
	◆ Nailing or physical impact
Deterioration	◆ Non provision of protective coating especially on the face in contact with masonry
	◆ Dry and wet rot in timber
	◆ Insect/termite attack
	◆ Corrosion on steel frames
	◆ Bimetallic action on aluminium frame exposed to environment.
Windows and doors	
Door sticking	◆ Warping of thin or insufficiently seasoned wood
	◆ Inadequate hinge support or loosening of hinges
Pin holes	◆ Beetle infestation and wet/ dry rot
	◆ Damp penetration under sill
	◆ Omission of DPC under sill
Decayed window boards	◆ Condensation
De-lamination of plywood panels	◆ Inadequate protection against moisture
	◆ Inappropriate specifications

TABLE 5.9: DECORATIVE/PROTECTIVE FINISHES

Elements and Defects	Causes
Paint films	
Blistering	◆ Entrapped moisture forces the paint into little bubbles or blisters
Bleeding	◆ The resinous deposits of the sub-surface coming to the surface of paint cause bleeding
Blooming	◆ Mist haze or milkiness due to presence of moisture or chilling of surface glassy coat
Chalking	◆ Powdering of the surface paint due to external exposure and lack of paint binder
Crazing	◆ Irregular cracking of surface due to use of hard drying paint with thick consistency
Curtaining and sagging	◆ Uneven application of paint
	◆ Too much application of paint
Flaking /peeling and cissing	◆ Lifting up and peeling away of the paint due to loss of adhesion
	◆ Incomplete preparation of substrate
	◆ Incomplete application of primer coat
Grinning	◆ Surface not cleaned
	◆ Paint partly applied
	◆ Poor capacity of the paint film
Gilling	◆ Formation of unstable gelly due to unstable thinners or long storage
Loss of gloss	◆ Porous under coat or moisture on the surface
	◆ Alkaline material left on wood
Mould	◆ Damp conditions
Uneven drying	◆ Surface not properly cleaned
	◆ Residues of paint removers
	◆ Finish applied before undercoat dried
Resin coming through Sticky film	◆ Knots not properly treated
	◆ Very resinous knots not cut out
	◆ Akali attack by salts in backing

TABLE 5.10: SERVICES

Material/Defects	Probable Causes
Water pipe	
Leaking pipes	◆ Corrosion
	◆ External damage due to vibrations
	◆ Freezing and thawing
Leaking joints	◆ Wearing out of gland
	◆ External damage
	◆ Poor installation
	◆ Bimetallic corrosion when dissimilar metals used
	◆ Chemical action of fluxes
	◆ Thermal expansion/contraction due to hot and cold water circulation
Water hammer	◆ Sudden increase in pressure in the System
	◆ Water hammer in cold pipes caused by loose washer plates in valves
	◆ 'Knocking' in hot water pipes caused by sudden obstructions in the flow in return pipes
	◆ Sudden closing of taps
Rupture of pipe	◆ Freezing of water in pipes resulting in volume increase leading to expansion and opening of joints
Taps and Stopcocks	
Water leaks around top of spindle	◆ Faulty gland packing
	◆ Defective handles or stem
	◆ The washer has deteriorated or broken
	◆ The seat is worn and pitted
	◆ Loosening of washer from the seat
Drips when closed or	◆ Faulty washer or worn valve seating
vibrates when opened	◆ Threads of spindle worn out
Ball Valves	
Discharge from overflow	◆ Valve fails to close properly due to faulty washer, or eroded seating or presence of grit or lime deposits
Sticking in closed position	◆ Usually follows a period of no use during which dirt and lime dry out in the working parts
Inadequate flow	◆ Use of high pressure instead of low pressure valve
Hissing noise	◆ High pressure feed

Material/Defects	Probable Causes
Storage Cistern	
Overflow or water flowing continuously	◆ Plunger washer worn out
	◆ Float valve broken or deteriorated
	◆ Float valve defective
	◆ Incorrect float adjustment
	◆ Flush valve defective
	◆ Deformation of rubber valve or valve not seating properly
No flushing	◆ Bell/syphon type arrangement out of alignment
	◆ Operating lever defective
Cistern fails to fill	◆ Defective float valve(Plunger jam)
Rust pitting of galvanised steel cisterns	◆ Electrolytic action due to iron filling from drilling holes left in tank or deposition of copper from water into sides of tank
Corrosion of cistern fittings	◆ Bimetallic action with fittings of brass or other copper bearing alloys
Distortion of PVC cistern	◆ Inadequate base support
Drainage Waste Pipe	
Softening and distortion	◆ Ageing
	◆ Reaction with chemical cleaners
Blocking	◆ Internal corrosion
	◆ Household waste
Traps	
Leakage	◆ Faulty plumbing
	◆ Body crack
Clogging	◆ Accumulation of grease, dirt and other materials i.e. hair and lint
Foul smell	◆ Water seal broken due to evaporation and drying of water seal
Underground Pipes	
Blockage	◆ Poor layout (pipe runs not in straight line between access points or laid with insufficient slope to clear soil)
	◆ Drain fractured by settlement, heavy traffic or tree roots

Material/Defects	Probable Causes
	◆ Accumulation of builders' rubbish during construction
	◆ Disposal of unsuitable materials through drains
	◆ Accumulation of grease from kitchen waste
Failure under tests	◆ Defective joints at:
	▸ Base of soil pipe
	▸ Entry to manhole
	▸ Intermediate points
	◆ Drain fractured.
	◆ Cracked flaunting in manholes
Tanks Overhead/Underground	
Overflow/no flow	◆ Float not functioning
Leakage	◆ Faulty plumbing
	◆ Cracks in tank body
	◆ Alternate wetting and drying
	◆ Faulty design and rebar placement
Septic Tanks	
Floating sludge	◆ Sludge decomposition due to long travel
Foul smell	◆ Septicisation of sewage due to long travel time in sewer line
	◆ Presence of sewage solids and plant growths
Froathing	◆ Presence of sewage solids and plant growth detergents used in the household washing causes frothing
	◆ Bathroom water mixed with w.c. output

TABLE 5.11: MATERIALS

Material/Defects	Probable Causes
Timber	
Deformation	◆ Seasonal moisture movement of timber at the end
Rot in structural timber	◆ Dry rot due to ingress of water through defective coverings or condensation
	◆ Wet rot when moisture content of timber is more than 30%
Decay/deterioration	◆ Insect/termite attack
	◆ Continuous dampness

Material/Defects	Probable Causes
Bricks	
White powdering deposit	♦ Efflorescence caused by crystalisation of soluble salts in brick or Mortar
	♦ Sulphate attack
Staining of bricks	♦ Efflorescence
	♦ Dampness
	♦ Absorption of chemical salts
Decay(Disintegration)	♦ Crystallisation of salts formed below the surface in the brick Crevices
	♦ surface erosion known as Cryptoflorescence. When Cryptoflorescence continues indefinitely it leads to thedecay of bricks
Stone	
Staining (organic)	♦ Presence of alkali in mortar, cement and hydrated lime
Staining (metallic)	♦ Metals attached to stone.
Pitting and staining of limestone	♦ Continuous dampness
Decay of sandstone	♦ Continuous dampness
Delamination of surface layers	♦ Penetration of rain water with subsequent disruption of surface on Freezing
	♦ Sulphate attack from soluble salts in backing or bedding mortar
Erosion of surface	♦ Weathering effect (stone soluble in rain water containing carbon Dioxide and sulphur dioxide)
Contour scaling	♦ Repeated alternate wetting and drying of the stone which leads to differential conditions in outer and inner layers
Blistering of surface	
Concrete	♦ Atmospheric pollution-sulphur dioxide absorbed in the rain water to form a weak acid solution which attacks the natural binding medium of the stone
Cracking	♦ Chemical reaction
	♦ Shrinkage, thermal expansion and contraction
	♦ Overloading or overstressing
	♦ Corrosion
Popout	♦ Excessive presence of soluble salts.

Material/Defects	Probable Causes
Permeability	◆ Poor mix design, higher W/C ratio and improper curing
Weathering of concrete	◆ Atmospheric pollution rains and acid rains
	◆ Physical abrasion
	◆ Chemical attack
	◆ Solar radiation
Carbonation	◆ Environment pollution
	◆ Porous concrete
	◆ Non provision of sealer coat
Discoloration	◆ Environmental pollution
	◆ Wind driven rains
Surface Damage	◆ Concrete surface can be damaged by
	▸ Mechanical impact or abrasion
	▸ Unusual condition of exposure
	▸ Improper design or construction
	▸ Unsuitable materials
Deterioration	◆ Sulphate attack
	◆ Presence of destructive bacteria
	◆ Presence of waste water pollutants
	◆ Absence of protective lining material
Honey combing	◆ Insecure form work which allows the leakage or bleeding of Plastic concrete, especially the cement content
	◆ Improper compaction and insufficient consolidation
	◆ Over compaction or excess vibration
Spalling	◆ Frost damage
	◆ Fire damage
	◆ Corrosion
	◆ Sulphate attack
Dusting/delamination	◆ Build up of water and air beneath a dense layer of surface Mortar resulting in de-lamination
Disintegration	◆ Chemical reaction.
	◆ Sulfhate attack.
	◆ Faulty mix design.

Material/Defects	Probable Causes
	◆ Impurities in materials.
	◆ Incorrect construction practices.

R.C.C
Failure of cover concrete
	◆ Cover concrete weak and porous-unable to resist expansion forces caused by
	  ▸ Corrosion of rebars.
	  ▸ Differential thermal expansion.
	  ▸ Moisture ingress
	  ▸ Frost action.
	◆ Exposure to atmosphere.
	◆ Exposed to moist polluted air

Mild Steel Rusting/Corrosion/Pitting	◆ Stress corrosion ◆ Attack by flue gases containing sulphur dioxide ◆ Contact of carbon with copper
High Strength Steel Corrosion	◆ Rain water draining from copper covered roofs ◆ The presence of copper together with damp conditions caused by heavy and repeated condensation leading to serious corrosion
Copper Pitting Corrosion	◆ Unprotected aluminium embedded in cementmortar or concrete exposed to salt spray will also leading to corrosion
Aluminum Pitting Corrosion	◆ Anodised layer disfigured by splashes of cement or lime
Bimetallic corrosion	◆ Attack from Lead ◆ Contact with Copper/Steel and presence of dampness
Lead Corrosion	◆ Attack by free alkali present in concrete/mortar
Zinc Corrosion	◆ Exposure to sulphur gases and compounds
Asbestos Sheets Cracking and breaking	◆ Subjected to more pressure due to overloading ◆ Impact damage ◆ Damage during erection

Material/Defects	Probable Causes
Deterioration	◆ Moss and lichen growth cause surface deterioration
	◆ Atmospheric pollution
Galvanized Iron Sheets	
Deterioration	◆ Corrosion of basic material
	◆ Ageing and atmospheric pollution
	◆ Corrosion of nails and wind action
Thatch	
Deterioration of thatch	◆ Ageing and atmospheric pollution
	◆ Heavy rains and winds
	◆ Rat eating and bird habitation
	◆ Termite attack
Ripping off	◆ Heavy wind/storm
	◆ Improper securing of thatch
Fibre Reinforced Plastics Sheets	
Cracks near J bolts	◆ Less tearing strengh
	◆ Improper fixing
Deterioration	◆ Ageing and effect of Ultra Violet effect
Glass	
Cracking/breakage	◆ Vandalism
	◆ Thermal stresses (Different coefficient of thermal expansion of glass and frame material)

TABLE 5.12: DAMPNESS IN BUILDINGS

Material/Defects	Probable Causes
Foundation	
Damp patches near plinth below DPC visible on external walls during rainy season	◆ Absence/failure of vertical water proofing system around foundation
	◆ Collection of water adjoining the foundation areas and impervious soil strata
	◆ Failure water proofing qualities of foundations
Basements	
Dampness is visible on walls and floors; Water may be leaking or seeping through walls or floors, or may be collecting in low level areas in severe cases	◆ Penetration of water under pressure through construction joints/Cracks/Porous Structures
	◆ Moisture penetration from subsoil through cracked basement walls/ raft slab
	◆ Inadequate water proofing system

Material/Defects	*Probable Causes*

	◆ Failure of water proofing system due to
	  (i) deterioration with age
	  (ii) lack of continuity of water proofing system e.g. under stanchions and other structural members which may penetrate the water proofing system
	  (iii) damage occurring during construction
	◆ Non provision of perforated pipe drainage system around the perimeter of the basement
	◆ Condensation. This is likely to arise when heating is used intermittently without adequate ventilation

Walls

Dampness at or near ground level

Semi permanent dampness seen on wall surfaces from ground level up to approximately 750 mm or even higher in severe cases; Decorations may be damp, blistered and discolored or dried out and pushed off the wall; There may be rot in floor and skirting boards adjacent to the deteriorated wall.	◆ This dampness is due to one or more of the following:
	  (i) Lack of a damp proof course
	  (ii) Bridging of the damp proof course
	  (iii) Failure of the damp proof course

External Solid Walls

Visible damp stains and mould growth on decorative internal face	◆ The wall cannot resist direct penetration of driving rain of high intensity ◆ Deteriorating brickwork joints or rendering. ◆ Defective gutter or rain water pipe ◆ Carelessly made construction and lift joints ◆ Loss of cement paste due to poorly formed joints resulting in honey combing and continuous capillary paths

Cavity walls

Dampness on internal decorative surface near DPC, or lintel level	◆ Excess mortar build up at various joints in the Cavity walls. ◆ Wall ties wrongly placed with slope down towards inner wall ◆ Cavity bridged by injected foam ◆ Absence or damage of DPC over lintel

Internal walls

Patchy damp areas on plaster accompanied by peeling of wall paper, discoloration or, surface stains or mould growth.	◆ Leaking or back filling of overflow pipes ◆ Leaking pipes embedded in the wall causing frost damage or corrosion ◆ Localised chemical action ◆ Localised condensation over buried rising main
Patches of dampness particularly in kitchens and cold bathrooms.	◆ Condensation of water vapour from air in contact with cold surfaces

Material/Defects	Probable Causes
Moisture beads appear on the surface of dense materials	◆ Condensation of vapour within the body of the wall ◆ Residual/construction water
Openings in walls	
Damp patch over head of opening	◆ Inadequate design (Drip/throating missing or partly constructed). ◆ Absence of gola/fillet at the intersection of wall and linke ◆ Omission of cavity tray over opening in cavity wall or tray too short causing dampness in corners
Damp patches around window/door frames	◆ Omission of vertical DPC at junction of inner and outer leaves of cavity walls or window/door frame
Damp stain at sill level	◆ Condensation running down on inner face of glass ◆ Gap between frame and glass ◆ Defective sealant incapable of accommodating differential movement of different materials
Chimney	
Dampness in upper part of stack	◆ Defective damp proofing of stack and gutter ◆ Absence or defective damp proof course or flashing in the chimney stacks ◆ Defective chimney and/ or bedding of the stack allowing penetration of water
Dampness in chimney breast or flue and discolouration of decoration	◆ Chemical action due to flue gases ◆ Condensation in unlined flue
Innerside below parapets Dampness visible on the internal plaster of the upper parts of external walls. Initially dampness appears at the wall/ ceiling junction and then spreads downwards, pronounced after heavy rain or snow fall.	◆ Absence of defective damp proofing in the parapet ◆ Split at the junction of the roof finish and parapet due to structural deformations ◆ Condensation ◆ Cracked, leaking or blocked gutter or spout behind parapet
Underside of flat roofs Damp patches visible after rain and in winters around the perimeter of the roof.	◆ Condensation of water vapour passing from the room below the ceiling into the structural roof slab ◆ Residual construction water entrapped during roof laying which drips slowly ◆ Direct rain penetration travelling horizontally between roof slab and felt ◆ Water dripping from cracks in ceiling or around electric fittings
Pitched Roofs Damp patches on false ceiling seen after rain or snow fall	◆ Inadequate overlaps of sheets or nailed through troughs instead of crowns of corrugations ◆ Defective tiles or slates

Material/Defects	*Probable Causes*
	◆ Defective battens affected by woodworm or fungal attack
	◆ Defective parapet or valley gutters
	◆ Leaking water tank
Solid Ground Floors Persistently damp floor except in very dry weather. Mould growth on the underside of carpet.	◆ Absence or ineffective damp proof membrane. ◆ Base concrete and floor material porous/permeable ◆ Carpeting or linoleum may retain base dampness
Dampness restricted to perimeter of floor	◆ Non linking of the floor damp proof membrane and wall DPC
Ceiling and Floors Damp patches on ceilings or floors.	◆ Leak from broken appliances ◆ Leaks from defective joints and pipes of services ◆ Leaks due to inadequate overlaps in case of pitched roofs
External Joinery (Door/windows) Water penetration from external walls	◆ Worn threshold or absence of water bar and doors weather fillet by the sides of the frame

Note: Dampness resulting from defective appliances, joints and pipes may well show itself some distance away from the source since small leaks will run down along pipes and ductwork before dropping onto other surfaces and becoming apparent.

5.4 SUMMARY

Defects and failures reflect status of performance efficiency and serviceability of building structures and its maintenance. Most of the building defects are avoided by effective repair and maintenance. Effective repair and maintenance requires analysis of defects for the correct diagnosis of root causes. Correct identification of root causes requires detailed information to be collected by observation of specific problem and site testing.

Defects may be grouped for various elements or according to characteristics of defects. Correct diagnosis of defects is essential before attempting effective repair and maintenance.

Defects are basically caused due to water movement and environmental factors. Cause of dampness may be due to construction, intruding, condensation and occupational water. The source and route of water in building element must be determined as accurately as possible for planning effective repair and maintenance. Most of the damages are caused due to combination of various factors.

Different tables may be referred for probable defects and their root causes.

QUESTIONS

5.1 Differentiate between building defects and failures.
5.2 Describe classification of defects based on characteristic approach.
5.3 List 3 main causes of building defects.

5.4   Explain importance of diagnosis of defects.

5.5   Explain different type of waters causing maintenance problems.

5.6   List factors causing dampness movement in building elements.

5.7   List environmental factors affecting building elements.

5.8   Describe not more than 10 most probable causes of defects in:

   (a) Building foundations

   (b) DPC

   (c) Building walls

   (d) Wall plaster

   (e) Wall pointing and rendering

   (f) Chimney stacks

   (g) RCC column and beams

   (h) Flat RCC roof slab

   (i) Damp proofing system of roof terraces

   (j) Finishes in floor slab on grade

   (k) Concrete terrazzo floors

   (l) Wooden joinery work

   (m) Paint films

   (n) Domestic water supply lines

   (o) Domestic sanitary lines

   (p) Rainwater drainage

   (q) Septic tanks

5.9   Describe not more than five most probable causes for defects in:

   (a) Bricks

   (b) Stones

   (c) Timber

   (d) Concrete (plain)

   (e) RCC

   (f) Plastics

   (g) Aluminium

5.10 Describe not more than 5 most probable causes for dampness in:

   (a) Foundations and basements

   (b) Walls

   (c) Flat roof slabs

   (d) Floors on ground level

UNIT–II

COMMON MATERIALS AND TECHNIQUES FOR REPAIR AND MAINTENANCE

UNIT-II

Chapter 6

MATERIAL FOR REPAIR, MAINTENANCE AND PROTECTION

Single Component Elastomeric Membrane Coatings

6

MATERIALS FOR REPAIR, MAINTENANCE AND PROTECTION

LEARNING OBJECTIVES

After studying this chapter, the learner understands the basic characteristics of modern repair materials and will be able to:

- **State** importance of basic characteristics of repair materials;
- **List** factors affecting durability of repair and construction materials;
- **Explain** compatibility aspects of repair and construction materials;
- **List** type of repair materials;
- **State** characteristics of

  - Anticorrosive coatings,
  - Bonding/adhesive agents,
  - Repair mortars,
  - Curing compounds,
  - Sealants,
  - Grouts,
  - Waterproofing materials,
  - Special concretes,
  - Protective coatings,

- **Explain** selection criteria of materials for a specific repair job;
- **List** the common commercial repair materials.

6.1 INTRODUCTION

Deterioration of building/structures is not uncommon in India. The building/ structure once built is subjected to resist weathering action and gradually looses its serviceability. Generally deterioration takes place due to failure in ensuring proper repair and maintenance. Most of structures today are constructed by using cement in one or the other way. Due to mismatching of multiple service requirements along with adverse climatic conditions and numerous specifications, the long-term durability of the structure gets affected. Factors like environmental pollution, thermal stresses, chemical and biological effects tend to lower the life of a structure. The important factor that has been hastening the deterioration phenomenon specially in India is the **lack of maintenance culture**. The dominant philosophy amongst a vast majority of construction industry was to construct and forget. This philosophy has inflicted a severe damage upon the structures and resulted in heavy national losses. Economic considerations and time constraints have made it obligatory to go for rehabilitation and repair of damaged structures rather than their replacement or total reconstruction wherever possible.

The repair materials are intended to protect and maintain **original form and serviceability** for a longer period. Selection and use of correct repair materials at proper time brings maximum economic benefit by enhancing serviceable life. Absence of proper repair and maintenance leads to total unserviceability necessitating demolition and reconstruction.

The need to repair is dictated by the severity and root causes of the deterioration as determined from the diagnosis. The success of the repair of damage will depend upon the correct assessment of the root cause, the right choice of repair materials and the quality of workmanship in its execution. The damages in many cases can be effectively repaired and rectified with the use of new materials in conformity with modern construction technology.

Although liquid plastics like epoxy, polyurethane based materials are expensive but if applied properly, solve some of the repair problems even with considerable economy. Repair and rectification of damages with liquid plastics are expensive and impractical in certain cases. Now a days, many ready to use polymer based and chemical materials are available for **preventing deterioration** and for carrying out repair of damaged structures effectively. Chemical materials can be used to provide a protective coating to safeguard structures from adverse environment. The correct selection of such materials for repair is of great importance.

6.2 DURABILITY

Thousands of buildings are constructed every year. Though initially these buildings are built stronger and carefully, but maintenance is necessary to sustain serviceability. Serviceability depends on durability. Durability of building structures is one of the greatest challenges that building technologists face today. One ton of steel is lost every 14 seconds due to corrosion alone in RCC in India. **Durability** is the property of a structure to give a satisfactory performance and service for the design life with minimum maintenance. American Concrete Institute (ACI) defines durability of concrete as its **ability to resist**

weathering, chemical attack, abrasion or any other **process of deterioration** under service conditions.

Normally, concrete provides adequate protection to embedded steel, provided there is enough cover to steel and the concrete is also sufficiently impermeable. Durability of concrete or RCC is affected by its strength, and other properties such as air content, permeability, water tightness, and volumetric stability. Type of concrete materials and their proportions, construction practices and exposure conditions also affect durability of the structures.

6.2.1 Factors Affecting the Durability

(a) **Physical and mechanical:** These comprise of internal and external factors.

   (i) **Internal**
   - Compaction
   - Porosity and permeability of Material
   - Surface finish
   - Duration of curing
   - Cover to the reinforcement and
   - Surface cracking

   (ii) **External**
   - Abrasion and erosion (wear and tear)
   - Exposure to moisture wetting and drying, and
   - Freezing and thawing

(b) **Chemical**
   - Chemical composition of Aggregate
   - Cement and water used
   - Exposure to Acid
   - Sulphate attack
   - Internal Aggregate alkaline reaction
   - Other chemical aggressive actions

(c) **Biological and environmental factors**
   - Biological growth, such as moss, algae.

(d) **Non structural/Construction workmanship**
   - Inadequate Drainage
   - Inappropriate Joints (element inter-connections, construction, expansion)
   - Inserts , Bearings ,Railings, Anchors and Fixtures

There is certain degree of interdependence of the above factors and deterioration occurs as a result of combination of these factors.

6.3 COMPATIBILITY

The inadequate performance of structures built over the past few decades focused the attention of researchers to look for newer modified materials and construction methods that can withstand the present day industrial environmental forces reasonably over the entire life span of the structure.

Compatibility means consistent with and capable of harmonious union. **Compatibility** refers to consistent structural behaviour of repaired element or material in union with the original material. They undergo deformations as monolithic unit without bond failure under variety of forces and weather conditions. Compatibility of repair materials with original construction materials plays very important role in maintenance of serviceability of structure.

Replacement of all or even a few of the already existing constructed elements of a distressed structure is near impossible, due to the considerable expenses in terms of the capital cost, manpower as well as time and safety involved.

This made engineering community to look for specific materials and methods of repair to keep the structures in operation. Thus the specialised field related to the repair and maintenance comprise of the following aspects:

(a) Understand the behaviour of the existing elements and basic materials (Deterioration and corrosion aspects)

(b) Improve resistance of existing construction materials; (Prevention of ingress of environmental forces)

(c) Develop newer basic construction materials for environmental compatibility

(d) Formulate newer repair and maintenance materials with compatibility of the existing materials

(e) Formulate special reinforcing processes for compatible repair
(Fibrous materials, plate liners, latexes, welded-mesh etc.)

(f) Formulate newer bonding agents and methods for stable repair
(Polymers, plate laminates, stitching etc.)

There are several materials available for the repair and maintenance of concrete structures. It is often seen that some of the repair strategies do not perform as expected. The primary reason for such a performance is probably due to inadeequate investigation· of the compatibility requirements of these repair materials. The main thrust in repair and maintenance process should be to achieve the required **integral action** of the old and new materials and structural forms.

6.3.1 Compatibility of Concrete Repair Materials

In order to design structural repair concrete or even to specify the characteristics of the repair materials which satisfy the requirements, we need to study the **material-environment** interaction. Depending on physical and chemical characteristics of the environment the repair concrete should be designed. We need to consider the following **parameters**.

(i) **Strength:** The design of a concrete structure is generally based on an assumed 28 days unconfined compressive strength. Though there are many other parameters like the tensile strength, modulus of elasticity and stress-strain relation, the compressive strength alone is taken as the main design criteria. In most cases of overlays or rebuilding the following important parameters are assessed :

● Strengths-for compatibility
● Heat of hydration - to reduce the micro cracking
● Shrinkage - both during set and while drying
● Modulus - to achieve compatible deformations
● Fatigue - to sustain the integrity over the service life
● Creep - under operating loads and temperatures
● Plasticisability - to define the compactability

(ii) **Porosity:** Porosity, pore size, pore distribution, and pore volume are some of the parameters that are directly responsible for many of the deterioration and corrosion problems in reinforced concrete. The determination of these is not only complex but also suffers from substantial variation in the matrix. In concrete most investigators preferred to establish the over all behaviour through parameters like permeability, diffusion (chloride, sulphate, oxygen), pH and their influence on other parameters. The characteristics of repair materials are influenced by the porosity parameter as under:

● Porosity -to allow breathing
● Impermeability-to restrain water permeation or moisture movement
● Carbonation -to contain the alkalinity reduction
● Diffusivity-to restrain chloride and oxygen ingress

(iii) **Durability:** The durability of concrete, particularly in marine environment is influenced by many physicsal and chemical parameters. The presence of sulphates in sea water can cause extensive damage to concrete due to chemical attack. Earlier researchers have pointed out that there are apparently two chemical reactions involved in sulphate attack on concrete. First relating to conversion of calcium hydroxide (liberated during the hydration of cement) to form calcium sulphate (gypsum) leading to softening and spalling, second relating to the formation of calcium sulphoaluminate (ettringite) due to the reaction of gypsum and hydrated tricalcium aluminate, leading to expansion and cracking.

Another consideration in evaluating the behaviour of concrete is the assessment of durability over a very long period of time . Further accelerated laboratory test results have to be validated using field results. The durability of concrete is directly influenced by porosity parameters (porosity, permeability, carbonation, diffusivity) apart from the following considerations.

- Chemical resistance to agents like sulphates and acids
- Microstructure to understand the chemical reactions.
- Compatibility to chemicals and mineral admixtures
- Filler effects to define the pore structure
- Pozzolanic effects-to define chemical reactivity

(iv) **Corrosion:** The corrosion effects are obvious only after a long period of exposure. Field exposure studies along with a number of accelerated studies in the laboratory have also been undertaken. The parameters like resistivity, diffusion of chlorides and sulphates and variation of pH have influence on corrosion potential and corrosion rate.

(v) **Other Parameters and Aspects:** It is to be recognised that the above overview is in no way comprehensive. The parameters of relevance for a comprehensive understanding of any material are far too many and will be decided by the experience of application and environment in question. This problem of understanding the behaviour of material gets aggravated in the case of repair materials as they have to perform in combination with the parent material. The primary aim should be to obtain a synergy in the system rather than using a repair material that was the best individually without considering the interaction effects.

6.4 TYPES OF REPAIR MATERIALS

The basic materials of construction (such as steel, cement, concrete, bricks, stones, timber, paints, etc.) can all be used for repair work. The present text describes mainly repair materials other than basic construction materials.

Depending upon requirements, properties and use, the repair materials can be broadly categorized into following groups:

- (i) Anti corrosion coatings
- (ii) Adhesives/Bonding Aids
- (iii) Repair Mortars
- (iv) Curing compounds
- (v) Joints Sealants
- (vi) Grouts
- (vii) Waterproofing Systems for Roofs
- (viii) Special Concretes
- (ix) Migratory Corrosion Inhabitor
- (x) Protective coatings.
- (xi) Materials for special applications

6.5 CHARACTERISTICS AND PROPERTIES

The basic principle of repair is that the repair medium should be as close as possible in all physical characteristics (Elastic modulus, co-efficient of expansion, strength), to the base material. This is aimed at ensuring that the properties of the old and new work are similar to facilitate maintaining a good bond by limiting the boundry stresses.

A good repair material should have the best combination of the following properties:

- Mechanical properties as close to the base materials as possible
- Good adhesion in dry, damp or wet conditions
- Low shrinkage (during curing and long term)

It is not always possible to have all these properties in one material, therefore, there are differing solutions which dictate selection of repair materials from each of the three basic groups.

Specifications for repair materials are becoming increasingly sophisticated and comprehensive. To ensure durability of repaired elements, repair materials must exhibit specific properties as described subsequently.

6.5.1 Anti Corrosive Coatings (Rebar Primer)

Anticorrosive coatings are provided to steel reinforcement and structural steel to enhance resistance against corrosion. The ideal characteristics for a rebar primer are:

- Resistance to ingress of moisture by film formaion
- Resistence to under cutting (i.e. progressive rust creep under the primer)
- Good bond to steel and subsequent repair layers
- Resist adverse effect on the adjacent steel
- Easy to apply

Following are examples of rebar primers in use today :

 (i) Cement mortar slurry
 (ii) Polymer modified Cement slurry
(iii) Non passivating epoxy
(iv) Passivating epoxy
 (v) Zinc rich epoxy
(vi) IPN (Interpenetrating Polymer Network System) coating

Interpenetrating Polymer Network System(IPN) are relatively novel types of polymer alloys consisting of two or more polymers in net-work form. One of the polymer synthesised and/or cross linked in the immediate presence of the other. IPN have better mechanical properties like Tensile Strength, Flexibility, Hardness, Shear Strength, Chemical and Weather resistance in cmparison to the independent polymers.The morphology studies conducted by C.B.R.I. Roorkee using scanning electron microscopy and differential scanning calorimetry have proved the efficacy of **IPN Polymer** System as primers.

Lot of studies revealed that **zinc rich epoxy** gives the best anti corrosion quality because of cathodic protection offered to the reinforcement and adjacent concrete.

6.5.2 Bonding Aids

Application to conventional concrete, sprayed concrete or sand - cement repair mortars, bond is often a problem. When the repairs are to be carried out at a high temperature, water loss at the interface between the repair material and the prepared concrete element surface may prevent proper hydration of hydraulic matrices along the interface resulting in poor bond. Hence, under such conditions, it is necessary to provide a bonding coat prior to the application of repair material.

The major advantage of **adhesive bonding** is that it allows distribution of an applied load over much larger area thus reducing the unit stress on the repaired elements. It allows adhesion without changing the shape of the element to be attached. The adhesive bond line also acts as moisture barrier. There are two main categories of adhesives:

(a) **Solvent free** adhesives and

(b) **Water borne** adhesive (e.g. latex and latex powder)

(a) Solvent Free Adhesives (e.g. Polymeric adhesives)

Solvent free adhesives cure by polymerisation of monomeric resins. Solvent free adhesives are Epoxies, Polyesters, Acrylics, Polysulphides, Polyurethanes, and Silicones .

(i) **Epoxy Adhesives:** Epoxy adhesives are generally composed of an epoxy resin, an amine or polyamide curing agent, reactive diluents and, in some cases, inorganic fillers and thixotroping agents. These are the most commonly used polymeric adhesives. These are very tolerant of the alkalinity of concrete.

(ii) **Polyester Adhesives:** These consists of unsaturated **polyester resins** dissolved in styrene monomer. They have relatively high shrinkage, thus have only limited use as adhesives. Most polyesters do not bond well to damp or wet substrates. Generally polyesters have excellent resistance to acidic environments.

(iii) **Acrylic Adhesives:** Methyl methacrylate and high molecular weight **methacrylate** monomers of the acrylic family are used as **solvent-free** adhesives for concrete. These adhesives generally share the same characteristics as polyester adhesives . These are most commonly used by mixing with fine aggregate to form an easily flowable adhesive mortar.

(iv) **Polysulphide Adhesives:** Polysulfides are most commonly used as flexibilizers in epoxy resin formulations . These formulations are sometimes referred to as "polysulfide adhesives" but these fall basically in "epoxy adhesive" category. Polysulfide materials are primarily joint sealants and can be used to bond glass to concrete.

(v) Polyurethane Adhesives: Polyurethane adhesives are available as both rigid and flexible materials . When combined with an aromatic amine, the urethane forms a rigid polymer similar to epoxy adhesives. When combined with a polyol, they form an elastomer. They have limited use with concrete because of their low bond strength. The flexible types have been used in membrane systems and for bonding ceramic tile to concrete where impact resistance is required.

(vi) Silicone Adhesives: Silicones that have the ability to cure in a wide temperature range are exclusively used as flexible joint sealant. These are used to bond window elements to concrete where a highly flexible adhesive is required to minimize concentration of stresses. Silicone should not be used in situations requiring resistance to sustained loads.

(b) Water Borne Adhesives

The only water borne adhesives currently used to bond concrete are **latex and latex powder.** There are two types of latex and latex powder adhesives. Type-1, which is used without further formulation, and Type-II, which is used in slurry form with a hydraulic cement, usually portland cement. For Type II adhesives, the ratio is about one part latex solids to four parts of cement by weight.

Both types of adhesives are generally used for bonding fresh unhardened concrete to hardened concrete. However, Type II adhesives have occasionally been used for bonding hardened concrete to hardened concrete. Latexes and latex powder are generally made by emulsion polymerization techniques. These adhesives include the following:

(i) Polyvinyl acetate (PVA)

(ii) Vinyl acetate copolymers (VAC)

(iii) Poly acrylic esters (PAE)

(iv) Styrene-butadiene copolymers (SB)

Type-I latex and latex-powder adhesives are generally made using a polyvinyl alcohol (PVOH) surfactant system. This type of adhesive gives a dried film that is redispersible upon application of water. This category includes polyvinyl acetate and vinyl acetate copolymers. The co-monomers are ethylene, butyl acrylate, and the vinyl ester of versatic acid.

Type-II latex adhesive are usually made with non ionic surfactant systems such as alkyl phenols reacted with various levels of ethylene oxide. Often, low levels of ionic surfactant are incorporated to assist in polymerization and specific latex properties. This type of latex gives a dried film that is not redispersible. Polyacrylic esters and styrene-butadiene co-polymers are included in this category.

Some of the commonly used Water Borne adhesives are discussed below:

(i) Polyvinyl Acetate: Polyvinyl acetate latexes are Type I adhesive and are usually formulated with a plasticizer such as dibutyl phthalate or di-propyl glycol di-benzoate. The plasticizers are added to decrease the minimum film-forming temperature. This type of adhesive is usually made in a polyvinyl alcohol surfactant system and is available both in the latex form and as a redispersible powder. Water resistance of such adhesives is suspect because of hydrolysis of the polyvinyl acetate. Films of the latex are redispersible.

(ii) **Vinyl Acetate Copolymers:** Copolymers of vinyl acetate with such materials as butyl acrylate, ethylene and the vinyl ester of versatic acid are Type I adhesives but can also be used as Type II adhesives. They are generally made in polyvinyl alcohol surfactant systems and are available in latex and redispersible powder forms. Their water resistance is much better than that of polyvinyl acetate, because the comonomer reduces the hydrolysis of the vinyl acetate grouping and also the resultant product is not as water soluble as polyvinyl alcohol. The water resistance of such polymers will depend on the type and ratio of comonomer to vinyl acetate. The comonomer also causes a reduction in the minimum film forming temperature, which eliminates the need for addition of plasticizers. When used as Type II adhesives, bond strengths usually exceed **6.9 MPa.**

(iii) **Polyacrylic Esters and Acrylic Co-polymers:** Polyacrylic ester latexes, such as polyethyl acrylate, and acrylic co-polymer latexes are Type II latex adhesives. They are generally made using primarily a non-ionic surfactant system. If the latex dries before placement of the fresh concrete, the dried film can act as a bond breaker rather than as an adhesive. These groups can improve adhesion by ionic reaction with metallic radicals in the surface of the fresh concrete. However, it has been observed that such groups may retard the initial hydration of the hydraulic cement.

(iv) **Styrene-butadiene Co-polymers:** Styrene- butadiene copolymer latexes are Type II adhesives. They could be used as Type I adhesive but are not recommended for this category, because their films are not redispersible. In addition, their surfactant system is primarily of the nonionic type. Such groups can improve adhesion and latex stability, but may also retard the initial hydration of the hydraulic cement.

Various Applications of Adhesives:

- Bonding of **hardened concrete** to **hardened concrete**
- Bonding of **plastic concrete** to **hardened concrete**
- Repair of **cracks in concrete**
- Bonding of **inserts into concrete**
- Bonding of **concrete** and **other materials**

6.5.3 Repair Mortars and Concrete

There are many reasons for the deterioration of the reinforced concrete elements and there are different ways in which these can be repaired. Deterioration mechanisms caused by corrosion of reinforcement, sulphate attack, or alkali-silica reaction, result in extensive cracking and/or spalling. In such situations, it is necessary to remove the damaged concrete and refill it with repair mortar or concrete.

The repair mortars commonly used can be broadly categorised as:

(a) Cementitious mortars/concrete
(b) Polymer modified cementitious mortars
(c) Resin mortars

(a) Cementitious Mortars/Concretes

These consist of cement as hydraulic binder and can further be subdivided into two categories:

(i) **Conventional Cement Concretes/Mortars:** Portland cement concrete or mortar offer a number of advantages as repair materials. Thermal properties are similar to existing concrete, similar in appearance and comparatively of low cost. It has ready availability and ease in applying. Concrete is often preferred for complete replacement of sections and deep cavities extending beyond reinforcing bars. Mortars can be used for small cavities/spalls (as small as 10-15 mm). Portland cement mortar with water content low enough to render it no-slump concrete and referred to as 'dry pack'. It finds its application as a good repair material. Low water contents reduces shrinkage of the repaired portion while the effectiveness gets increased.

The properties of cement mortar/concrete can be further modified with the addition of suitable **admixtures** like superplasticizers, hydrophobic agents, pozzolanas, flyash, silica fume, volcanic ash, diatomaceous earth, etc.

(ii) **Gunite/Shotcrete:** Gunite is a cementitious repair material that is pneumatically applied to concrete surfaces. **Shotcrete or gunite** is mortar or concrete conveyed through a pressure line and pneumatically at a high velocity on to a damaged or worn out prepared surface. It has the advantages of requiring no formwork, self-compacting and monolithic joints. Using specially graded aggregates and guniting equipment, 28-day equivalent cube strengths in excess of 50 N/mm^2 can be obtained. Reinforced gunite can be obtained by pinning the reinforcement to old surfaces and then guniting. Gunite linings are usually reinforced with steel bar fabric (welded mesh) and the distribution steel may be more than that provided in the original concrete. Normally, gunite would have a w/c ratio of about 0.5, though w/c ratio up to 0.33 can be used. Several types of fibres and additives are added to the gunites to improve its properties and performance.

(b) Polymer Modified Cementitious Mortars/Concretes

Since the early 1950s, it has been known that certain polymers can be added to cementitious mortars to help over come many of the problems of using unmodified mortar or concrete as repair materials. The polymers used as admixtures for cementitious systems are normally supplied as milky-white dispersions (latex) in water and are used to gauge the cementitious mortar as a whole or, as partial replacement of the mixing water. Such mortars offer the same alkaline passivation protection of steel as conventional cementitious materials. They can be placed in a single application of 12-15 mm thickness which gives adequate protective cover. The advantages of using polymer latex are :

- It functions as a **water-reducing plasticizer,** producing a mortar with good workability and lower shrinkage at lower water/cement ratios.
- It improves the bond between the repair mortar and the concrete being repaired, provided it is applied and used properly.
- It reduces the permeability of the repair mortar to water, carbon dioxide and oils and also increases its resistance to some chemicals.
- It acts as an integral curing aid, but careful curing is generally essential.
- It increases the tensile and flexural strength of the mortar.

Styrene butadiene rubber (SBR), acrylic and modified acrylic latexes are most commonly used as modifiers in repair mortars and base concrete. When they are properly formulated for compatibility with cement, there are no significant differences in the long term performance of the repair mortars. The latex modified concrete has the advantage of being able to be designed for adequate workability and placed in a manner similar to ordinary concrete by conventional methods. However, site batching and mixing often results in production of unsatisfactory mortars due to lack of adequate quality control. To overcome this problem, there are now complete prepacked sets of latex and preblended sand and cement available which simply require mixing at site.

Polymer modified cementitious mortars are mainly used for the repair of reinforced concrete, where the cover to be replaced is more than **12 mm** and less than **30 mm** in thickness. Above 25-30 mm of cover, conventional cementitious repair materials are commonly used.

(c) Resin Mortars

Where the cover to be replaced is less than 12 mm and areas to be repaired are relatively small, resin mortars are being used. When resin mortars are used, the protection of steel reinforcement depends wholly upon the impermeability of the repair mortar. This requires very careful application including surface preparation of the steel reinforcement to a very high standard. Resin mortars are based on reactive resins filled with carefully graded aggregates. Epoxy resins are most commonly used, but polyester and acrylic resins are also used. Epoxies, polyester and acrylic resins are classified as thermosetting materials because, when cured, the molecular chains are locked permanently together. They are generally supplied as, 2-3 component system (resin, hardener and filler).

(i) **Epoxy Resins:** Epoxy resin consists of a reactive resin which is cured with a curing agent called 'hardener'. Concrete proportioning and correct mixing are essential operations when using epoxy resin. The curing of epoxy resin system is an exothermic reaction and the rate of curing is temperature dependent. As a rule, the rate of curing doubles as the temperature increases by 100C. Epoxy-based formulations possess high chemical resistance but low elasticity. These are also suitable for filling joints subjected to low movements. These are used to repair concrete or masonry damaged due to chemical attack and areas subjected to high impact or tensile loads, leakage of chemical tanks, cracks of dams and spillways, to fill cracks and honeycombs. The basic characteristics and properties of these resins include the following:

- **High strength**: compressive strengths of **60 to 80 N/mm²** and very high tensile and flexural strengths (usually 5 to 10 times greater than concrete).
- **Thin sections**: Sections as thin as **5 mm** can safely be applied.
- **Adhesion**: Exceptional adhesion to concrete; repaired area never fail along correctly prepared bond line.
- **Rapid cure**: Most of the strength gain occurs on first day in normal atmospheric conditions
- **Chemical resistance**: Excellent resistance to alkalis; very good resistance to most acids and solvents; excellent resistance to most of minerals.
- **No maintenance**: Correctly applied repairs need no replacement .
- **Impermeable**: Properly compacted mortar is impermeable to water, water - borne contaminants and airborne gases.
- **Limitation**: These are not suitable above 100 °C temperature.

Most epoxy resins must be applied on dry surface. Ambient air temperature and relative humidity has to be kept within a fairly narrow limits .

(ii) Unsaturated Reactive Polyester Resins: Polyester resin systems are chemically much more simple. Both the reactive components are present in the resin. The hardener is a catalyst which is required only to initiate the reaction. Mixing and proportioning of hardener component is, therefore, less critical than for epoxy resins. There is a reduction in volume in the set polymer and so, polyester resin formulations must be limited to applications in relatively smaller areas at one time.

Advantages

- Very **rapid cure**: Polyesters develop strength extremely quickly; repaired areas can be reused within one to three hours.
- **Part mixing**: Part packs may be used .
- **No priming**: Polyester tends to rely on mechanical rather than chemical bonding.
- Independent primer is not necessary .However in specific conditions priming may be helpful.

Limitations

- Areas greater than approximately 30cm square will be prone to shrinkage. Repair sizes are limited. Deep repairs need to be applied in layers in order to avoid heat build-up.

(iii) Unsaturated Acrylic Resins: Acrylic resin systems form high strength materials by chemical cure mechanism. In general, acrylic resins are based on monomers of very low viscosity or blends of monomers with methyl methacrylate monomer. Acrylic resins permit higher proportions of fillers because they have lower viscosity and so the mortars exhibit less shrinkage upon curing.

6.5.4 Curing Compounds

Curing is defined as methods employed to ensure hydration such that concrete maintains sufficient free water within its structure to allow **cement hydration** to continue. Curing affects the concrete in the cover. The concrete in cover protects the reinforcement from corrosion due to ingress of aggressive agents. Adequate thickness and quality of cover is necessary to transfer the forces in the reinforcement, to provide fire resistance to the steel, and to provide an alkaline environment at the surface of the steel.

Methods of Curing

- **Water Immersion** - The ideal solution but impractical
- **Water Spray** - Expensive to maintain consistently and can be affected by wind
- **Wet Hessian** - ideal when covered with polythene
  - Can be 100%efficient
  - Difficult to keep wet
  - May shuck out water if it dries
  - Difficult on large structures
  - Affects surface finish
- **Wet Sand** - Similar to wet hessian
- **Tenting** with polythene
  - Similar to wet hessian to some extent
  - Risk of wind tunnel effect if it becomes loose
- **Keeping form and shutters** in place-Expensive and slow
- **Curing membranes**-Maintain internal moisture for hydration

Wet curing is specified in majority of construction projects but it is rarely achieved. Thus we should adopt the next best possible option of using curing compounds forming membrane with the following distinct advantages :

- Easy to apply with spray application and reduced labour cost
- Single application and no repeat maintenance required.
- Reliable with no risk of erratic or poor curing
- Positive water loss
- 90%curing efficiency can be achieved
- Ideal for overhead and vertical locations

Curing compounds are more relevant for repair jobs which are spread over large areas and in small patches. There are various types of curing membranes available. Some of these are:

- Light reflective and aluminised
- Degradable or permanent
- Solvent or water based
- Wax or resin based.

6.5.5 Sealants

As the name suggests, it is something that seals the pores and gaps. There are many places in a building, both internally and externally where joints require filling especially where the two elements of the joints are subject to relative movement. Building and construction industry the term sealant is specially used for sealing of expansion **joints and gaps** between dissimilar building materials. In present construction practices, many diverse materials like steel, concrete, plastics, aluminium and glass are used which essentially requires joints. These joints or cracks make the building vulnerable to influences of external environment. These cracks and joints get widened due to differential thermal coefficients of the materials used. The sealant is required to seal the inner portion from the external environment like wind, rain, pollution, air borne dust, micro-organisms, smells, insects, small animals, water, heat and sunlight.

(a) Functions of a Joint

A joint in a building or structure may be defined as a **discontinuity** in the component located in a predetermined position between either similar or dissimilar materials, and capable of:

- Allowing for **shrinkage, contraction** and other **movements** without causing excessive tensile stresses in the components.
- Allowing for expansion and other movements without causing severe compressive stresses in the components.
- Accommodating shear movements.
- Accommodating the allowable deviations from design in the dimensions of the building components.
- Allowing for temporary interruption in the progress of construction.

Joints may be either movement (flexible) joints, or non-movement (rigid) joints.

(b) Types Of Joints

The main types of joints are expansion, contraction, construction, building and fillet joints. These joints are briefly described in subsequent paras as per their functions.

(i) **Expansion Joints:** Expansion joints are designed in buildings to cope with the dimensional changes imposed on the building substrate by climatic changes such as temperature, humidity and movements caused by settlement of foundation.

(ii) **Contraction Joints:** Contraction joints, also known as control joints are designed to regulate cracking that might be due to unpredictable and unavoidable contraction or initial shrinkage of concrete. These joints are used to divide large but relatively thin concrete structure such as floors and retaining walls.

(iii) **Construction Joints:** Construction joints are designed specially to perform certain specific functions. For example, special joints are designed to accommodate rotational movement or sliding movement. Similarly when two different building materials such as brick and concrete, are used in conjunction with each other, a joint is created which needs to be weather proofed. When large slabs are casted at different times, a social joint is planned along the line of least shear force toward weak zones.

(iv) **Building Joints:** A Building requires a large number of minor joints such as around window and door frames and around various pipe penetrations. These are also called Door/Window Joints.

(v) **Fillet Joints:** The **narrow gap** provided for accommodating sufficient movement of pipes are called fillet joints. These are also called Pipe and Sanitary joints.

(c) Functions of Sealant

A joint sealant is any material intended to maintain a seal or act as barrier between two sides of a joint which may be subjected to some degree of movement.

A sealant does not contribute to the structural properties of the joint but has to perform some or all of the following functions:

- Accommodate continuing changes in the size of a joint
- Accommodate movement of various elements
- Exclude rain, snow and wind from the interior of a building
- Resist freeze/thaw influences
- Resist abrasion and water flow through the joint
- Exclusion or retention of water or liquid
- Prevent the ingress of anything which would interfere with movement in horizontal joints (e.g. in floors, debris) or cause damage to the building components
- Exclude chemical or biological contaminants from the joint to preserve the background or interior from chemical attack , or to maintain hygienic conditions.
- Provide sound insulation.
- Prevent heat losses arising from movement of air through the joint.

(d) Properties of a Sealant

An ideal sealant must have the following properties:

- **Stability** in storage
- **Ease of mixing** and ease of preparation for application.
- **Ease of application** over a practical temperature range
- **Adhesion** to faces of the joint
- Freedom from **creep**, slump or cold flow
- No undue **shrinkage**
- Freedom from **bleeding** or staining
- Adequate **cohesive properties** to accommodate movement without splitting
- Resistance to chemicals encountered in services
- **Compatibility** when used in combination with other materials
- Resistance to **weathering** and ageing
- Adequate **hardness or abrasion resistance** to fulfil its functions
- Retention of physical properties
- Freedom from regular maintenance
- Capability of being repaired or resealed

Different materials exhibit these properties to varying degrees, and some compromise may have to be made in balancing the requirements of individual joints. It is essential, therefore, to establish the functions required, and the degree of performance of each, in order to select the most suitable sealant for a particular joint.

Key feature of any sealant is its specification and ability to cope with movement during its service life both in compression or in tension. The ability is expressed by Movement Accommodation Factor(MAF), which is given by:

$$MAF = \frac{Joint\ Movement}{Joint\ width}$$

(e) Sealant Types

Sealants can be classified under a range of options based on polymer chemistry, typical applications or performance and market trade names. We would examine the sealants in terms of performance characteristics as this is usually the most relevant for the users. The key groupings are described in subsequent paras.

(i) **Oleo Resinous Mastics:** These are materials which form the surface skin after application, thereby protecting the main body of material underneath. These are commonly referred to as mastics. Both oil- based and butyl-rubber reinforced materials are available. Oleo-resinous mastics are suitable for joints where very little movement is expected. These materials have a limited life, but this can be extended by regular painting. They have good adhesion. No chemical bonding takes place with the substrate.

Typical applications include window perimeter pointing, but they should not be used on microporous finished timber or PVC. The buytl-rubber reinforced materials are commonly used for sealing lap joints in metallic building panels.

(ii) Bitumen and Rubber/Bitumen-based Sealants: Bitumen -based sealants are thermoplastic, and they retain a degree of flexibility. Enhanced elasticity and flexibility can be obtained by modifying the bitumen with rubber. Rubber/bitumen sealants are essentially plasto-elastic in nature and can develop a non-tacky surface within 24 to 48 hours after application.

Bitumen and rubber modified products are used in roofing, water-retaining structures and areas where compatibility with other bituminous materials is desirable .

(iii) Acrylic resin Sealants: Basically, two types of acrylic resin sealants are in common usage: solvent and water based. Solvent based acrylic sealants are primarily used for refurbishment work where their tenacious adhesion to surface which are difficult to clean is a distinct advantage. They are thermoplastic materials with plasto-elastic properties. They are used externally in the vast majority of cases. Appearance and durability can be improved by painting. Water-or emulsion based acrylic materials are widely employed as sealant for painting of window and door perimeters particularly internally. If used externally, care should be taken to avoid early contact with water or rain, because this can lead to washing of the sealant.

(iv) Flexible Epoxide Sealants: These are based on epoxy resins, and varying degrees of flexibility can be achieved by the addition of other polymers or extenders. Epoxide sealants are normally available as multi-component products which, when mixed, get cured at ambient temperature. Epoxide sealants are predominantly plastic in behaviour. Although they show a degree of **flexibility at room temperature** and above, they become **rigid at low** temperatures.

(v) Polysulphide Sealants: These are available in one and two component versions. The single part materials cure on exposure to atmospheric moisture. They are essentially elasto-plastic in nature. The polysulphide sealants require site mixing and they do not rely on atmospheric moisture for curing . Curing takes place uniformly throughout the body of the material.

Polysulphide sealants have been widely used in normal building and civil engineering applications over many years.

Two component polysulphides are available in gun applied and flow grades.

(vi) Polyurethane Sealants: Polyurethane sealants are also available in gun applied and flow grades as one or two component systems. These can exhibit a wide range of properties and normally their cure rate is rapid. When cured they have elastic properties. Two component polyurethane sealants are chemically curing products not relying on atmospheric moisture to effect curing.Some polyurethane sealants are **tar** or **pitch** modified. Polyurethane sealants are used both in building and civil engineering applications.

(vii) **Sealing Strips:** These are preformed extruded materials based on non-drying oils, polymer modified bitumens, synthetic polymers, or resins. These may be load bearing or non load bearing, and applications include lap joints in roofing, joints in tanks, joints between concrete wall panels, etc.

(viii) **Silicone Sealants:** These sealants are commonly available as single component materials with a variety of cure systems. Curing takes place by reaction with atmospheric moisture. Skin formation is generally rapid and the acid-curing version cure quickly in depth. Recent advances have produced a wide variety of cure systems. Specific information regarding these materials and their intended applications should be obtained from the relevant sealant manufacturers.

Silicones are versatile enough to produce a range of properties with respect to modulus and elongation. Applications include sealing of curtain wall panels, pointing of PVC components, glazing and sealing of joints in road pavements. Two component silicones have recently been introduced for insulating glass.

(ix) **Hot Poured Sealants**

These comprise of bitumen, rubber with bitumen and pitch with polymer combinations. These are primarily used in horizontal joints in road pavements, water retaining structures, and subways. It is essential that the correct equipment should be used to apply these materials. Temperature of the sealant is a critical parameter, and needs to be accurately controlled.

Sealant classification alongwith **Movement Accommodation factor** and life expectancy is given in Table 6.1.

6.5.6 Grouts

(a) Definition

A **grout** is a fluid material which is designed to be introduced into the cavity for the purpose of filling and subsequently hardening to give specific physical properties. The process of grouting is to fill voids, gaps or cavities with a material having certain specific properties. Grouts are most common materials for repair of building elements.

(b) Properties

- The materials of grout should **not shrink**, but should **expand to fill** gaps;
- Grout must have **ease of filling** with enough **fluidity**;

- Grout must have capacity to bear loads by having compressive and tensile strength and vibratory load resistance.

(c) *Types*

The grouts comprise of **cementitious** or **epoxy based resins**.

TABLE 6.1: SEALANT CLASSIFICATION

Method of Application	Chemical Types	Character Accommodation	Movement	Life expectancy (year)
Hot-applied	Bitumen	Elastic	Low	Up to 10
Hot non sag	Bitumen	Plastic	Low	Up to 10
	Bitumen/rubber	Plastoelastic	Medium	Up to 10
	Pitch/polymer	Elastoplastic	Medium	Up to 10
Cold poured	Polysulphide	Elastoplastsic	Medium-High	Up to 20
Two component	Polyurethane	Elastic	Medium-High	Up to 20
One-component	Polyurethane	Elastic	Medium-High	Up to 20
Chemically cured	Epoxy	Elastoplastsic	Medium-High	Up to 20
Gun-applied	Oil	Plastic	Low	Up to 10
Non-cured	Butyl	Plastic	Low	Up to 15
	Acrylic	Plastoelastic	Low-Medium	Up to 20
Gun-applied,	Polysulphide	Elastoplastic	Medium-High	Up to 25
Chemically cured	Silicone	Elastic	Medium-High	Up to 20
Strip	Nondry oil	Plastic	Non applicable	Up to 15

Cementitious

- Cement grout
- Cement sand grout
- Cement sand grout with additive
- Polymer modified cement grout

Epoxy/Polyester Resins

- Normal Epoxies
- Low viscosity Epoxies
- Low Exothermic Epoxies
- Polyester Resin Grouts

Grouts commonly used for repairs are described in detail along with their method of application subsequently..

Cement Grouts

Cracks wider than 1mm can be sealed by brushing in dry cement followed by spraying lightly with water. For cracks wider than 2 mm, it is preferable to use cement water slurry grout. Alternatively, cracks can be chased out to a width of 5-10 mm and pointed with cement and sand mortar.

Polymer Sealing

Low viscosity liquid polymers can be used in a similar way as cement grout for repair of cracks. Cracks may be filled by brush application or by temporary ponding with liquid polymers. When no further materials will penetrate the crack, the surplus materials are removed by wiping.

Resin Injection of Cracks under Pressure

Cracks in reinforced concrete greater than 0.3 mm may require sealing by resin injection to prevent ingress of moisture, oxygen and harmful agents. Before deciding upon the most appropriate method and material for repairing the crack, it is imperative to establish the nature and cause of the cracking. Cracking is caused normally by tensile stresses and if these stresses re-occur after the crack repair, the concrete may crack again. Low viscosity epoxy resin system with a viscosity below 6 stokes at 20° C are generally used for repair of cracks. Low viscosity acrylic or polyester resins are found to give lower bond strength, especially under damp conditions. The pressure of injection depends upon the width and depth of crack. When repairing cracks of 2 mm or more, it is not always possible to seal the outlets. In such cases, low viscosity resins are not preferred and thixotropic resins are used. These modified resins flow readily into relatively fine cracks under low pressures, but stop flowing immediately when the pressure is released.

Repair of live Cracks with Mastics

Normal grout injection will not be effective when repairing live crack, as the crack further widens or new crack opens at the interface of existing hardened grout and concrete. In such situations, materials that can accommodate considerable strains against cracking are to be selected.

When the anticipated future movement of the crack is small (less than 15% of its width) apply "mastics"-viscous fluids such as **non-drying oils** or **low melting soft asphalts** along with **fillers/fibres**, to the cracks. When the movement is more (greater than 25%) apply "thermoplastics" such as asphalt, coal tar or pitches. However, these have limited durability under ultra-violet light rays. Elastomers have advantage over mastics and thermoplastics and are more commonly used in recent times. Elastomers include polysulphides, polyurethanes, certain silicones and acrylics.

6.5.7 Waterproofing Materials For Roofs

(a) **Butyl Rubber (BR) Sheeting:** It is used for water proofing of roof slabs and treating leaking roofs of buildings. It is tough, black and flexible sheeting with considerable abrasion resistance. It requires careful laying as it can be punctured by sharp tools or edges of concrete surface while laying. Sheets joined together by surface adhesives result in unsatisfactory performance and hence, hot vulcanising of sheet joints is preferred. The sheets are of thickness 0.5 mm to 2.0 mm. These have good abrasion resistance and durability.

(b) Polyisobutylene (PIB) Sheeting: It is black, flexible-waterproofing sheeting similar to the BR sheet with improved properties. It has a low restitution, as a result of which when it is stretched, there is lesser tendency to regain its original shape and residual stresses are reduced. The PIB sheets have increased life in ultraviolet light, ozone and a wide range of chemical environments. PIB Sheets do not support fungi growth. These sheets have lower abrasion resistance compared to BR sheets and are not recommended for waterproofing of heavy foot traffic roofs.

The sheets are available in 0.5 to 2.0 mm thickness. For general purpose, 0.8 to 1.0 mm thickness is adequate. The joints are solvent welded. The solvent softens and activates the sheets at the joint and completely evaporates leaving strong and durable joint.

(c) Glass Fibre Reinforced Plastics(GRP): It is a composite material of polyester resin and glass fibre, usually referred to as GRP. GRP is applied on liquid retaining structures to prevent leakage and to protect against chemical attack. Usually, GRP is spray applied, with the resin, hardner and glass fibre through a 3-point nozzle gun. Special technique is employed to ensure that the fibres are completely covered by the resin.

(d) Bitumen and Bituminous Emulsions: Bitumen is frequently used in protection of leaking or damaged concrete roof slab against ingress of water and water borne aggressive agents. Minimum 2 coats are to be applied, the second coat at right angles to the first to help eliminate pinholes. It is cost effective and easy to apply. For drinking water retaining structures, special grade bitumen which is non-toxic and taint free is used.

(e) Latex-Cement Coatings: For water retaining structures of low porosity, the inside can be treated with a coat of 2 parts of cement and 1 part latex by weight. It gives a coverage of 0.8 m2/litre of the paste. The cleaned concrete surface is given a brush application of the latex -cement rendering. On drying, second coat is applied at right angle. It is simple to use. It is relatively cost effective.

(f) Epoxy and Polyurethane (PU) Resin Coatings: These are basically two or more component organic polymers and can be formulated to give coatings of wide range of characteristics and uses such as water-proofing, chemical resistance, bond, viscosity, pot life, curing and colouring.

Concrete surface should be **free of dust and oil** to ensure bond with repair material. For water proofing a minimum thickness of 0.5 mm in 3 coats is recommended. It can be applied by brush or spray.

(g) Acrylic Polymer Modified Cementitious Coating: These are designed to resurface and evenout variations in concrete surface. These provides a long lasting barrier to waterborne corrosive salts and atmospheric gases providing resistance to concrete decay. These provide a seamless, waterproof coating suitable for use in water tanks, reservoirs, fountains, roofs and sealing tie bar holes to ensure water tightness. These effectively seal concrete masonry walls, bridges and static shrinkage cracks. These provide a tough and durable coating which cannot be easily damaged or worn away. These have excellent weather resistance properties and are suitable for exterior applications.

Advantages

- Minimum surface preparation needed
- Applied directly to the concrete and masonry
- Excellent adhesion and bonding to porous and nonporous surfaces

Surfaces

- Non-toxic hence ideal for potable water tanks
- Excellent for damp-proofing basements
- Breathable-allows transmission of water vapour from interior of building
- Eliminates hand rubbing of concrete surfaces
- Good resistance to carbon dioxide and chloride ion diffusion
- Combines beauty with strength and durability

(h) Polymeric Modified-Bituminous Membranes: These membranes consist of a centre core of High Molecular High Density Polyethylene film (90 micron), which is the heart of the membrane which forms complete barrier against water and moisture. The centre core is protected on both sides with a high quality polymeric asphaltic mix with properties of high penetration high heat resistance and high softening point to make it ideal for waterproofing purposes. It also achieves high bonding strength with substrate and in overlaps. The polymeric asphaltic coating film is protected on bottom side with thermofusible High Molecular High Density Poly-Ethlene (HMHDPE). This aids the thermofusing process and also reinforces the non permeable quality of the entire membrane including overlaps and joints. The membrane is finished on top with Nitrocellulose Lacquered Embossed Foil of 75 micron thickness. The resultant membrane has elongation exceeding 200% to absorb any structural movement. It also has high tensile strength and is extremely flexible and pliable to adopt to contours. It is lightweight and is ideal for precast structures. It is also highly **Sound absorbent** on top of light CGI roofing. It is available in rolls of 1m width and 15-20 m length.

Special Features

- Polymer modified membranes have high softening point (>150°C for APPmodified membrane)

- Low cold flexibility (upto-25°C for SBS modified membranes)
- High tensile strength and absolute water tightness
- Membranes can be overlapped to form a continuous blanket over the whole surface (overlap strength being equal to or more than the membranes)
- Membranes have good Ultraviolet (U / V) resistance (as tested for 2000 hours of accelerated weathering test equivalent to ten years of natural weathering)
- Membranes bond well to different type of surfaces such as RCC, asbestos, or metal decks.

Application

Polymer modified bituminous membranes are laid on the pre-primed surface by use of LPG/propane gas torches resulting in thermofusion bonding

Generally there are two types of polymers which are used for modifying the bitumen. They are: APP - Atactic Polypropylene; SBS - Butadiene Styrene.

The two polymers differ fundamentally in their chemical nature. APP is a plastomer, whereas SBS is an elastomer. This chemical difference manifests itself physically in much greater elasticity for SBS -based modified bitumen, with more nearly uniform properties under a wide range of temperature and greater flexibility at low temperatures. APP modified bitumen are generally stronger and stiffer than SBS. APP also have greater resistance to high temperatures.

Special Uses

- Waterproofing of precast roof structures and where light weight waterproofing is required
- Waterproofing of CGI / AC sheet roofing
- Soundproofing of CGI sheet roofing

(i) **Polymer Based Waterproof Coating:** These are single component based on modified polymers and are free from tar bitumen and solvents . The product is available in paste like consistency. The product is applied on cementitious substrates by means of brushes or roller after one coat of primer. The dried coat forms a seamless membrane which is flexible and elastic in nature having a breaking elongation of 82%. The coating has the breathing properties. The coating has adequate resistance to U/V rays and does not crack or flake after multiple thermal cycles. These coatings have special feature of:

- Excellent Adhesion to substrate due to polymer base
- Application for water proofing of old and new terraces, roof slabs, balconies and parapets
- Suitability for structures of complicated geometry like domes, arches, shells, folded plates , paraboloids and corrugated sheets flexibility.

6.5.8 Special Concretes

These are used for repair and restoration of large damaged areas or for strengthening of an existing structure or to provide some special property to the structure. Details of special concretes are described at appropriate place. Some of the special concretes used for repairs are:

- Polymer concrete
- Polymer modified concrete
- Polymer impregnated concrete
- Shrinkage compensated concrete
- Underwater concrete
- Heat resistant concrete
- Free flow microconcrete
- Epoxy concrete
- Sulphur concrete
- Slurry infiltrated concrete
- Fibre reinforced concrete

6.5.9 Migrating Corrosion Inhibitors (MCI)

These are mostly Amine based compounds having optimum vapour pressure at ambient temperature for effective migration and are capable of forming a stable bond with the metal surface. These water based migrating corrosion inhibitors (MCI), when sprayed or brushed on concrete migrate through the hardened pore structure of concrete by diffusion. Upon contact with reinforcing steel, the MCI compounds form a mono-molecular protective layer. MCI compounds are currently being used in the developed nations. Recently Indian construction industry hus started using these materials for repair. The use of MCI may be very useful as it can protect the structure from further deterioration by simple coatings or injection into concrete.

6.5.10 Protective Coatings

Protective coatings are special products which represent the most widely used materials of corrosion control. These coatings are based on epoxies, polyurethanes and acrylics. These provide long term protection under a broad range of corrosion conditions (atmospheric exposure to full immersion in strong corrosive solutions). The use of **protective coatings** to save concrete from deterioration is fairly new idea which has been derived from the established paint technology. The potential of coatings to improve long-term performance of concrete is well recognised. This has triggered the development and manufacture of special coatings for concrete structures. Different types of coatings are available for the protection of concrete structures.

The quality of various manufactured construction units continues to improve but despite of these quality improvements, the problems of efflorescence, leakage, moisture vapour movement, spalling, cracking and staining of exterior surface still occur. One way to reduce these problems is to treat such surfaces with protective coatings. These coatings can be applied on the surface to be protected by brush or spray. The coatings are suitable for common building materials such as bricks, stones, concrete and various renderings.

Following type of coatings are normally used for protection of concrete surfaces under different conditions:

- Conventional coatings-Oleoresin, paints, alloys and phenolic modified alkyls;
- Bituminous coatings;
- Vinyl coatings
- Chlorinated rubber coatings
- Epoxy coatings
- Coal tar epoxies
- Polyurethane/resins
- Inorganic zinc rich coatings
- Silicones
- Acrylics and methacrylic resins

The special advantages and disadvantages of some of the **generic types of coatings are described below:**

(i) **Acrylics:** These form the base for a large number of coatings available in the market viz., methacrylates, styrene methacrylate, styrene acrylics, and synthetic rubber latex. Some excellent coatings have been formulated based on acrylic polymer.

(ii) **Urethanes:** These can form an ideal coating for concrete. **It is likely that single component, moisture-curing polyurethane will gain much popularity for concrete protection in future.**

(iii) **Chlorinated Rubber:** Coatings which were traditionally used for concrete were based on chlorinated rubber. While they have excellent diffusion resistance to carbon dioxide, they also retain the water vapour inside the concrete. This vapour barrier effect has resulted in many failures of this type of coating. Additionally, they are not permanently elastomeric and subsequently crack and debond.

(iv) **Epoxy:** These can provide excellent protection to concrete in extremely aggressive environments where high chemical resistance is the main parameter. However, they may not be cost effective under other conditions.

(v) **Cementitious-traditional Cement-based Masonry Paints: The traditional cement paint does not give the desired protection.** However the cement based paints modified with polymer, such as **Flexicrete** offer excellent resistance to carbon dioxide and water but permit diffusion of water vapour. Some of the products may be so impermeable that they are equivalent to 1m of concrete, when tested under a 100m head.

(vi) Silanes/Siloxanes: These are gaining popularity as surface impregnants. They have high penetrating capabilities and their **hydrophobic nature makes the surface water repellent.** When concrete is impregnated with a silane, not only does it react with itself to form a short chain polysiloxane, but also reacts with the silicate found in the cement hydrate. These consequently **cover the surfaces of pores and capillaries with a water repellent substance.** However, there are several drawbacks to their use on reinforced concrete. **Their hydrophobic nature expels water from any pore or capillary into which they penetrate.** This opens the surface of the substrate and allows gases to pass in more easily. This action has been shown to increase the rate of carbonation.

Amongst the various types of coatings, only acrylics and urethanes, which, when properly formulated, can fulfill all the general performance requirements.

Protective coatings can also be specified depending on its specific purpose/performance. Some of these group of coatings are described in subsequent paragraphs.

(a) Anti Carbonation Coating

Carbonation occurs because carbon dioxide defuses into concrete and reacts with calcium hydroxide. This results in loss of alkalinity and makes the reinforcement susceptible to corrosion. Coatings can be applied to concrete surface to arrest diffusion of CO_2, from the atmosphere. Commonly used anti-carbonation coatings are, chlorinated rubber, polyurethane resins and acrylic emulsion. In addition to reducing diffusion of CO_2, these coatings also limit the penetration of chlorides and other harmful chemicals into concrete. Anti-carbonation coatings allow free passage of water vapour. This prevents building up of vapour pressure behind the paint-film and hence, avoids blistering.

(b) Coatings to Protect from Acidic Environment

In certain industrial environments, concrete structures are subjected to attack of acidic compounds often due to combustion of fossil fuels, sulphur and other chemical plants. Effect of 'acid rain' due to SO_2 and other gases causes direct etching/pitting of concrete leaving exposed aggregate finish on vertical concrete surfaces. This kind of prolonged exposure with poor drainage may lead to severe damage or complete disintegration of concrete. Under these circumstances, **polyurethane coatings are provided.** In tougher environments, it is preferable to **use a high thickness epoxy paint to achieve the necessary protection.**

(c) Surface Impregnation Coatings

These materials penetrate the surface of the concrete and act as hydrophobic agent or completely block the pores. Silicon products, viz., silanes and siloxanes are commonly available as water repellent sealants. Polymer impregnates are chemically resistant, but has limited applicability in-situ.

Silanes are more correctly described as alkylalkoxy-silanes. The most widely used **"monomeric silane"** for the protection of concrete is isobutyl trimethoxy silane. Silanes become reactive in the presence of moisture and the speed of reaction is governed by the pH of concrete. Thus in normal alkaline concrete, the silane will react with the pore surface quite rapidly. In deteriorated concrete, normally the pH is low and the reaction time is more. Since silane is a volatile material and therefore, very high concentrations (upto 100%) are necessary for coating.

Siloxanes are more correctly described as **oligomeric alkylalkoxy siloxanes**. They have properties similar to the silanes and in addition, have low vapour pressure. They are extensively used for protection of concrete highways and structures.

It is recommended that silane and siloxane impregnant coatings be protected with an additional anti-carbonation coating to avoid diffusion of CO_2 and other gases.

Silicon resins which have much higher molecular weight as compared to silanes or siloxanes. These are also used as penetrant coatings for concrete. Their penetration is normally not more than 0.1 mm under site conditions. Because of their limited penetrability, they are less durable and not commonly used.

(d) Surface Coatings for Concrete Protection

Surface coatings are applied to concrete surface to prevent ingress of moisture and provide protection against aggressive environment. Preparation of the substrate prior to the use of surface coatings is necessary. All concrete surfaces whether new or old will have weak cement laitance, oil, grease, dust and other contaminations. If these are not removed, the coating may fail in bond.

The function of surface treatment is two-fold (i) to act as a barrier against aggressive environmental attack, (ii) to improve the visual appearance in addition to giving the surface self cleaning properties. Surface treatment can be used for:

- Enhancing appearance (colour, texture, opacity, cleanability and reflectance)
- Improving chemical resistance to sulphates, acids, brewery and dairy products
- Controlling ingress of chlorides, oxygen, carbon-dioxide, water vapour and moisture
- Improving mechanical and physical endurance such as resistance to abrasion, impact and skidding

Surface treatments are expected to perform satisfactorily under widely variable and extreme service conditions such as hot and humid, cold wet or dry cycles, covered and underwater. They often have to fulfill several distinct requirements at the same time. Depending upon the condition of the concrete, requirements of the structure and the environment, the factors to be considered for concrete coating selection are as:

- Resistance to water ingress
- Resistance to carbon Dioxide diffusion resistance
- Resistance to chloride ingress

- Water vapur diffusion resistance
- Ultra violet light resistance
- Elastic/crack bridging abilities
- Chemical resistance
- Abrasion resistance
- Ease of application
- Service life required
- Ease of over coating
- Aesthetic appearance

The coating of elements normally include floors, buried pipes, exteriors of structures and tanks, tank interiors, marine structures, bridge piers and superstructure and nuclear power contaminant vessels. These structures need protection against a variety of conditions such as food chemicals, fats, oils, corrosive soils, chemical plant wastes and fumes, fungus and even bacteria. Common coatings used for these purpose are:

(i) **Bituminous Cutbacks:** These coatings are solvent solutions of coal tar or asphalt. The coal tar cutbacks have better chemical resistance and better water impermeability than the asphalts. On the other hand, the asphalt cutbacks have better weather resistance and sunlight resistance. A thin coat is generally applied as primer, followed by heavier coats over the surface to protect water penetration. Some bituminous cutbacks are used as concrete penetrates. These are applied as thin material coat and allowed to soak into the concrete, increasing the surface density and thereby minimising water penetration and concrete spalling. Bituminous coatings are also available as water emulsions which cure by water evaporation and resin coalescence.

(ii) **Chlorinated Rubber Coating:** Chlorinated rubber coatings have water, chemical and alkali resistance. They are lacquer type and hence dry rapidly to form a good resistant film. They have very good adhesion to concrete. They are extensively used as coating for concrete water tanks, swimming pools and are specially formulated for tough, abrasion-resistant concrete floor enamels. They perform well under high humidity conditions and can be easily recoated. Chlorinated rubber coatings are vulnerable to rodent attacks, vegetable oils and greases. Chlorinated rubber coatings get softened in presence of oils/greases and hence these coatings are not recommended for sewer or linings.

(iii) **Vinyl Coatings:** Vinyl coatings have been used for many years as a coating for different concrete structures, from tank lining to nuclear power plant installations. They have good chemical resistance to both acids and alkalies. In view of their high molecular weight, their application needs to be preceded by a thin primer for maximum penetration into concrete surfaces. Vinyl coatings are lacquer type which dry very fast, hence it is advisable to coat them in cooler temperatures. Vinyl coatings generally have relatively low solid content so multiple coats are needed to build the required thickness. For exterior applications, vinyl acrylic top coatings give better weather resistance.

(iv) **Epoxy Coatings:** Epoxy coatings are formulated with liquid epoxy resin, liquid curing agent and highly penetrating solvent for good performance. These coatings allow the epoxy resin to penetrate deeply into the concrete surface, react to increase the density as well as the strength of the concrete surface providing excellent adhesion and chemical resistance.

Epoxy coatings are extensively used for a variety for structures such as tank lining, floor, pump base an nuclear fuel storage areas etc. These are several types of epoxy formulations used in coating concrete surfaces. These could be grouped into the following three types.

- Thin **epoxy coating** which can be applied over sand-blasted concrete surface. These are usually solvent based epoxies with relatively high molecular weight resins. They are similar to many epoxies used for steel coating.

- Thin **thixotropic liquid epoxy**-based primers can be applied to sand blasted surface. They provide both, a good surface and a base for other epoxy topcoat.

- Thick epoxy coating can be applied by trowel or by spray. This can be applied directly to a clean unprepared concrete surface. This fills the concrete surface imperfections and can be used alone or with additional coats of conventional epoxy topcoats.

Epoxy coatings can be cured by either amines or by polyamides and should be applied over smooth concrete which has been etched or sandblasted to a fine sand paper finish. The thickness of the coating usually ranges between 200 to 375 microns.

(v) **Coal Tar Epoxies:** Coal tar epoxies can be classified as the fourth type of epoxy coating. They have both properties of coal tar and epoxy. They have good adhesion to concrete and better chemical and abrasion resistance. These are easy to apply either by conventional roller or by airless spray. These are normally applied in two coats, having coating thickness of 375-500 microns. Amine-cured coal tar epoxies can effectively withstand the action of severe corrosion, bacteria, H2S and acid condensate.

(vi) **Integrated Four Coat System:** Central Electro Chemical Research Institute (CECRI) has developed an integrated four coat system for coating of concrete structures consisting of **epoxy polyamide ion oxide** primer, **epoxy polyamide MiO** undercoat, **epoxy polyamide TiO_2 top coat** and **aliphatic polyurethane sealer coat** to prevent corrosive attack from aggressive marine and industrial environment. Epoxy-based coatings are known for their ability to penetrate deeply into the concrete surface, thereby increasing the strength of the concrete surface. They are also well known for their **high alkali resistance** and have **good adhesion** and **compatibility** with other epoxy coatings as well as polyurethane coatings. Water absorption studies carried out on bare concrete surface, general painted concrete structures and CECRI formulation indicated that the CECRI coat system has the minimum water uptake value and minimum permeation to ions.

(e) IPN Polymer coating

Interpenetrating polymer networks (IPN) is a new technology that combines incompatible polymers. It promises to provide a much better cost-effective performance. IPNs are relatively novel type of polymer alloys consisting of two or more polymers in network form, atleast one of which is synthesised and/or cross -linked in the immediate presence of the other polymer. The polymer phases are devoid of chemical linking between them, interwoven to each other and held together by permanent entanglements.

Central Building Research Institute (CBRI), Roorkee, which undertook elaborate research work to assess the performance of protective coating systems have ultimately recommended the application of IPN polymer coating systems for vital concrete structures located in aggressive surroundings.

IPN polymer system satisfies most of the required properties of concrete protection and serves as an appropriate protective coating.

The adhesion, tensile strength and moisture vapour permeability of IPN polymer film are satisfactory

Elongation of polymer system is adequate to take care of micro cracks which might occur due to dynamic loading. Salt spray and weather resistance of the film is observed to be good. The life of IPN Polymer system can be estimated to be beyond 7 years.

TABLE 6.2: PROPERTIES OF POLYMERS-USED AS PROTECTIVE COATINGS.

Property Materials	Alkali resistant	Water resistant	Flexible	Breathable	UV resistant
Alkyd	No	Yes	Yes	Yes	No
Vinyl	No	No	Yes	Yes	No
StyreneCopolymer	Yes	Yes	No	Yes	No
Chlorinated Rubber	Yes	Yes	No	No	No
Pure Aliphatic Acrylic	Yes	Yes	Yes	Yes	Yes
Epoxy	Yes	Yes	No	No	No
Polyurethane	Yes	Yes	Yes	No	Yes

6.6 SELECTION OF MATERIALS FOR REPAIRS

Several repair materials are available commercially, their use depends upon nature and intensity of damage, type of environment, required life, importance of structure and available funds. Proper selection of repair materials is very essential for its effectiveness in repair.

Important factors which influence the selection of materials for repair are as follows:

(i) **Bonding** of repair materials to existing substrates

(ii) **Non shrink properties** of repair material (volumetric stability)

(iii) **Strength** of repair materials as compared to old materials of construction

(iv) **Resistance to shrinkage** crack formations

(v) Type and extent of damage

(vi) Type of structure

(vii) **Environmental** conditions

(viii) **Serviceability** conditions

(ix) Site conditions

(x) Time factor

(xi) **Temperature** at application

(xii) **Maintenance** of proper cover

(xiii) **Durability** of repair material

(xiv) **Curing** of newly laid repair materials

(xv) **Economical** considerations of repair

(xvi) **Matching** of repair mateials with substrate

The nature of repairs in question guides selection of materials depending on the basic cause of failure. Several repair materials are available commercially. Their use depends upon the nature and intensity of damage, type of environment, required life, importance of structure and available funds. The basic repair materials can be classified as grouts, **coatings, mortars, cement admixture and concrete.** The term concrete broadly includes polymer systems, epoxy resins, fiber-reinforced cement and other composites. There are many materials available for preventing moisture movement to **repair dampness damaged structures.** All these materials have certain advantages and limitations. No single material can be prescribed as a general solution to repair. It is, therefore, necessary to identify the basic cause of damage and then carry out remedial work using suitable repair materials.

The selection of appropriate material is one of the important step for the effective repair and rehabilitation of the damaged structure. The civil engineer is confronted with a large number of proprietary materials available in the market and is liable to err on this count. Commercial literature provides all the advantages of the materials but lacks in stating minimum necessary technical data and limitations. Table 6.3 gives the recommendations of the Working Group of Concrete Society which can serve as valuable guidelines for the selection of repair materials depending on nature of damage.

TABLE 6.3: SELECTION OF MATERIALS FOR CONCRETE REPAIR

Repair/Damage Type	Type of Material
1. Large Spalling	
(i) >25 mm	Plain Cement Concrete(PCC), Cement Mortar (CM), Polymer Modified Cement Mortar(PMCM)
(ii) 12-25 mm	Polymer Modified Cement Mortar(PMCM), Epoxy Resin Mortar(ERM)
(iii) 6-12 mm	Epoxy Resin Mortar(ERM), Polyester Resin Mortar(PRM)
2. Small Spalling	
(i) 12-25 mm	Polymer Modified Cement Mortar(PMCM), Epoxy Resin Mortar(ERM)
(ii) 6-12 mm	Epoxy Resin Mortar(ERM), Polyester Resin Mortar(PRM)
3. Crack Sealing	Styrene Butadiene Resin(SBR),Acrylic Resin Co-Polymer Latex(AR-CL), Low Viscosity Polyester Resin and Acrylic Resin (LV-PR&AR)
4. Structural Crack Repair	Low Viscosity Epoxy Resin(LV-ER)
5. Bonding Aids for Repair	Moisture Tolerant Epoxy Resin(MT-ER),Styrene Butadiene Resin(SBR),Acrylic Resin Co-polymer Latex(AR-CL)
6. Honey Comb Concrete	Low Viscosity Polyester Resin and Acrylic Resin (LV-PR&AR), Low Viscosity Epoxy Resin(LV-ER)
7. Permeable Concrete	Interpenetrating Polymer Network(IPN), Sealants , Protective Coatings (Penetrants)

6.7 COMMERCIALLY AVAILABLE REPAIR MATERIALS

Commercially available modern repair materials alongwith their special characteristics and trade names are specified in Table 6.4. This Table 6.4 facilitates in selecting the appropriate repair material for a particular situation.

TABLE 6.4: COMMERCIALLY AVAILABLE REPAIR MATERIALS

Type of Material	Description/Advantages (Special Features)	Commercial Trade Name
● Rebar Primer	● Anti corrosive steel primer ● Quick drying ● Two component mineral based corrosion inhibitor ● Can be welded through, forms conductive film ● Improves binding for further coats	(i) Nitozinc primer (ii) Colusal MK-25 (iii) IPNet-RB (iv) Rusticide SS (v) Sikatop Armatec 108 (vi) Cinter flux-20
● Epoxy based (Solvent free or Pollutionymer) Adhesives	● Bonding hardened concrete to fresh concrete ● Usable in wet conditions ● Develops High Strength ● Has long open life ● Suitable for chloride contaminated host concrete.	(i) Nitobond EP (ii) Epibond (iii) Sikadur 32 (iv) Araldide Systems (v) Fairbond EP (vi) Aqua EP

- Water borne Adhesive (Latex Adhesives)
  - Single component
  - Water resistant bonding agent for repair mortar
  - Polymer latex Additive for repair mortar/concrete (PMC &PMM)
  - Fully compatible with PCC repair systems
  - Universal carbonation resistance bond coat

  (i) MONOBOND
  (ii) Nitobond AR/SBR
  (iii) Hack-Aid Plast
  (iv) Nafufill BB2
  (v) Sika Latex
  (vi) High Bond-40

- Shotcrete Admixtures/ Sparayable repair mortar
  - Cementitious Sprayable repair mortars.
  - Develops Extra Strength
  - Applied Fast
  - High build/less rebound
  - High abrasion resistance
  - High chloride/carbon dioxide resistance

  (i) Sprayset SL
  (ii) Conplast-MS
  (iii) Microsilica-600
  (iv) ELKEM-Micro silica 920-D)
  (v) Rendroc SP range

- Prepacked Cement Mortars
  - They contain polymer to improve low permeability qualities and to enhance adhesion, heat development is low
  - Suitable Mortars for concrete repair
  - Ready to use polymer based mortar for repair
  - Provides increased bend strength
  - Improves/provides resistance to carbonation

  (i) RendrocS2/HB2/RG/ UW
  (ii) Zentrifix/Nafuquick
  (iii) Cico ExcemV
  (iv) Sikatop 121/122/123
  (v) Patchroc/Superpatch

- Epoxy resin mortar
  - Resin mortars suitable for small/ shallow repairs
  - High performance/Exceptional adhesion to concrete
  - Better chemical resistance
  - Rapid Cure/Strength gain
  - High strength-Abrasion impact, tensile strength

  (i) Nitomortar-S/30
  (ii) Sikadur-42/43
  (iii) Epoxy mortar KP-HP-350
  (iv) Araldite Systems
  (v) Cico Resicrete
  (vi) Dobeckot Systems
  (vii) Cinter SL Screed J

- Polyeste Resin Mortar
  - Rapid setting-harder than concrete within 2 hours
  - No primer coat
  - Good resistance towards bleach liquids
  - High chemical resistance

  (i) Nitomortar PE
  (ii) Corocem P103/Z7L/ B/O
  (iii) Lokset S/L/P
  (iv) Anchorgrout
  (v) Poly-V

Curing Compounds	• Single application, spray applied • Quick drying • 90% curing efficiency achieved • Easy to apply • Spray applied after disappearance of water from the surface • Can be done on horizontal as well as vertical surface after removal of form work	(i) Cuncure WB (ii) Antisol E/(WP) (iii) Polycure (iv) Faircure C/WC/RA (v) Emcoril White/AC
	• Rapid foaming for sealing water leaks	
Injection systems	• Solvent free, two-part polyurethane for sealing cracks • Low Viscosity, High molecular weight polymer for grouting • High bond strength • Effective for structural repairs	**(a) Cement/Polymer Grouts** (i) Conbex100 (ii) Sifumex100 (iii) MC-Einpresshife EH (iv) Intraplast (v) Fairadd (S) (vi) Roft Hygrout (vii) B-Crete
		(b) Low Viscosity Epoxy Grouts (i) Conbextra EP10 (ii) Epoxy EIJ-KP-HP250 Monopole LV (iii) Pidifil EP-200 (iv) Cico Resiscrete 212 (v) Sidadur 52/53 (FLV&UF) (vi) Flow grout EPLV
		(c) Aqua reactive grouts (i) Polygrout (ii) Hydro Active Flex LV/SLV (iii) Purseal (iv) Purflex
Grouts	• Innovative Repair Mortar • Free flowing • Non shrink • High strength	**Resin Grouts** (i) Conbextra EP65/120 (ii) MC-DUR GROUT (iii) Araldite System (iv) Debeckot systems (v) Flowgrout EP **Cementitious Grouts** (i) Conbextra GP/HF (ii) Shrinkcon (iii) Emcwkrete (iv) Sikagrout 214 (v) Flow Grout 40/60/AG (vi) Core fit H

Joint Sealants

- Elastomeric sealent for construction
- Excellent adhesion
- Accommodates high and low temperature.
- Accommodates high movement

Silicone Sealants
(i) Nitoseal 125
(ii) Elastosil
(iii) Tech seal-RTV
(iv) Wilsil
(v) Pideseal N-22
(vi) Crack Seal RTVA

Polysulphide Sealants
(i) Thioflex-600
(ii) Pidiseal PS
(iii) Techseal RDL-940.
(iv) Sikalastic
(v) Tuffseal

Polyurethanes/Acrylic/ Mastic/Epoxy Sealants
(i) Pidiseal A-11/B-11
(ii) Techmast-71
(iii) Techseal RDL-600
(iv) EPCO-100
(v) Nitoseal-100/280
(vi) Elastocil - Wacker

Special Sealants
(i) Colpor 200
(ii) Nitoseal 525
(iii) Elastosil 355,
(iv) Non-staining
(v) Nitoseal 200
(vi) Euco-700

Protective Coatings
(a) For concrete and steel

(i) Corroseal
(ii) Polycot
(iii) Nitocote EP-140/405/ 410
(iv) Sikaguard 67/Sealoflex
(v) Safe core SF/AR/ ultracoat

(b) Water Repllents

(i) Nitocote SN 522
(ii) Nisiwa SH
(iii) Repellin Super
(iv) DF-105
(v) Techsilicone

(c) Decorative/Anti carbonation

(i) Dekguard S
(ii) Fair coat A
(iii) Emcecolor-flex
(iv) Nitocote CR
(v) B-Guard

Water proofing
(a) **Membranes**

(i) Multiplas
(ii) Mak-polyplast

		(iii) Tera Roof
		(iv) Bituseal
		(v) Bitumat APP
		(vi) Thermolay
		(vii) Tarfelt
		(viii) Plastofelt
(b) Acrylic Polymer Modified Cementitious Coatings		(i) Brushbond RFX
		(ii) Zentrafix Elastic
		(iii) Roof Hyguard
		(iv) Sika Top seal 107
		(v) Water guard A
		(vi) Pidiflex WPM
		(vii) Bond fit 5000
(c) Polymer Coatings		(i) Roofex 200
		(ii) Sealer Coat
		(iii) Aquatight
		(iv) Sealoflex
		(v) Water guard AC
		(vi) Walcrete
(d) Miscellaneous		(i) Waterseal
		(ii) Waterguard PU
		(iii) Nitocote CR
		(iv) Tuffcoat CRP
		(v) Tapecrete
		(vi) Isothane EMA
Migratory Corrosion Inhabitor (MCI)	• Protects and controls reinforcement corrosion • Makes concrete durable • Does not affect the corrosive strength • Surface Applied	(i) MCI-2000
		(ii) MC-Corrodur

6.8 SUMMARY

Serviceable life of any building structure depends on its materials, environmental conditions, workmanship, protection and preventive maintenance. All building materials deteriorate by weathering actions at different rates and ultimately loose their serviceability. Repair materials are intended to protect and maintain original form, quality and serviceability for a longer period. Lack of proper repair and maintenance leads to total unserviceability necessitating demolition and reconstruction. Many new repair materials such as polymer based chemicals and liquid plastics (epoxy, polyurethane) are available for treating many of the building defects quite effectively to retard the process of deterioration.

Durability of repair and construction materials play most important role in serviceability of buildings. A large number of modern buildings use cement concrete as major construction material and hence durability of concrete elements plays a critical role in building serviceability. Concrete durability is affected by its grade, W/C ratio, permeability, chemical and biological resistance and construction practices (batching, mixing, compaction, curing and finishing).

Compatibility of repair material with original construction material refers to harmonious monolithic structural behaviour without loss of bond when subjected to different kind of forces and weather conditions. Compatability of repair materials, thus, plays an important role in maintenance of serviceability of building elements. Apart from compatibility, repair materials must reduce or prevent the ingress of environmental forces and maintain bond with the original construction material.

Repair materials can be grouped based on desired characteristics and properties in to anticorrosion coatings, bonding aids, repair mortars, curing compounds, sealants, grouts, water proofing systems, special concretes, migratory corrosion inhabitor, protective coatings and special materials. A good repair material should have the best combination of low shrinkage, mechanical properties and adhesion both in dry and wet conditions.

Repair concrete include cementitious mortars or concretes, polymer modified cementitious mortars, and resin mortars. These mortars and concretes are used depending on the condition of deteriorated elements.

Curing compounds are used to ensure adequate free water inside repair concrete for hydration process to continue for a longer period. Adequate curing makes cover concrete effective to resist ingress of water or dampness.

Sealants are used to create barrier to passage of water or liquid in joints which are subjected to relative movement. Sealants seal the innerside from external environmental forces like wind, rain, pollution, dust, micro-organisms, insects, water and heat. Sealant can expand or contract and occupies required shape as per joint function and movement (Movement accommodation factor - MAF). Sealants are grouped according to their chemistry, performance or application, and trade names.

Fluid materials used as grouts for filling cavities, voids or cracks must have enough fluidity, must not shrink on hardening and must be capable of bearing compressive, tensile and vibratory loads. Grouts consist of cementitious or epoxy resins.

Water proofing system include Butyl rubber sheeting (BR), polyisobutylene (PIB) sheeting, glass fibre reinforced plastics (GRP), Bitumen and Bituminous emulsions, latex-cement coat, Epoxy and polyurethane (PU) resin coatings, Acrylic Polymer modified cementitious coatings, polymer modified Bituminous membranes, and polymer based water proofing coatings.

Special concretes are also used for repair jobs. Special concretes include: polymer concrete, polymer modified concrete, polymer impregnated concrete, shrinkage compensation concrete, under water concrete, heat resistant concrete, free flow microconcrete, epoxy concrete, sulphur concrete, and fibre reinforced concrete.

Migratory Corrosion Inhibitors (MCI) are mostly Amine based compounds having optimum vapour pressure for effective migration and stable bond formation with metal. Sprayed or brushed MCI migrate through the concrete pores by diffusion and form mono-molecular protective layer.

Protective coatings are representing special corrosion or dampness control products. These coatings are based on epoxies polyurathanes and acrylics. These coatings are also used on surfaces of concrete or masonry to minimise efflorescence, leaking moisture movement, staining and spalling. These coatings include **oleoresin, paints, phenolic modified Alkyls, bituminous layers,** vinyl **chlorinated rubber, epoxy resins, coaltar epoxy, polyurethanes, zincrich, silicones, Acrylics** and **methacrylic resins**. Surface coatings protect the element from wide range of conditions such as hot and humid, cold wet and dry cycles and under water. Important factors considered in protective coating selection are resistance to water ingress, CO_2 and water vapour diffusion, chloride, ultra voilet rays, **crack bridging, chemicals, abrasion** and **ease of application, service life,** and aesthetic **appearance**.

Selection of repair material depends on availability, nature and extent of damage, environmental conditions, required life, importance of element and available funds. Selection of repair material is influenced by factors such as: Bond, non shrink, strength durability, nature of damage, environment, service condition, time, temperature at application, curing and economic aspects.

QUESTIONS

6.1 Describe importance of repair materials for effective maintenance of buildings.

6.2 Describe factors influencing durability of repaired elements.

6.3 Explain importance of compatibility of repair matrials with original construction.

6.4 List 10 important type of repair materials.

6.5 Describe 3 most important characteristics of any repair material.

6.6 Describe briefly important characteristics of the following bonding aids:

    (i) Epoxy resins

    (ii) Polyester resins

    (iv) Polysulphide

    (v) Polyurathane

    (vi) Silicone

    (vii) Polyvinyl acetate

    (viii) Styrene Butadiene copolymers

6.7 List 3 type of repair mortars/concrete.

6.8 Describe briefly important properties of following repair mortar/concrete.

    (i) Gunite/shotcrete

    (ii) Polymer modified cementitious mortars

    (iii) Epoxy resin mortar

    (v) Polyester resin mortar

    (vi) Acrylic resin mortar

6.9 Describe briefly special properties of curing compounds.

6.10 Describe briefly properties of sealant.

6.11 List 5 important type of sealants.

6.12 List most important characteristics of grouts.

6.13 Describe main properties of:

    (i) Polymer modified cement grout

    (ii) Cement grout

    (iii) Polyester resin grout

6.14 List 5 type of repair water proofing systems for roofs.

6.15 List 10 types of special concretes used for repair.

6.16 Explain the protection mechanism using migrating corrosion inhibitor (MCI)

6.17 List 10 type of protective coatings.

6.18 Describe in the properties of the following coatings:

    (i) Anticarbonation

    (ii) Silanes/siloxanes

    (iii) Chlorinated rubber

    (iv) Interpenetrating Polymer Network (IPN)

6.19 Describe the 10 most important factors which influence the selection of repair materials.

6.20 List the 5 most common commercial repair materials for treating dampness along with their basic characteristics.

6.21 List the commercial trade names of following repair materials:

Rebar Primer

Epoxy based adhesives

Shotcrete admixtures

Silicone sealants

Polysulphide sealants

Protective coatings

Water proofing polymer coatings

Curing compounds

Cement-polymer grouts

Chapter 7

PREVENTIVE MAINTENANCE AND SPECIAL PRECAUTIONS

7

PREVENTIVE MAINTENANCE AND SPECIAL PRECAUTIONS

LEARNING OBJECTIVES

After studying this chapter, the learner understands the preventive maintenance of buildings and will be able to:

- **Describe** importance of preventive maintenance;
- **List** main considerations for preventive maintenance;
- **Describe** importance of floor dusting for preventive maintenance;
- **Describe** the importance of washing in preventive maintenance;
- **Describe** main features of joint maintenance;
- **Explain** termite treatment in a existing building;
- **Describe** importance of repair and maintenance of damp proofing systems in roofs and wet areas;
- **Describe** special precautions required for repair and maintenance of buildings.

7.1 INTRODUCTION

All building components require periodic inspection for repair and maintenance for satisfactory long- term service. Periodic inspection, thus, forms part of maintenance of building structure. The inspector must be familiar with the construction materials as well as their structural behaviour. The inspector must be very careful not to alarm the occupants who may readily jump to the conclusion that their structure is in danger.

From inspection report, repair work may be identified for maintenance of building elements. **Repair** means making good, but prior to any repair work it is essential to establish the extent and root causes of the problem. Repair and maintenance go hand in hand. The consideration of repair method must take into account owner's expectations from the structure. It must also take into account serviceability and expected life of the structure. The availability of building for repairs must be ensured on the basis of convenience and economy. . There should be minimal interference with the use of a building structure which is occupied. Maintenance of building structure can be placed into three categories.

◆ Routine maintenance to ensure adequate performance and to prolong the service life of the structure;

◆ Specific maintenance to correct obvious problems;

◆ Preventive maintenance to eliminate major costly repair of anticipated problem.

The type and extent of maintenance is best determined by an ongoing building inspection programme. Building inspections help establish the need for maintenance, identify the causes of problems, and prescribe appropriate repair maintenance procedures.

To conduct a thorough inspection, an inspection sheet should be used to note specific problems and ensure that no areas are overlooked during inspection. This allows the **location**, **nature** and **extent** of all problems to be properly documented. Always note the precise location, size, and extent of all **cracks**, **spalls**, **stains**, and other evidence of structural **deterioration**.

The inspector should have access to design and construction documents that show original construction details and any modifications that have been made. This allows the inspector to locate **hidden** details.

Building inspections should be carried out systematically by concentrating on specific areas or features. For example, examine parapets, then check window and door openings, proceed to control joints, flashing and weep holes, examine wall surface and mortar joints. Inspection method may vary depending on the type and configuration of structure. Compare any deterioration noted during earlier inspections with that observed during the current inspection to establish the rate of deterioration and natural ageing process. Decisions required to be taken in maintenance are:

◆ Job scheduling-resource based

◆ Safe access

◆ Good and effective house keeping

◆ Safety and insurance provisions

- ◆ Environmental conditions (dust, noise, smell)
- ◆ Service equipment during repairs
- ◆ Monitoring frequency and norms.

Structures are constructed to minimise future maintenance. But maintenance needs still arise in building's life. It can be a challenging task to find an inexplicable cause of a difficult maintenance problem and to prescribe the correct solution. **Buildings which are not maintained are going to perish soon.**

7.2 PREVENTIVE MAINTENANCE CONSIDERATIONS

Preventive maintenance of a building components aims at nipping the problem before it becomes serious. A systematic approach has to be followed. It must start from top. An inventory with following details will facilitate in planning preventive maintenance schedule:

- ● Location
- ● Date of completion
- ● Construction cost
- ● Present condition
- ● Minor repairs and routine maintenance job (due on)
- ● Budget provision
- ● Check list for building inspection (with date and major items to be checked)
- ● Equipment inspection, etc.

Generally it is assumed that concrete does not require any preventive maintenance but preventive maintenance of concrete structure enhances its serviceability and useful life.

Every one knows steel must be painted to prevent rusting and wood must be treated to postpone rotting. But what about concrete members? It is an eloquent endorsement of our favourite construction material that preventive maintenance of concrete is a little known subject.

The usual primary requirement of a good concrete in its hardened state is a satisfactory compressive strength. It is assumed that **strong** concrete is also **durable**. We need to have higher quality of concrete for severe exposure conditions. A **good quality concrete** must be **dense** and **impermeable**. For severe conditions of exposure both strength and durability have to be considered explicitly at the design stage.

In order to ensure durability, it is not only necessarily to produce durable concrete, but also to put into place a system of regular maintenance procedures. Thus the concept of preventive maintenance is very important and it should be preferred to repairing at the later stages after substantial deterioration has already taken place.

Although concrete performs well even when left on its own. However, there are applications where a little periodic maintenance will enhance or extend the useful life of concrete. There are some basic activities, which must be carried out on a regular basis to prevent deterioration of concrete. These are:

- Sweeping/washing
- Joint maintenance
- Dusting of floors
- Termite control

7.3 SWEEPING / WASHING

A suitable type of cement and concrete must be used to match the environment. Knowledge of environmental conditions facilitate users to evaluate the suitability of the type of concrete for the proposed job. Concrete can withstand intimate contact of many materials. Some of the materials can coexist with concrete by proper house keeping procedures. The detrimental materials are required to be removed from the concrete surface at the earliest by sweeping and washing regularly to minimise contact time to enhance the concrete life.

Certain users of building structures such as dairy building should be especially vigilant for washing of floor instead of superficial cleanliness alone for its life. Milk spilled on concrete floor will not attack the floor immediately, but it is likely to sour soon and form lactic acid. If not washed away before it penetrates deep into the surface, the potential for deterioration exists.

If any plant floor is subjected to spillage of fine materials that could act as abrasives, a regular schedule of sweeping, washing or both is warranted. The same is true for spillage of fluids that are slippery or flammable. Protective coatings are available to prevent penetration and minimize chemical attack.

When designing a floor likely to be subjected to occasional chemical spillage, consideration should be given for achieving a highly reflective hard surface. For high traffic areas, heavy-duty floor can be achieved by use of dry shake non-metallic floor hardener when the floor is being finished. The light reflectance of floor hardener encourages good house keeping practices and reduces requirements of high intensity lighting. The dense surface resists infiltration of spilled materials. The high concentration of well graded natural aggregates (components of the dry shake) at the surface of the floor renders it much more durable under heavy traffic.

7.4 JOINT MAINTENANCE

Joints are vitally needed components of a concrete structure, but they are also potential problem spots. Most joints are incorporated into buildings for the purpose of accommodating differential movement in building elements. The joint might be used at any place where drying shrinkage, differential settlement or other change could set up horizontal or vertical stress greater than the tensile or shear strength of the concrete.

A concrete building can frequently represent one of the most monumental and assumed to be static man made edifice. In reality, however, concrete is continually moving, growing and shrinking in response to changes in temperature, load and other weather conditions. Thus when a crack forms it merely beguns to perform its function as joint. Joint must retain its ability to accommodate movement if it is to be successful. When the joint locks itself or becomes immobile, the concrete is subject to stresses and likely to adjust to repeated stresses by further cracking.

Unfilled joints, especially in floors, are likely to become clogged with dirt, aggregate particles and other materials that eventually cause them to lose their ability to move. To prevent this, maintenance crews must be especially vigilant to keep the joint clear of such debris. A more practical approach to joint maintenance in industrial buildings and warehouses is to fill them with **sealant** that resist infiltration of foreign matter and accommodate movement for long periods of time. Joints in vertical surfaces and most outdoor applications are best sealed with elastomeric sealant. These should not be applied to the full depth of the joint, but to a depth that is less than the joint width. The bottom of the joint is first filled with **nonbonding backing material** to support the sealant.

Interior floor movement due to temperature changes is likely to be small. After the floor has undergone most of its drying shrinkage the joint seldom undergoes much further movement (opening and closing). For this reason floor joints are usually sealed with less deformable sealant, particularly where they are called on to support the loads of wheeled traffic, specially in roads and industrial floors. Joints should be regularly inspected to ensure that they are functioning as designed.

7.5 DUSTING FLOORS

A concrete floor that is correctly designed for the service expected and is properly placed will not produce dust by normal usage. However, some industrial floors produce dust. It is possible to reduce the amount of dusting through proper maintenance procedures.

First determine the cause of dusting and then plan the treatment. If the dusting is because of **cracking and pulverising** of aggregates at the floor surface due to traffic, the alternatives are to **resurface the floor concrete** or **reduce the traffic**.

Dusting of the floor may be the result of inadequate curing and/or carbonation of the surface concrete. The problem can be temporarily relieved (alleviated) with the use of commercial fluosilicate, zinc fluosilicate or sodium silicate. The reaction from liquid treatment containing magnesium to form hard crystals to reduce the amount of dusting experienced in light or medium duty floors. These treatments must be reapplied periodically, the frequency of reapplication depending on the severity of dusting and the amount of traffic. Concentration of the active ingredient varies with the brand resurfacing material used.

7.6 TERMITE CONTROL

The preventive measures are generally adopted for controlling the termite menace in existing building components such as:

- Foundations
- Soil under floors
- Voids in masonry
- Contact points with woodwork
- Woodwork

7.6.1 Treatment Outside Foundations

The soil in contact with the external wall of the building is treated with chemical emulsion at the rate of 7.5 litres per sq. m of the vertical surface of the sub-structure by excavating a trench exposing the foundation wall to a depth of about 30 cm all along the periphery of the building.

7.6.2 Treatment of Soil Under Floors

The points where the termites are likely to seek entry through the floor are cracks at the following locations: (a) at the junction of the floor and walls as a result of shrinkage of concrete, (b) On the floor surface owing to construction defects, (c) at the construction joints in a concrete floor cast in sections and (d) Expansion joints in the floor.

Chemical treatment should be provided within the plinth area on the ground floor of the structure wherever such cracks are noticed, by drilling 12 mm vertical holes at the junction of floor and walls, construction and expansion joints at an interval of 300mm to reach the soil below. Chemical emulsion should be applied inside these holes using a hand operated pressure pump until refusal or to a maximum of one litre per hole. The holes should then be sealed with non shrinking cement or cement mortar..

7.6.3 Treatment of Voids in Masonry

Termite are known to seek entry through foundation masonry and work their way up through voids in the masonry and enter the building at ground and upper floors. The ingress and movement of termite through the masonry walls may be arrested by drilling holes in the masonry wall at the plinth level and squirting chemical solutions into the holes to be absorbed by the masonry and create a barrier of chemical film or coating near the plinth.

Insecticides recommended for termite control in the existing buildings are the same as those recommended for new buildings. It may be noted that these chemicals are highly poisnous and should be handled carefully by skilled workers under the supervision of experienced hands with suitable precautionary measures.

7.6.4 Treatment at Points of Contact with Wood Work

All existing woodworks in the building which are in contact with the floor or walls and infested by termites, should be treated by spraying at the point of contact with the adjoining masonry with any of the standard insecticides of the required concentration. This is done by drilling 6 mm holes at a downward angle of about 45° at the junction of the wood work and masonry and squirting the chemical emulsion into these holes till refusal or to a maximum of half a litre per hole. The treated holes are then sealed with cement.

7.6.5 Treatment of Wood Work

For the purpose of treatment, woodwork condition may be classified as below: (a) that which is damaged by termite beyond repair and needs replacement, and (b) that which is damaged slightly by termite and does not need replacement. The woodwork that has already been damaged beyond repair by termite should be rejected and replaced. The new timber should be dipped or liberally brushed at least twice with chemicals in oil or kerosene.

All partially damaged woodwork which do not need replacement should be treated with suitable chemicals. Infested woodwork in chaukats, shelves and joints in contract with the floor or the wall should be treated with protective chemicals by drilling slanting holes of about 3mm diameter to the core of the woodwork on the inconspicuous surface of the frame. These holes should be at least 150mm centre-to centre and should cover the entire framework. Any one of the available insecticides should be liberally infused into these holes. The holes are subsequently sealed with appropriate putty or sealant.

If wooden wall shelves are attached, it may be necessary to drill holes in the wall panels at intervals of about 300mm and drench the wall with insecticide solution. While replacing any member, either a liberal coat of insecticide solution is applied to the wall surface or only preservative treated timber is used.

7.6.6 Precautions

All the recommended insecticides are highly poisonous. There will be adverse effects on health of person if these chemicals are absorbed through skin, inhaled as vapours or swallowed. The precautions suggested in the insecticide containers should, therefore, be strictly followed.

All workers must use **masks, gloves and aprons** when working with these insecticides. Workers must wash their hands and body after closing the days work. The work place must also be properly ventilated. Children should be strictly kept away from the work place . Treated area should not be used or occupied for at least 3-4 days.

7.6.7 Latest Development

The chemicals used as pesticides namely **Aldrin, Chlordane** and **Heptachlor** have been found to be highly potent for anti termite treatment. Millions of dwellings have been treated with these chemicals during the last three decades with very successful results. However, during the past few years it has been found that these chemicals are highly toxic and persistent in environment. The US National Academy of Science has suggested a limit for long term continuous exposure of chlordane as 5hg, heptachlor 2hg, and aldrin as 1 per cum of air. Realizing the high toxicity of the chemicals used for termite control, use of such chemicals is increasingly being discouraged in India too.

In recent years, CBRI Roorkee has undertaken research in respect of development of alternative nontoxic methods for controlling termites in buildings. One such alternative for control of termites is the development of termite repellent surface coating from pine tree. The pine tree has been found to restrict the presence of termites around it. In order to develop nontoxic termite repellent surface coating for buildings, investigations were carried out to find the termite repellency of extract of leaves, barks and roots of various trees and horticulture plants. Out of nine viable extracts three were found to show 85-90% termite repellency.

When other alternatives being explored at CBRI, the safer limits of pesticides for controlling termites shall be used. **Chlorphyrifos,** an organophosphorus compound has been tried out in USA and Japan with better results. Its worthiness in Indian context has been examined in detail by CBRI. Four years of investigation regarding effectiveness of chlorphyrifos carried out at Roorkee, Pune and Jorhat reveal that **Chlorphyrifos is 100%** effective in controlling termites in buildings. Various alternatives to the present practices of termite control in building being presently explored by CBRI Roorkee have now been included in the national standards.

7.7 DAMP PROOFING OF EXISTING ROOF AND WET AREAS

Repair and maintenance of damp proofing system of existing roof and wet areas is very critical as most of the problems of leakage and seepage starts from these areas. It is essential to prevent failure of these systems. Most of the leakage problems occur in roof terrace, parapet-walls, and joints of external walls with roof slabs. Comprehensive treatment of damp proofing problems of roof and wet areas is dealt with in Chapter 9.

To prevent frequent failures of total damp proofing systems, the roof terraces should be regularly inspected and checked for drainage and obstructions to rain water spouts specially before rainy season to clear rainwater drains. Terrace surface of tiles or any other finishing material must also be checked for cracks and depression if any. The depressions must be repaired immediately after removing tiles to provide correct slope towards rain water spout to avoid leakage on the structural slab. After rectifying the main depressions, all cracks in tile joints and gola should be cleaned and filled with liquid epoxy sealants or bituminous emulsions. The terrace surface must be treated with protective coatings from time to time as required to prevent damp proofing system failure. Wider cracks may be filled with bonding epoxy resin mortars and non shrink cementitious mortar mixed with water proofing admixtures.

Check terrace tile joints and continuity of water proofing layer specially, near the joints of roof and parapet wall including gola expansion joints and khura around rain water spout. In case of wet areas of toilets and kitchens check water proofing failure around floor traps and drains etc. Any leaks or cracks around floor traps and drains must be sealed with appropriate sealing and damp proofing material which penetrates to the bottom of the leak or crack. These minor repairs should be carried out on a regular basis till major repair jobs are undertaken for treating bigger failures.

7.8 WATER SUPPLY AND SANITARY SYSTEMS

It is desirable to prevent failure of water supply and sanitary drainage system to maintain these services without interruptions. The objectives of preventive maintenance is to keep the systems functioning as required by the user. Maintenance of these services can be divided into following:

- Roof drainage
- Foul or sewage drainage
- Surface water from paved areas
- Subsoil drainage

It is necessary to prepare a checklist for maintenance programme of these systems. Regular inspection of water closets, wash basins, sinks, C.I soil and waste pipes, cisterns, traps, etc. should be carried out once in three months. Any problem or cause of trouble should be removed. Outlets should be cleaned at regular intervals and not allowed to become blocked by leaves, dust, debris, etc.

There is no point in maintaining a system that is not functioning as per requirements of the user and such a system should be replaced.

All water supply and sanitary lines must be cleared of blockage. Any leakage due to cracks, holes or joint failure must be immediately rectified. These minor repairs and regular inspections of WS and sanitary lines prolong their useful life. Major repair of existing water supply and sanitary lines is also dealt in detail in Chapter-14.

7.9 SPECIAL PRECAUTIONS FOR REPAIR OF BUILDINGS

Adequate planning and safety considerations are of prime importance before undertaking repair of any building structure. For any repair work to be successful, correct investigations of the problem must be carried out to determine the root causes. To achieve effective long-term solutions to problems, it is necessary to select correct repair materials, techniques and team of skilled workers.

Planning includes the decision making on appropriate sequence of various subtasks and main repair tasks from safety and other considerations. Proper planning brings down the layoff period for a building to a minimum. Sometimes layoff period of building can result into much bigger loss of production in comparison to extra cost for faster repair methods and materials. Planning must therefore, consider all these factors when deciding the technology, materials and working team for repairs. Cost and effectiveness of repair methods must be compared by also considering indirect factors of loss of production, quality and loss of profit due to non-occupation of buildings under repair. Thus repair tasks must be planned in advance considering all direct or indirect factors such as comparative cost, availability of skilled workers and technology, time of completion and reoccupation, loss of production and profits, etc.

Before starting repair of any building, its safety must be examined in the present condition, during repairing phase and after the completion of repair work. Repairing may cause redistribution of loads and stresses and hence safety of structure must be assessed with reference to repaired structure. **Safety** must also be ensured for the **workers, occupants** and **public** in general during the repair process. Any **hazardous** repair material or process must be well notified to all concerned with appropriate **warning signs**.

Success of repair job depends on correct investigation of root causes and eliminating these causes through repair. Superficial repair, without detailed investigation, may not last long and result in waste of repair efforts. Correct investigations facilitate selection of appropriate techniques and suitable materials for effective repair. There are very wide range of materials and techniques available in the market and the selection must be done scientifically. Scientific considerations are effectiveness for the specific purpose, economic considerations, facilitate long life of repaired elements, time and expertise available for repair. Repair must be planned at macro and micro level. Special care must be taken for surface preparation prior to repair work to ensure proper bond of new material and old construction. Specific instructions on the job must be given to all workers regarding correct techniques, safety and other site considerations of the job. All important repair tasks must be supervised by qualified engineer or supervisor to ensure quality of the repair work.

7.10 SUMMARY

Periodic inspection for repair and maintenance are essential acts for satisfactory long term service life of buildings and prevent major problems in functioning. Building maintenance includes routine repair and maintenance to ensure adequate performance of structure, specific repair to correct a problem which might have arisen during use. **Preventive** maintenance is to eliminate minor problems and keep the building operational. Repair **job scheduling** providing access, good and effective **house-keeping** ensuring safety and insurance, maintaining **service equipment** required, are some off the major activities which should be undertaken for building maintenance.

Preventive maintenance is also important in buildings constructed with concrete elements. To ensure usability of concrete elements, it is not only necessary to produce good concrete, but it requires preventive maintenance and protection for long life.

Regular sweeping and washing of concrete floor surfaces facilitate removal of many detrimental materials resulting in longer service life. Floor hardness can be used in areas where occasional spillage is expected during service of building.

Joints are incorporated in buildings for the purpose of accommodating differential movement in components. Maintenance staff must be especially vigilant to keep the joint clear of clogging. Joints should be regularly inspected to ensure that they are functioning as per required specifications.

Maintenance of floors free from dusting problem requires application of commercial fluosilicate, zinc fluosilicate or sodium silicate. These must be applied periodically with appropriate coating depending on the usage.

Antitermite treatment for foundations, soil under floors, voids in masonary and wood work should be done regularly for controlling the menace of termite in existing buildings. Chemicals such Aldrin chlordane and Heptachor have been found to be highly potent for termite control but should be used after taking proper precautions under experienced supervisor.

Roofs and other wet areas in building require special attention during regular repair jobs. It is important not to allow water to pond in any part of the roof and damage damp proofing system. Water drainage reduces the problem of leakage and dampness in the building elements specially roof structure.

QUESTIONS

7.1   Describe in not more than 100 words the importance of preventive maintenance of buildings for their satisfactory long-term service.

7.2   List 5 most important factors to be considered while planning maintenance jobs.

7.3   List 4 main considerations in preventive maintenance of buildings.

7.4   Explain the importance of following to prevent deterioration of concrete elements:

    (i) Sweeping and washing

    (ii) Joint of floors

    (iii) Dusting of floors

7.5   Explain briefly the measures for controlling the termite menace for following existing elements:

    (i) Foundations

    (ii) Soil under floors

    (iii) Masonry

    (iv) Wood work

7.6   Describe briefly latest materials available for antitermite treatment.

7.7   Explain the importance of preventive maintenance in about 100 words for roofs and other wet areas in buiildings.

7.8   Describe 3 most important aspects considered in preventive maintenance of WS and sanitary lines.

7.9   State 5 critical precautionary measures necessary for repair jobs.

Chapter 8

COMMON TECHNIQUES OF BUILDING REPAIR

8

COMMON TECHNIQUES OF BUILDING REPAIR

LEARNING OBJECTIVES

After studying this chapter, the learner understands common techniques of building repairs and will be able to:

- **Describe** the importance of surface preparation for repair jobs;
- **Explain** common methods of surface preparations;
- **List** advantages of different methods of surface preparations;
- **Explain** common repair techniques;
- **Explain** common methods of crack repair.

8.1 INTRODUCTION

It is very important to ascertain the main cause of defects prior to carrying out repair and remedial works. Once the causes are known, the most suitable method and materials of repair are selected. Generally the damaged part may require replacement, strengthening and/or protection. The repaired component must have adequate desired **strength**, **adhesion** between new and old construction, **reduced permeability** and **enhanced resistance** to environmental forces. Cementitious materials and repair mortars are required to show no or very low degree of shrinkage. These repair materials must have low permeability alongwith good compatibility with the substrata. Repair materials must also provide good protective coating for reinforcement in case of RCC elements.

Successful repair work requires knowledge and understanding of surface preparation using appropriate techniques. There is need to understand proper application of common technique to suit the specific repair job. Maintenance engineer must give due consideration to surface preparation and selection of method prior to undertaking any repair. The deteriorated or damaged area must be prepared to receive repair material for exhibiting compatibility, bond, and adequate strength as per desired specifications.

8.2 SURFACE PREPARATION

Surface preparation is critical to repair, resurfacing of topping or coating for long-term performance of the substrate. Best quality material systems applied correctly may fail in the absence of adequate and proper surface preparation. The technique of surface preparation varies from case to case depending on various factors, but it is important to understand the importance of surface preparation in general. Guidelines to carry out proper surface preparation before application of any repair materials must be clearly stated and understood by the Maintenance engineer and workers for its long-term performance. The overall success and performance of repair materials applied to substrate is highly dependent upon the quality of surface preparation carried out.

Surface contaminant may be defined as a material, either liquid or solid that has the potential to cause problems related to **adhesion, curing,** and/or application of repair materials to substrate. Surface contaminants must be completely removed by a synergistic combination of **cleaning and preparation**. The common examples of contaminants are dust, **efflorescence, laitance, form release agents, oil, grease, tar and gum**.

8.2.1 Methods

Various methods for preparation of surface prior to repair are:

(i) Chemical cleaning

(ii) Steam cleaning

(iii) Shot blasting

(iv) Sand blasting

  (v) Vacuum cleaning
  (vi) Acid etching

(i) Chemical Cleaning

Hot water solution of trisodium phosphate (TSP) or commercially produced detergents, cleaners and emulsifiers are generally used to remove contaminants from the surface. Thorough flushing with clean water is necessary to completely remove residue of cleaning chemicals from the substrate surface.

(ii) Steam Cleaning

Steam cleaning provides an effective method of removal of many forms of water-soluble contaminants that may be present on the surface. Steam cleaning machines produce wet or dry steam directed at the surface in high concentration and at high velocity sufficient to loosen, soften, and remove the contaminants. Detergents, degradations, and other chemicals are often added to water to increase effectiveness.

(iii) Shot Blasting

Centrifugal shot blasting is a very effective, clean and dust free method for removing hardened films of contaminants from textured horizontal surface. This process involves impacting the surface with high velocity abrasive steel shots. The shot blasting media available in a range of sizes and shapes is thrown against the concrete from an enclosed high velocity rotating paddle wheel. The abrasive, dust, and contaminants are then removed by a separate dust collector. The cleaned steel shots are then recycled to the blast wheel where the cycle repeats.

Shot blasting provides a clean, physically sound substrate, with a relatively uniform texture. The method is particularly useful and cost effective on large unobstructed horizontal surfaces.

(iv) Sand Blasting (Abrasive Blast Cleaning)

Sand blasting is a method for preparing any surface including reinforcement by impact with high velocity stream of fine mineral aggregate abrasive propelled by clean compressed air. The blasting medium usually consists of hard angular mineral aggregates. Sandblast cleaning produces a textured, physically sound substrate free of surface contaminants and fines.

Generally, larger sizes of abrasive are used for preparing concrete than steel. Sand abrasives having a diameter of 1.6-2.0 mm size are recommended for heavy cleaning. A diameter of 0.35 to 0.84 mm graded sand is sufficient for removal of laitence.

(v) Vacuum Cleaning / Air Blast Cleaning

Vacuum cleaning or air blast with oil free clean compressed air is a final step to remove loose dust or dirt on a prepared surface immediately prior to coating or patch repair application. Vacuum cleaning is preferable to air blast when the dispersion of dust must be controlled and limited.

(vi) Acid Etching

Acid etching is useful either for preparing concrete surface to receive a repair material or to roughen the flat smooth surface that has proven to be slippery. Muriatic acid usually diluted to the job requirements is poured onto the concrete surface vigorously and broomed until bubbling ceases. Then the surface is promptly and thoroughly flushed with clean water. The amount and concentration of the acid and to some extent the amount of brooming will determine how much of the concrete surface is removed. The acid fumes are noxious and the liquid can cause skin rashes and hence appropriate boots, gloves and goggles must be used while working. Suitable ventilation of the area should also be provided. Acid etching can be done along with other techniques depending on the field requirements.

8.2.2 Basic Considerations for Surface Preparation

Different techniques and equipment can be used to clean and prepare concrete surface for repair. Any shortcoming in use of appropriate technique and equipment may lead to ineffective performance of even the best of patch repair system. From the science of adhesion it is well known that surface contamination is a prime obstacle for effective contact between an adhesive coating and the substrate. For perfect adhesion, the coating must be able to 'wet the contact surface adequately and efficiently.

Splashes of paint, oil, plaster, mortar and dirt from atmosphere even on a new surface are common causes of contamination. On existing old surfaces oil and dirt are common in work areas and may have penetrated the surface over a long period of time. It is, therefore, necessary that any repair material is applied to the surface only after proper preparation of the substrate. The basic considerations in surface preparation are:

- Removal of any weak concrete from surface or body of substrate.
- Removal of any contaminated concrete, to an agreed depth as determined by chemical analysis.
- Removal of spalled concrete caused by corrosion of steel reinforcement.
- Removal of corrosion products from steel reinforcement.

Various methods of surface preparations, their specific purposes and advantages are given in Table 8.1.

8.3 COMMON REPAIR TECHNIQUES

The concept of repair and its strategy is developed by asessing the type of deterioration and main causes. If the deterioration is of a continuing nature, investigate the practicality of eliminating the causative factors. Check, if water eroding a concrete surface can be channelised in some other direction, or can further corrosive spillage be prevented or heavy traffic on a road be diverted. The standards of quality of any patch repair job must be equal or better than the quality of the original concrete and if needed match the appearance of the surrounding concrete. In case it is impossible or impractical to eliminate the causative conditions completely then there are three alternative choices for repair:

- Tear out the damaged concrete and replace it with suitable concrete to match exposure conditions;
- Increase the concrete strength if it is structurally inadequate (for example , by stressing);
- Protect the concrete if deterioration is caused by severe exposure conditions (for example, by coating jacketing or resurfacing)

TABLE 8.1: SURFACE PREPARATION METHODS AND ADVANTAGES

	Method	Purpose	Advantages
(i)	Acid etching	Removal of Light laitence in concrete, mortar and plaster stains, rust stains, mould oil/curing compound traces, light rust in steel	Inexpensive, simple application, leaves clean surface
(ii)	Wire brushing	Removal of loosely adhering and soft materials, curing compound residues in concrete, light rust in steel	Relatively inexpensive, hand or mechanical application, dry process
(iii)	Grit blasting	Removal of deep laitence, soft material, curing compounds and most contaminants except oil in concrete and heavy rust metal	Most effective method, Leave deep cut, clean surface to bright metal
(iv)	Mechanical scrubbing	Removal of deep penetration of surfaces in concrete	Most effective for actual removal of concrete surface
(v)	Degreasing	Removal of mineral oil and grease in concrete and metal	Inexpensive simple application
(vi)	Flame cleaning	Removal of mould oil/curing compound traces, Light oil deposits, light laitence in concrete	Clean surface and simple application by burning
(vii)	Water jetting	Removal of mould oil/curing light oil deposits, deep laitence in concrete	Simple application, uses clean water, very effective
(viii)	Wet sand blasting	Removal of rust, Salt contamination, mineral oil, grease and marine growths in steel	Cleaner than dry method, effective removal of soluble salts and marine growths.

Repair of concrete play an important role in repair of total building. There are many common techniques for carrying out repair of concrete elements. These techniques are:

- Caulking
- Coating
- Replacement by plastic concrete
- Grinding or grooving
- Injection
- Jacketing
- Prepacked concrete application
- Thin or regular resurfacing
- Bonded or unbonded resurfacing
- Stressing
- Shotcreting

The various techniques are explained as follows:

8.3.1 Caulking

Caulking by definition is arbitrarily restricted to the filling of narrow ruptures in concrete with plastic material that is neither flowable like grout nor stiff like dry pack mortar. Concrete with small or medium size ruptures and not requiring replacement may be repaired by caulking. If the cracks are dormant, Portland cement mortar or expanding mortar can be suitable as caulking material. If the cracks are active, an elastomeric caulking material should be used.

8.3.2 Coating

Coatings are defined as materials of **liquid** or **plastic consistency** applied directly over the surface to protect it by developing desired characteristics in the existing surface. This is the painting forming the plastic or liquid coating over damaged surface. The coating must be carefully selected for the needed characteristics. Coatings can be applied by brushing, rolling on or spraying. Their permanence varies greatly. Some common applications of coatings are for waterproofing, protection from aggressive chemicals or other injurious pollutants or providing longer life against heavy traffic. Some of the coating materials used for concrete are epoxy resins, bituminous compounds, linseed oil, fluosilicate compounds and other silicone preparations. These coating materials partially permeate the concrete and also provide a thin film over the surface. Coatings can be secured to concrete by bond or chemical reaction with concrete. Opaque coatings hide discoloration and also protect the surface from the abrasive or aggressive forces.

8.3.3 Replacement by Plastic Concrete

Unsound concrete can be removed and replaced with concrete of plastic consistency which may either be conventional cement concrete or other patching material. Complete or partial removal of existing concrete depends on the extent and nature of the deterioration. This is one of the most commonly used repair techniques and is appropriate for applications where the cause of deterioration is non-repetitive or root cause has been eliminated. It is better to first determine the cost of replacement before taking decision to replace. In case of fire or other severe damages, it is necessary to replace the whole concrete. Total replacement is advisable when the partial replacement cost exceeds the total cost or partial replacement is not long lasting.

8.3.4 Replacement by Dry Pack

This procedure is similar to the plastic replacement technique with the exception that a no slump concrete or stiff mortar is used rather than a plastic repair material and it is rammed into the place. The method has the advantage that the repair material has a **low pressed or water content** and therefore **low shrinkage** but it requires greater skill on the part of workman and increased vigilance by inspectors to minimize voids. It is appropriate for use in deeper cavities which have good accessibility. Since formwork is not required, dry pack is especially convenient for narrow-deep cavities in vertical members. It is not suitable for replacement of extensive, wide or shallow areas or for applications requiring compaction behind reinforcing bars.

8.3.5 Grinding or Grooving

This technique is used to improve bonding characteristics of the surface to be repaired. When a slab has irregularities and varies from a plane or has shallow cracks or unsound pitted concrete, grinding is necessary part of repair technique. Grinding can also be used incase of **pitted** and **stained** concrete. However, unless modern heavy-duty equipment is used it can be a slow, expensive method of repair. Consideration should also be given to acid etching or chipping (by hand or with chipper). Grooving is the process of **cutting grooves** into flatwork to reduce slipperiness. Before a slab is grooved, other ways of roughening the surface, such as acid etching, should also be investigated.

8.3.6 Injection

Narrow cracks can be repaired by epoxy resins. Methyl methacrylate resins have also been recently introduced for the purpose. Injection ports spaced at short intervals are drilled and the concrete surface is sealed with the resin to prevent leakage from the crack during injection. Two component resins or catalyzed resin is injected in a port unit. The material is pumped successively through all the ports in series. Evidence that the area up to the next port has been filled must be ensured. Injection technique for crack repair is described in detail under crack repair method.

8.3.7 Jacketing

Jacketing involves fastening of a repair material to enhance the resistance to the threatening environment that is causing deterioration. The material can be metal, rubber, plastic or high strength concrete. It can be secured to the existing element by bolts, nails, screws, bonding adhesives, straps or gravity. The material and method of jacketing depends on the exposure conditions and locations. There are a number of proprietary mortars and grouts used for this work. Common applications of jacketing include tanks, spillways, piers and other concrete elements that are exposed to corrosive materials or erosive force of rapidly flowing water.

8.3.8 Prepacked Concrete

Prepacked concrete is also called **preplaced aggregate concrete.** It is a technique in which gap graded aggregates are packed into a cavity and inundated with water to saturate the aggregate. Then mortar or grout is pumped in from the bottom, displacing the water. The method is most suitable for inaccessible applications, such as submerged concrete or deteriorated concrete that is being jacketed. Prepacked concrete provides low shrinkage and good bonding qualities but it can leave voids. Cores are taken out after the job is completed to ascertain the complete filling of voids. Prepacked concrete work should usually be carried out by a specialist as it requires special skills and equipment.

8.3.9 Regular Resurfacing

Regular resurfacing is the frequent application of a uniform thin layer of repair material, over an existing surface. It is often used to repair floors and pavements that are basically sound structurally but whose surfaces have deteriorated from heavy traffic or other aggressive environmental exposures. No standards have been promulgated but frequently topping of 50mm or less is categorized as thin and those over 50 mm are considered thick resurfacing. The thickness of the resurfacing will usually hinge on the amount of increase in slab thickness or level that can be tolerated. If no increase in level is permissible it will be necessary to remove enough loose and weak concrete from the existing slab to accommodate the layer of resurfacing material. Normally, this entails greatly increased cost and time.

8.3.10 Bonded or Unbonded Resurfacing

A decision is made regarding thin or thick resurfacing and whether or not it will be bonded. If the deterioration is a surface phenomenon (such as spalling or scaling), it is usually advised that it should be bonded. If the problem involves cracking or structural movement, it may be desirable not to bond the toping so that resurface will not transfer and reflect the distress of the substrate. In this case provision should be made to see that the topping is distinctly separated from the substrate by use of a sand layer, polyethylene sheet, or both so that it can move independently. Unbonded resurfacing is sometimes used on floors subject to severe chemical attack. When the replaced surface becomes sufficiently corroded it can be removed more easily than if bonded. Whether the toping is bonded or unbonded care should be taken to prevent partially bonded job.

8.3.11 Stressing

Stressing becomes necessary when the cracked area is too large for stitching and the cracks require to be closed. This involves embedding rods or cables in the distressed concrete, stressing them to a predetermined tension and anchoring them. The stressing must be done after proper structural design to prevent the distress from reappearance somewhere else in the structure. For jobs requiring minor repair, cracks can be injected with an expansive material.

8.3.12 Shotcreting

Shotcreting process is sometimes called guniting or pneumatic application of concrete (including mortar). Application of shotcrete involves shooting concrete under pressure onto the prepared surface of the deteriorated concrete. Shotcrete may require pumping of completely mixed material through the hose pipe or blowing the dry constituents through the hose pipe and mixing them with water at the nozzle.

The later method requires an experienced and competent operator but offers the capability of customizing the shotcrete with reference to water content and consistency according to the specific needs of areas of the repair job. Shotcrete is practicalble for large jobs, both on vertical and horizontal surfaces. The surface profile should be somewhat irregular and rough textured. Curing is especially critical in shotcrete work.

Special methods for repairing cracks are also described separately in subsequent paragraphs. Some of crack repair methods can also be used for common repairs.

8.4 COMMON METHODS OF CRACK REPAIR

Repair of cracks is one of common problems in buildings and requires high degree of skill. If the cracking is primarily due to drying shrinkage in concrete, then it is likely that the cracks will stabilize. However, if the cracks are due to foundation settlement or other structural changes, they will continue to grow. Following repair procedures work quite effectively for live cracks while others are recommended only for dormant cracks:

- Epoxy injection
- Grooving and sealing
- Stitching
- Adding reinforcement
- Grouting
- Flexible sealing
- Dry packing
- Polymer impregnation
- Overlays and surface treatment
- Autogenous healing

These methods are described in subsequent sections.

8.4.1 Epoxy Injection

Epoxy injection is a proven technique for bonding cracked concrete sections. A successful epoxy injection requires evaluation, preparation, and planning. Epoxy injection forms the part of the structural repair. Epoxy injection can be used to restore structural soundness of buildings, bridges and dams where cracks are dormant or can be prevented from moving further. Cracks as narrow as 0.05 mm can be bonded by the injection of epoxy. The technique involves drilling holes at close intervals along the crack and then injecting the epoxy under pressure.

Except for certain specialized epoxies, the method can not be used if the cracks are actively leaking. Although moist cracks can be injected, but water or other contaminants in the crack reduce the effectiveness of the epoxy repair.

Success in epoxy injection requires correct understanding of injection materials, equipment, techniques and sequence of steps.

The basic steps and sequence of work needed in epoxy injection are described in subsequent paras.

I. Steps for Epoxy Injection

- Preparation of the crack;
- Drilling holes;
- Surface sealing of crack;
- Fixing of injection ports or nipples in the holes;
- Mixing of resin;
- Injection of resin;
- Removal of nipples and plugging the holes;
- Removal of sealing material and finishing the surface;

The preparation of crack is required to ensure perfect bonding of the injection material with the crack surface. The preparation of crack should aim at complete removal of dirt, loose materials and moisture (if the crack sealing system chosen is not compatible with moisture). This can be done with oil free compressed air or solvent depending upon the width of the crack and type of contamination. Remove contamination by flushing with water or effective solvent. Then blow out the solvent with compressed air or allow adequate time for air-drying.

Port holes are located along the crack. Holes are drilled for fixing port nipples of a suitable size and depth so as to avoid leakage of epoxy during injection.

Seal the surface to keep the epoxy from leaking out before it gets gelled. The surface can be sealed by brushing the epoxy over the surface of the crack and allowing it to harden. Cut the crack in a V-shape and fill with the epoxy and strike off flush with the surface for high pressure injection.

Install the entry ports with metal or plastic nipples specially formed for injecting epoxy. The key feature of nipples is its direct connectivity to the injection nozzle of the pump and allowing no loss of pressure. These nipples can be stuck to the surface or can go deep into the holes.

The spacing of injection ports depends upon the width of crack as well as the porosity of concrete. As a thumb rule, the spacing should be about half of concrete cross section thickness in case of adhesion nipples. Adhesion nipples should not be used for high-pressure injection exceeding 6 N/mm². Entry port spacing along the crack also depends on crack width, nature of the epoxy and thickness of the concrete element injected.

Mix the epoxy just before injection. Epoxies are multi component materials that react when mixed. In batch mixing the adhesive components are premixed according to the manufacturer's instructions usually with a mechanical stirrer like a paint-mixing paddle. In the continuous mixing system the two liquid adhesive components pass through metering and driving pumps before passing through an automatic mixing head. The continuous method allows the use of fast setting adhesives that have a short working life. Most injection applicators or contractors prefer continuous mixers.

Inject the epoxy through the ports with adequate pressure by simple hand operated caulking guns or hydraulic pumps. Control of injection pressure is important. Select the pressure carefully because too much pressure can extend the existing cracks and cause more damage. In most of cases technicians begin injecting at one end of a crack and proceed from port to port to the other end. For vertical cracks start by pumping epoxy into the lowest entry port until the epoxy level reaches the entry port above. Then cap the lower injection port and repeat the process at successively higher ports until the crack has been completely filled. For horizontal crack go from one end of the crack to the other in the same way. Crack assumed full when the pressure is maintained.

Remove the surface seal after the injected epoxy get cured. Remove the surface seal by grinding or some other appropriate means. Fillings and holes at entry ports should be painted with an epoxy-patching compound and smoothened by grinding.

Remove nipples and plug the holes next day after completing injection of epoxy. The sequence of work described is of utmost importance in case of vertical cracks and the injection should start from the **bottom most port** and continued until the resin flows out of immediate top port. Then the lower port should be sealed.

II. Injection Techniques

The simplest of the injection methods is the **brush injection**. The resin is brushed on the non-moving surface cracks and is absorbed in by capillary action. In case of pressureless injection, the material is poured into the **nipples** especially in case of pipes acting as nipples. The use of such injection depends largely on the dimensions of the crack. In case of structural cracks of width 0.2 to 1.0 mm, it is advisable to resort to **low pressure injection**. This low-pressure can either be created with **hand guns** (or grease guns) and normal diaphram pumps. The pressure developed is around 6-10 bars. Depending upon the crack width and depth, **high-pressure injection** can be resorted to for structural crack repair. It is possible to develop pressure to the tune of **50 N/mm²** using **mechanical or pneumatic pumps.** The injection method should be clearly specified prior to the commencement of the work and should be supervised by competent person to ensure suitable specifications.

Remove nipples after the injection resin has hardened. The surface sealing material (which is normally quick setting hydraulic system or thermoplastic resins) should be scrapped off completely and the surface should be prepared for further cosmetic fine finishing or strengthening treatment.

III. Materials and Equipments for Injection

Several proprietary **materials** and **machinery** are available for treating the cracks by injection system. Injection materials are mostly **synthetic resin** based or cement bound. The synthetic resins are usually two components comprising of epoxies and polyurethanes. The cement-based materials are invariably modified with polymers to impart flowability, non-shrinking characteristics and better bonding.

(a) Crack Repair Materials

(i) **Solvent free unfilled epoxy** are suited for injecting cracks of more than 1mm width. The viscosity as well as strengths are very high in case of solvent free epoxy.

(ii) **Solvent free epoxy modified** with fillers is suitable for cracks of about 1 mm. It has lower viscosity enabling better flowability than the unfilled.

(iii) **Epoxy injection resin** with a very low viscosity is suitable for injection of cracks wider than 0.2 mm for structural repairs.

(iv) **Epoxy injection resin** for structural repairs of cracks having still lower viscosity is suitable for injection of fine cracks.

(v) **Polyurethane** based two component materials, which form gel within seconds when comes in contact with water is suitable for crack repair where water is also present. This should be used as a primary injection to stop water and then normal polyurethane injection can be used for sealing.

(vi) **Water compatible,** two component polyurethane injection resin is suitable for non-rigid and elastic sealing of cracks. It is suitable for sealing cracks subjected to differential movements and standing water.

(vii) **Non shrink grout** is used to develop high strengths and flowability at very low w/c ratios. They are suitable for injecting dry or wet over 15 mm wide joints or cracks. This has the following charactristics:

Water/Powder ratio = 0.15, Compressive strength = 75 N/mm^2
Flexure Strength = 10 N/mm^2, Working time = 30 minutes.

(viii) **Polymer modified ready to use grouts** remain in suspension at a very low water cement ratio and are suitable for wet or dry joints aboout 2 mm width. The strength developed is also quite high.

(ix) **Polymer modified cement grout** is suitable both under wet and dry conditions. This has following characteristics:

Water/Powder ratio = 0.50, Compressive strength = 35 N/mm^2
Flexural strength = 3.5 N/mm^2, Working time = 25 minutes.

(b) Injection Equipment

The equipment required for crack injection ranges from a simple bucket with an outlet to most sophisticated pneumatically compressed machines capable of producing about 50 N/mm^2 pressure. These machines have hand-controlled nozzles and a mixing assembly to mix the two components at the point of injection. The sophisticated machines are designed to provide adequate working pressures, better nozzle nipple combinations to take care of pot life and proper crack sealing. The machines for injection can be assembled to provide desirable pressure controlled by attached gauges and nozzle nipple assemblies. Following equipments are normally used for crack injection repair:

(i) **Hand guns, sealant guns or grease guns** in which two components of epoxy are mixed and filled with a pressure of about 6-10 bars*. (0.6 to 1.0 N/mm^2).

(ii) **Foot pumps** can be employed to create a pressure up to 400 bars (40 N/mm^2). This is attached to single vessel container for adding premixed two components. These are normally suitable for small quantities of material.

(iii) **Automatic mixing** machines are also available in which the two components are separately introduced in two containers. Automatically controlled quantities can be mixed at mixing assembly near the nozzles before injection. This arrangement solves the pot life problems. These machines are connected to pneumatic machines for mechanical transmission and for creating a pressure to the tune of 500 bars (50 N/mm^2). Since the exact quantity of different components can be preset, these machines are very suitable for continuous injection operations avoiding wastage.

The type of material chosen depends upon crack width, movement and presence of moisture inside the cracks. Some of the commonly used epoxy materials along with their mode of application is listed in Table 8.2.

8.4.2 Grooving and Sealing

The simplest and most common crack repair method is routing and sealing. It is useful for cracks that are dormant and are of no structural significance. This method involves enlarging the crack along its exposed face for filling and sealing with suitable material. This method is suitable for sealing both fine pattern cracks and larger isolated cracks and other defects.

Do not attempt grooving and sealing work on active cracks or cracks subjected to strong hydrostatic pressure except when sealing the pressure face.

Clean the surface of the grooved joint with blow of air. Allow the joint to dry before placing the sealant. The purpose of the sealant is to create barrier to water from reaching the reinforcing steel or to stop hydrostatic pressure from developing within the joint to prevent staining the concrete surface. This helps in getting rid of moisture problems on the far side of the member.

Choice of sealant depends on how tight or permanent a seal is desired. Epoxy compounds are often used as sealants. Hot poured joint sealant works well when water tightness and appearance of the joint are not important. Urethanes, which remain flexible through large temperature variations, have been used successfully in cracks up to 3-4 mm width and of considerable depth.

TABLE 8.2: SELECTION OF INJECTION MATERIAL FOR CRACKS

TYPE OF CRACK	TYPE OF MATERIAL	MODE OF APPLICATION
1. Shrinkage cracks in concrete. Width = < 0.2 mm Movement = No Water = No	Two component injection Epoxy.	Surface treatment, Based on capillary action.
2. Shrinkage cracks in plaster. Width = < 0.2 mm Movement = No Water = Generally no	One component acrylic base flexible Paint.	Coat with roller brush.
3. Shrinkage cracks in concrete, Brickwork Width = 0.2-1 mm Movement = No Water = No	Two component Low viscosity injection epoxy.	Low-pressure injection, shorter cracks with high-pressure injection.
4. Shrinkage cracks in concrete, Brickwork Width = 1-2 mm Movement = No Water = No	Two component injection Epoxy and Solvent free Epoxy.	Low pressure injection
5. Shrinkage cracks in concrete, Brickwork Width = 2-5 mm Movement = No Water = No	Solvent free epoxy Thixsotropic.	Low pressure injection with hand pump.
6. Shrinkage cracks in concrete, Brickwork Width = >5 mm Movement = No Water = Dry/wet	Polymer modified cement-based grout.	Grout with injection pump or by gravity or hand pump.
7. Shrinkage cracks in concrete, Brickwork Width = >15 mm Movement = No Water = Dry/wet	Non Shrink Grout Mortar.	Cut and fill Non Shrink
8. Shrinkage cracks in concrete, Brick work Width = 0.2 to 1 mm Movement = Due to Temperature changes Water = Dry/Wet	Two Component Polyurethane injections and flexible paints. Primary injection with Polyurethane forming gel.	High or low pressure injection Then coat with roller/brush.
9. Butt joints In pre stressed concrete Width = 0.2 to 2 mm Movement = Vibration Water = Dry/Wet	Two component polyurethane injection and joint sealant/for wet joints. Primary injection with polyurethane forming gel.	For joints pressure injection, for floor joint seal with sealant using gun or spatula.
10. Moving cracks in concrete, Brickwork and floors. Width = > 2 mm Movement = Vibration Water = Dry/Wet	Sealants of different grades including flowable grades. For horizontal surface flowable grade of joint sealant.	Sealant gun or spatula

8.4.3 Stitching

When there are major cracks across which structural continuity must be reestablished, stitching may be the answer. This technique uses stitched dogs positioned across the cracks anchored with reinforcement fixed in holes by means of non-shrink grout. Dogs of several different sizes are positioned along varying planes to avoid stress concentration. It must be kept in mind that more the cracks are stiffened, the greater will be the tendency for the concrete to crack elsewhere. To combat this an overlay with welded wiremesh reinforcement can be applied to the area most likely to crack. An experienced structural engineer should be consulted if stitching is to be done.

Stitching is a simple cost effective technique for restoring the tensile strength of a cracked concrete element. It also can increase the shear capacity of flexural members. The most common stitching methods use either stitching dogs (U-shaped metal units), thin metal interlocking plates or dowel bars for reinforcement. In each method the reinforcement is installed across the crack and is bonded to each side of the crack with epoxy or cementitious materials. The amount of reinforcement is varied to achieve the desired strength restoration as per structural design.

(a) Stitching dogs

Unlike interlocking plates, dowel bars are embedded in the concrete. Stitching dogs are surface mounted. Concrete in axial tension, therefore, requires stitching dogs on both faces of the crack. Stitching dogs are most effective for restoring tension in bending members by placing them at the critical location on the tension face.

To install stitching dogs, drill holes on both sides of the crack and clean the holes then anchor the legs of the dog in the holes with non-shrink grout or epoxy. Vary the length, orientation and location of the stitching dogs so that the tension is transmitted across an area not across a single plane within the section.

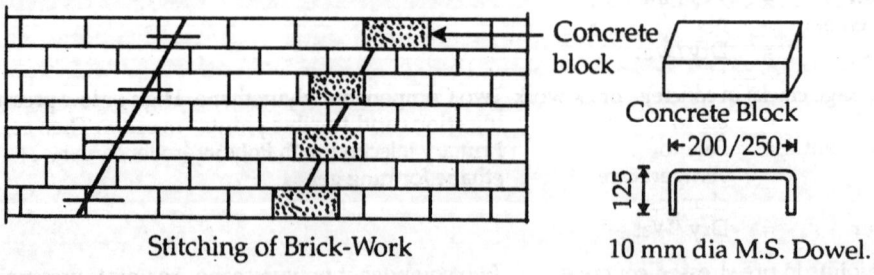

Stitching of Brick-Work   10 mm dia M.S. Dowel.

Fig: 8.1

Because dogs are thin and long and are not supported laterally they can not take much compressive force. If the crack closes and opens, stiffen and strengthen the dogs to prevent buckling. One method to prevent buckling of stitching dogs is to embed the dogs in an overlay. If an overlay is unsuitable stitching dogs can be placed in slots created by saw cut into the concrete. After the dogs are anchored in place use epoxy or cementitious materials to fill the slots.

Regarding the stitching method used, reduce the spacing of the reinforcing near the ends of the crack. If possible, drill a small hole at each end of the crack to stop the crack from propagating and to relieve the stress concentration.

Stitching a crack stiffens the local area increasing the restraint such that a crack may form elsewhere. If necessary, strengthen the adjacent area using external reinforcement such as bonded plates or a welded mesh reinforced **overlay**.

Stitching does not close a crack; it only prevents it from opening wider. If water seeps through the crack then seal the crack before stitching to protect the reinforcement from corrosion. The crack is easier to route and seal before the reinforcement is in place. For severely corrosive environments consider using epoxy coated or stainless steel reinforcement for dowels.

Most cracks can be stitched but a few cases avoid stitching cracks near control points of expansion joints. If the crack is too close to joint (within 30 mm), there is not enough concrete on one side to anchor the reinforcement and the crack will continue to widen. Also don't stitch a crack by embedding the reinforcement across a control or expansion joint. This prevents the joints from working as originally intended.

This method involves drilling holes on both sides of the crack and grouting in U-shaped metal units with short legs (called stitching dog) that span the crack. Stitching is suitable when tensile strength must be re-established across major cracks. Use either a non-shrink grout or an epoxy resin based bonding system to anchor the legs of the dogs.

(b) Interlocking Plates

To install interlocking plates make two saw cuts parallel and one saw cut perpendicular to the crack. Fill the saw cuts with cementitious grout then lightly hammer three interlocking plates into the saw cuts. The interlocking plates form an H-shape. The flanges act as anchor to transmit tension through the web. The plates can be embedded to any depth within the concrete section but typically are within **10-15 mm** of the top surface. Unlike surface mounted stitching dogs, the embedded plates won't buckle if the crack closes.

(c) Dowel Bars

To install dowel bars, drill two diagonal holes through the crack one on each side. Fill the holes with non-shrink cementitious or epoxy materials then drive a dowel bar into each hole. The bonded dowel bars transmit force across the crack face.

The angled dowel bars restore shear transfer and transmit axial tension but are not effective for restoring tension in flexural members.

8.4.4 Adding Reinforcement

Cracked reinforced concrete have been successfully repaired by inserting additional reinforcing bars to supplement epoxy injection. First the crack is sealed then holes are drilled across the crack plane at about 90 degree. Both the holes and crack plane are filled with epoxy pumped under low pressure (0.36 to 0.60 MPa) and reinforcing bars are inserted in the drilled holes. Silicone rubber gap sealants also work well and are especially suitable in cold weather or when time is short.

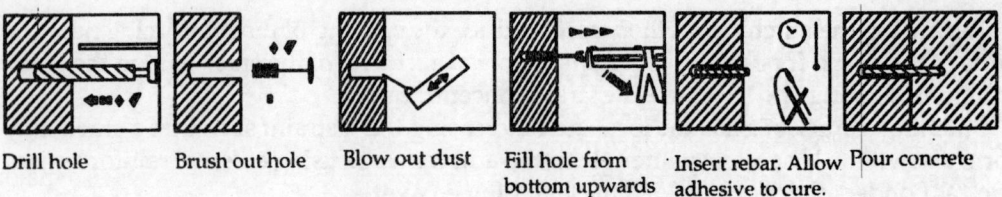

Drill hole   Brush out hole   Blow out dust   Fill hole from   Insert rebar. Allow   Pour concrete
                                                  bottom upwards   adhesive to cure.

Fig: 8.2: Adding Reinforcement

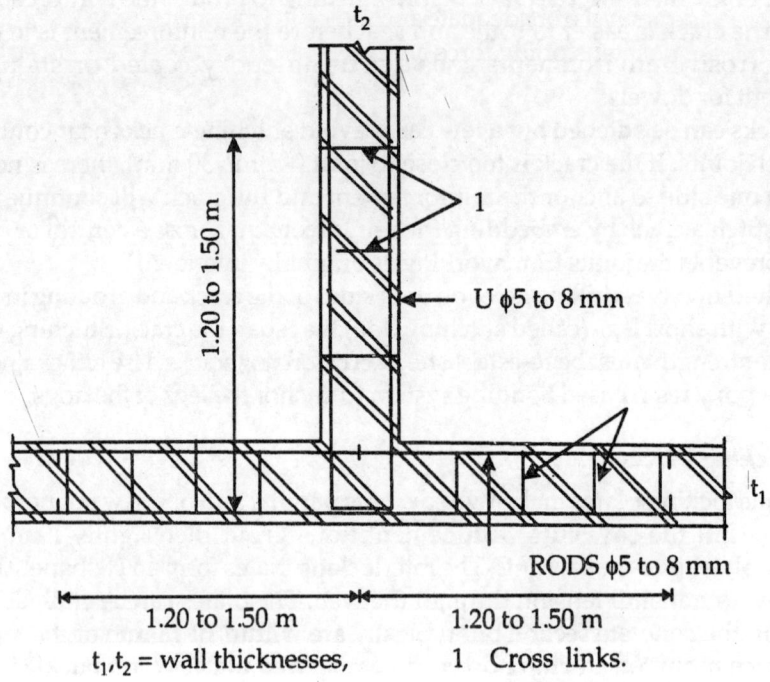

t_1, t_2 = wall thicknesses,   1   Cross links.

Fig. 8.3 (a): T-Junction-Strengthening by Dowel Reinforcements

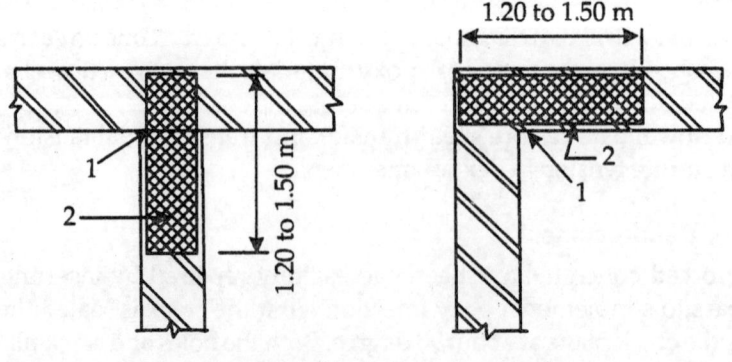

1 construction joint,   2 wire mesh.

Fig. 8.3 (b): Strengthening by Wire Fabric at Junction and Corner

The epoxy used to rebond the crack should have a very low viscosity and modulus of elasticity. The epoxy should be able to bond to concrete surface even in the presence of moisture and should be 100 per cent reactive to fill the crack. Appropriate reinforcement is laid and anchored across the crack. Sometimes the damaged portion is further reinforced with welded mesh near the surface using suitable epoxy concrete for embedment.

External pre-stressing and Post tensioning often provide a good solution when a major portion of a member requires to be strengthened or when a crack requires to be closed. Prestressing strands or bars are used to apply compressive force to the ailing member. This calls for design analysis for adequate anchorage for the prestressing steel, and analysis of the effect of the post tensioning force and eccentricity on the stresses in structure.

8.4.5 Portland Cement or Chemical Grouting

Wide cracks particularly in gravity dams and thick concrete walls may be filled with Portland cement grout. The procedure consists of cleaning the concrete along the crack by **compressed air and water under pressure**. Grout nipples are installed at intervals and the cracks are sealed with cement paint, sealant or grout. Flush the crack with water to clean it. Test the crack seal and then pump the grout. The grout is mixture of cement, admixture and water or cement plus admixture, sand and water depending on the width of the crack. Plasticisers, water reducers or other admixtures are also used to improve the properties of the grout. After the crack is filled the pump pressure should be maintained for several minutes to ensure good penetration.

Narrow cracks may be filled with chemical grout consisting of solution of chemicals which react with water to form gel, or solid precipitate. Concrete cracks as narrow as 0.05 mm have been treated successfully with chemical grout.

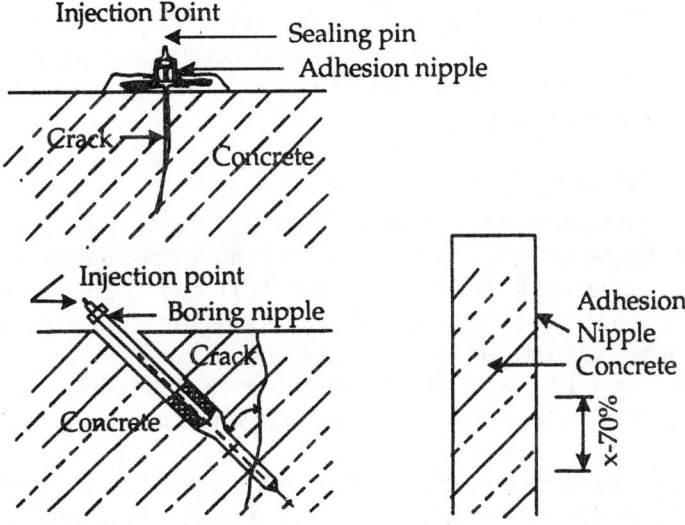

Fig. 8.4: Crack Grouting

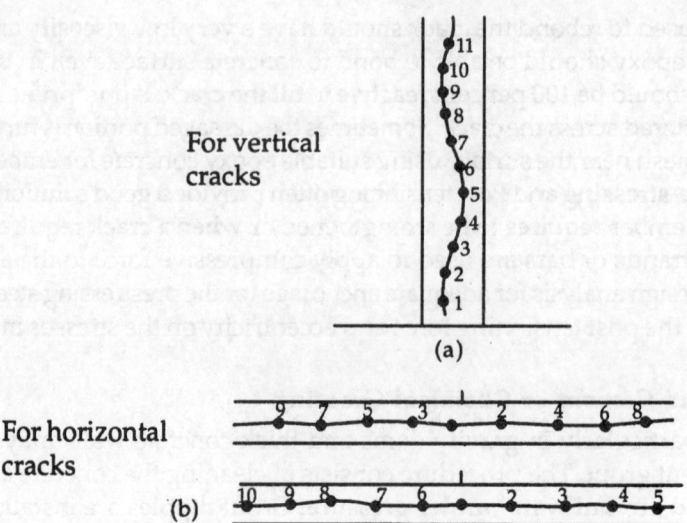

For vertical cracks

(a)

For horizontal cracks

(b)

Fig. 8.5: Showing the Pattern of crack injection

Chemical grouts can be used in very fine fractures in moist environments. Although there are wide limits of control of gel formation time, however, they have disadvantages of lack of strength and requiring high degree of skill for satisfactory use. The chemical grouts does not dry out in service.

8.4.6 Flexible Sealing

Active cracks can be cleaned by sandblast or water jet, or both and filled with a suitable field moulded flexible sealant. The slot or groove for the sealant should have a suitable width and shape factor for the expected movement.

The active cracks may be sealed with a flexible surface seal where the appearance is not important and the active cracks are not subjected to traffic or mechanical abuse. Using a bond breaker over the crack permits movement and the flexible joint sealant is troweled on top for bonding to adjoining concrete.

8.4.7 Dry Pack Mortar

Dry pack is the hand placement of low water content mortar followed by tamping or ramming of the mortar into place producing tight contact between the mortar and the existing concrete. There is little shrinkage and the patch remains tight with good durability, strength and water tightness. Drypack is used for the repair of dormant cracks but is not recommended for filling or repairing active cracks.

Before attempting a dry pack repair, widen the crack at the surface to a slot about **25 mm wide** and **25 mm deep**. A power driven saw tooth bit is a good tool for cutting this groove.

Clean and dry the slot thoroughly and then apply a bond coat consisting of cement slurry or equal quantities of cement and fine sand mixed with water to a fluid paste consistency. Bonding admixture can also be added to the dry pack mortar. Place the dry pack mortar immediately.

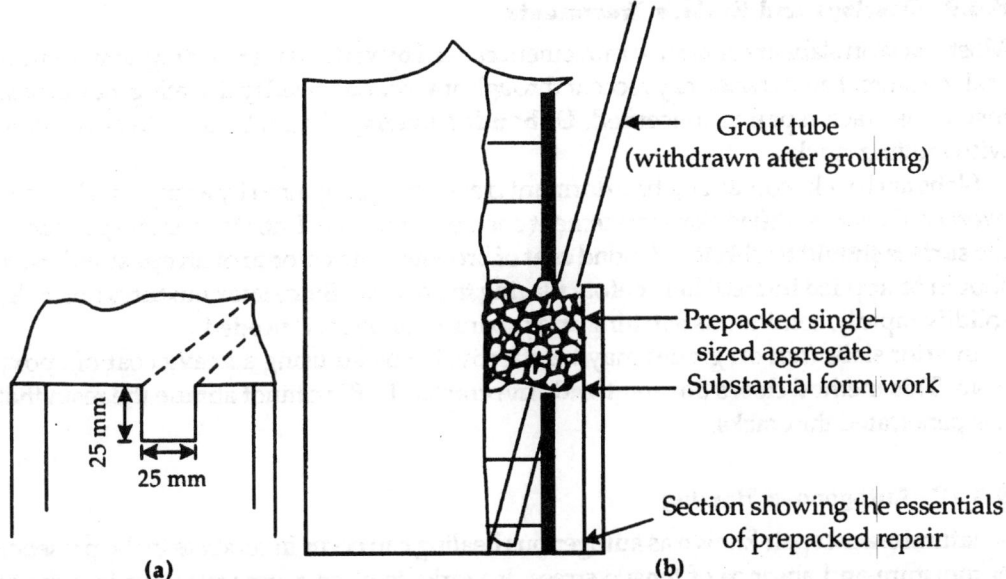

Fig. 8.6: (a) Dry Pack Method, (b) Repair of a Column: Section Showing Detailed Formwork with Grout Tube in Position

Place the flat side of a hardwood piece against the mortar to finish and strike several times with a hammer. Surface appearance may be improved by a few light strokes with a rag or sponge float. Cure by applying a curing compound or supporting a strip of folded wet burlap along the length of the crack.

8.4.8 Polymer Impregnation

A monomer system is a liquid that consists of small organic molecules capable of combining to form a solid plastic. These monomers vary in volatility, toxicity and flammability and do not mix with water. These are fluid and can be soaked into dry concrete filling the cracks much the same way as water.

Monomer systems used for impregnation also contain a **catalyst** or initiator and the basic monomer (or combination of monomers). When heated the monomers join together or polymerize becoming a tough strong durable plastic that greatly enhances a number of concrete properties.

If a cracked concrete surface is dried, flooded with the monomer and polymerized in place, the cracks get filled and structurally repaired. However, if the cracks contain moisture the monomers shall not soak into the concrete **at the moist crack face resulting in unsatisfactory repair**. If a volatile monomer evaporates before polymerization it will be ineffective. Polymer impregnation has not been successful in repairing fine cracks.

Badly fractured beams have been repaired using polymer impregnation by drying the fracture temporarily encasing it in a tight monomer proof band of sheet metal soaking the fractures with monomer and polymerizing the monomer with heat. Large voids or broken areas in compression zones can be filled with fine and coarse aggregate before doing a polymer repair.

8.4.9 Overlays and Surface Treatments

Most cracks in slabs are subject to movement caused by variations in loading, temperature and moisture. These cracks may reoccur through any **bonded overlay** defeating the purpose insofar as crack repair is concerned. **Unbonded overlay** should be used to cover slabs with moving cracks.

Slabs and decks containing fine dormant cracks can be repaired by applying a bonded overlay of latex modified Portland cement concrete or mortar. Prior to overlay application the surface should be cleaned. A bond coat of broomed latex mortar or an epoxy adhesive should be applied immediately before placing the overlay. Since latex mixtures normally solidify rapidly, continuous batching and mixing equipment is needed.

Interior slabs (not on grade) may be effectively coated using a heavy coat of epoxy resin. This method closes **dormant and fine cracks**. Traffic cannot abrade the resin that has penetrated the cracks.

8.4.10 Autogenous Healing

A natural crack repair known as autogenous healing can occur in concrete in the presence of moisture and absence of tensile stress. It works to close dormant cracks in a moist environment due to continuous hardening process.

This healing depends on the carbonation of calcium hydroxide in the cement paste by carbon dioxide, which is present in the surrounding air and water. Calcium carbonate and calcium hydroxide crystals form and grow within the cracks. The resulting chemical and mechanical bonding between the crystals and the surfaces of the mortar and the aggregate restores some of the tensile strength of the concrete across the cracked section and the crack may get sealed.

Healing will not occur when the crack is subjected to movement. Healing will also not occur if there is a positive flow of water through the cracks. Water dissolves and washes away the lime deposit unless the flow of water is so slow that complete evapouration occurs at the exposed face causing redeposit of the dissolved salts.

Saturation of the crack and adjoining concrete with water during the healing process is essential for development of any substantial strength. If the saturation is not continuous for the entire period of healing the regained strength may be lost again.

8.5 SUMMARY

Surface preparation is critical to repair, resurfacing of topping or coating for long term performance of the substrata. Surface contaminants either solid or liquid that causes problem related to adhesion, curing, or application of repair mortar, must be properly removed. The various methods of surface preparation includes chemical cleaning, steam cleaning, shot blasting, sand blasting and vacuum cleaning.

Chemical cleaning is done by using hot water solution of trisodium phosphate. Steam cleaning is done by producing wet or dry steam directed at the surface at a high velocity to loosen, soften and remove contaminants. **Shot blasting** involves impacting the surface with high velocity abrasive steel shots. Sand blasting is method of preparing any surface including reinforcement by impact with a high velocity stream of fine mineral abrasive aggregate propelled by clean compressed air. Acid etching is done using muriatic acid diluted to required strength and is poured on the concrete, which is then flushed with clean water.

Vacuum Cleaning is done with oil free compressed air and is prepared when the dispersion of dust is to be controlled.

The main considerations in surface preparation are removal of weak concrete, contaminated concrete, spalled concrete and also removal of corrosion products from steel reinforcement. Common repair techniques for concrete surfaces include **caulking, coating, replacement by plastic concrete, replacement by dry pack, grinding, injection, jacketing, replacement by prepacked concrete,** thin or **bonded** or **unbonded** surfacing, **stressing** and **shotcreting**.

Caulking is the filling of narrow ruptures in concrete by plastic material. **Coating** is the painting of a film forming plastic or liquid coating over the distressed surface. Repair of badly damaged surface is done by replacing the surface with plastic concrete. Cavities which are **deep and not very wide are repaired by dry pack** having low water content. **Grinding** or **grooving** is done for concrete surface which have shallow cracks or irregularities to enhance the bonding characteristic. **Epoxy resin injection** is used to repair narrow cracks.

Jacketing is done to make the surface more resistant to the environmental forces that cause deterioration. **Prepacked concrete** is used to **fill cavities** by graded aggregates which are **inundated** with water to saturate. Thin resurfacing is carriedout by the application of a uniform layer of repair material over an existing area of concrete. For defects such as **spalling or scaling,** the repair is done by bonded resurfacing. In case of surfaces subjected to chemical attack unbonded surfacing is done by separating the substrata fromthe resurfacing layer. **Stressing** involves embedding rods or cables in the distressed concrete. **Shotcreting** is the process of shooting concrete or mortar under pressure on the prepared deteriorated area for strong adhesion of repair material.

Live cracks due to foundation settlement or other structural changes must be repaired. **Epoxy injecting** technique for narrow cracks involves drilling of holes along the crack and injecting epoxy under pressure after sealing the crack. **Grooving and Sealing** is done for cracks that are dormant and are of no structural significance. **Stitching** is simple cost effective technique for restoring the tensile strength to a cracked concrete section. Cracked RCC is repaired by adding reinforcement bars in case of severe corrosion. Wide cracks in thick concrete walls can be repaired by **cement grouting** or **chemical grouting**.

Active cracks are repaired by cleaning with air, water jet and filling with flexible sealant. **Drypack** is used to repair dormant cracks. **Polymer impregnation** is used to repair badly fractured elements. Cracks subjected to movement caused by variation in loading, temperature and moisture are repaired by providing **unbonded overlay of latex modified portland cement concrete** or mortar. A natural crack repair (known as autogenous healing) occurs in concrete in the pressence of moisture and absence of tensile stresses. It works on dormant cracks in a moist environment due to continuous hardening process.

QUESTIONS

8.1 State objectives of surface preparation prior to repair of any building element.

8.2 List methods of preparing surface for repair.

8.3 Write short notes on

(a) Shot blasting

(b) Sand blasting

(c) Air blast

(d) Acid etching

8.4 Describe main principles involved in surface preparation for repair.

8.5 List most common techniques used in repairing building elements.

8.6 Write short notes on:

(a) Caulking

(b) Replacement by plastic concrete

(c) Injection

(d) Shotcreting

8.7 Differentiate between:

(a) Coating and resurfacing

(b) Drypack and prepacked concrete

(c) Bonded and Unbonded resurfacing

(d) Stressing and jacketing

8.8 Name common techniques of crack repair.

8.9 Explain briefly epoxy injection steps for crack repair.

8.10 Explain briefly epoxy injection materials and equipment.

8.11 Differentiate between:

(a) Grooving and grouting techniques of crack repair.

(b) Adding reinforcement and stitching dogs

(c) Flexible sealing and autogenous healing

8.12 Explain briefly the stitching technique of crack repairing with sketches.

8.13 Write short notes on:

(a) Dowelbars

(b) Chemical grouting

(c) Polymer impregnation

(d) Overlays

UNIT–III

REPAIR AND REMEDIAL MEASURES FOR BUILDINGS DEFECTS

11

Chapter 9

REPAIR OF EXISTING DAMP PROOFING SYSTEMS IN ROOFS AND WET AREAS

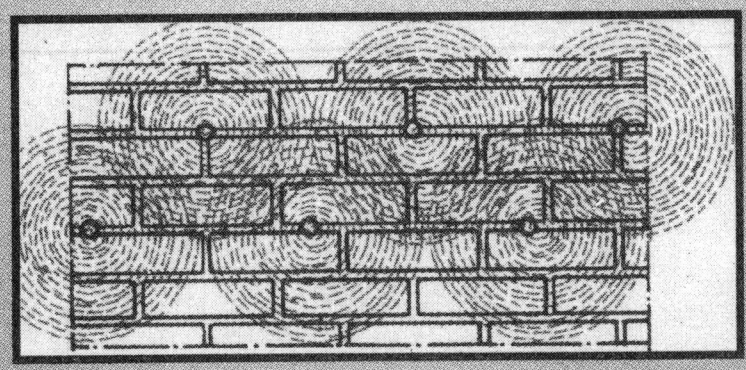

9

REPAIR OF EXISTING DAMP PROOFING SYSTEMS IN ROOFS AND WET AREAS

LEARNING OBJECTIVES

After studying this chapter, the learner understands repair of damp proofing systems in building elements and will be able to:

- **Describe** importance of damp proofing of building elements specially roof and wet areas;
- **Describe** sources of dampness;
- **Explain** the method of repairing existing water proofing system in wet areas;
- **Explain** the method of repairing existing damp proofing system of flat roofs;
- **Explain** the modern materials and techniques in repairing of existing water-proofing systems;
- **Explain** the protective coating to repair and enchance effectiveness of existing damp proofing systems.

9.1 INTRODUCTION

Water proofing is a critical element in design, construction and occupation of a building structure. Water infiltration and leakage severely damage a building structure and its elements. Water promotes growth of unwelcome life such as rot and insects. It consumes wood, erodes masonry, corrodes metals, peels paint, and expands building elements. It causes warping, swelling, discolouring, rusting, loosening, mildews growth and stinking. Because of the damaging effects of water, we must pay special attention to **selection of quality water proofing system which** provide proper in-place performance.

9.2 SOURCES OF WATER

Most buildings are constructed using a variety of materials, many of which are water absorbent. Dampness is a very common phenomenon in all types of building structures. Dampness **represents the water out of place** and is due to multiple sources. Basically water exists in the structure in liquid form or water vapour condensed from the atmosphere. The water ingress or existence of dampness in the building is by the following means:

- Construction water
- Intruding water
- Condensation water
- Occupational water

The most common sources of moisture movement or water leakage are due to structural defects such as cracks and voids, or construction and control joints. Below-grade areas are additionally susceptible to fluctuating water tables. The excess dampness in buildings can cause lot of deleterious effects.

Water proofing is the most important process in maintaining the beauty and increasing the life of the buildings. Water proofing provided in new buildings gives satisfactory performance for few years. But as the building becomes older, it settles unevenly, forms cracks in walls and roof slab. Water proofing system gets deteriorated or becomes ineffective and thus initiating leakage and seepage through cracks.

Dampness is caused due to inadequate drainage of rain water from roof slabs and obstructions in the drainage system. Leakage and dampness is also caused through the joints and cracks in embedded water supply or sanitary pipes. For eliminating dampness, these root causes must be set right before repairing damp proofing system.

Dampness brings a lot of avoidable unhappiness and inconvenience for the occupants and the builder. This can be avoided by taking proper care during water proofing installations. For ensuring the durability and proper functioning, the construction engineers must do the work **right at the first time**.

9.3 REPAIR OF WATER PROOFING SYSTEMS IN WET AREAS

The wet areas in the building comprise of areas where water can be spilled due to occupational W.C, bathroom and kitchen. The leakage through bathroom occurs at the junction of pipe and floor trap or floor and wall, P traps and joints between trap and pipe leading to the outside face of the building.

The sources of leakage through kitchen are the area surrounding the sink, joints between the sink and draining pipe, junction point between the kitchen platform and the wall, junction of wash basin and the floor trap lead joint between outlet and floor slab. Dampness arising out of plumbing defects are normally found in the service area. This is generally due to constructional defects and/or poor maintenance. Leakage from drainage or water supply pipes and fittings should be immediately attended to prevent dampness in walls.

Maximum water is used in wet areas. The problem of leakage through wet areas appears from joint between flushing cistern and flushing pipe, junction of flushing pipe with W.C. pan, junction of W.C. and traps with the branch pipe, depression of reinforced concrete slab to accommodate the pan and trap, joints between pan with the flooring, incorrect placement of overflow pipe and floor traps. The sunken slab is required to be completely waterproofed before and after levelling and installation of the fixtures. Major damp proofing system requires comprehensive steps to carry out repairs in totality. Patch repair work may only suffice for minor problems and for a short period.

9.3.1 Preparation for Repair of Wet Areas

The following work should be completed before starting repair of any water proofing work to prevent reoccurance of dampness in wet areas:

- Completion of internal plaster of walls before occupation
- Removal of existing defective and deteriorated water proofing system completely
- Completion of grooving for concealed water supply (G.I or electrical conduit) piping in wet areas
- Removal of all debris and chiselling of the extra mortar, if any, to expose the bare slab completely
- Completion of making holes in external walls for connecting various traps to external laying drainage line

9.3.2 Laying of Water Proofing System

Water proofing system should be laid as per the following steps:

1. Apply base coat of cement slurry mixed with bonding agent
2. Install water proofing system suitable for damp application. The best suitable materials are Acrylic Polymer Modified Cementitious coatings or liquid polymer membranes.
3. Provide protective rendering of cement mortar 1 : 4 admixed with integral water proofing compound.
4. Provide for water proofing of pipe to ensure elimination of trouble in the hidden wet areas.

Check possibility of leakage at the following points in wet areas:

- Joint between C.I/P.V.C pipe piece and P trap
- Joint between Tee of outer vertical stack and PVC/C.I. pipe piece
- Joint between P trap and W.C. pan
- Joint between flushing pipe and W.C. pan
- Gap between nahani trap and the floor
- Gap between floor tile joints
- Wash basin drain pipe connection with traps

All these points should be checked and the leaks in joints should be re-sealed with the proper sealant. The escape water pipe of flush tank should be cleaned of any blockade. After checking and sealing all the concealed pipeline if the leakage does not stop, apply cement slurry or aqua reactive grouts by pressure grouting the base slab and walls. Wet area floor is broken and removed completely for laying new water proofing system in case of failure of the piece meal repair system.

9.4 REPAIR OF EXISTING WATER PROOFING OF FLAT ROOFS

9.4.1 Introduction

Leakage from roofs is the **most common problem** in buildings. This problem may require for extensive repairs. Several methods of treating the roof damp proofing are practiced in different parts of the country. However, lack of proper detailing, use of inferior materials and /or poor workmanship often leads to problems of leakage in buildings. Hence there is a need to emphasise these aspects of construction and repair of water proofing treatment.

Unlike a sloping roof, a flat roof is more prone to leakage, stagnation of rain water for absorption or drain off at a slow rate. Reinforced concrete flat roofs, undergo movements due to variation in moisture, temperature, loading or other causes. Unless precautions are taken during design and construction to accommodate these movements, cracks may develop in the slab, leading to leakage. Permeability of concrete is another factor, which influences the behaviour of the slab against leakage. A water proofing treatment has to be provided over the roofs. The thermal resistance of 100 to 120 mm thick R.C.C slab is not adequate in tropical climate for comfort conditions. There is need for the provision of thermal insulation over the slab. It is therefore necessary to provide two treatments, one for thermal protection and the other for water proofing system. These two treatments are required to be properly integrated together. Ideally, the water proofing treatment should be placed on the exterior to prevent ingress of moisture into the thermal treatment. But it is usually not possible with conventional materials such as bitumen and polyethylene. Since the nature of these materials used for water proofing need their own protection against direct sun rays and wear and tear caused by usage, hence, the waterproof layer is generally provided below thermal treatment. Alternatively, the water proofing treatment is covered with a protective layer, making it a three-tier treatment.

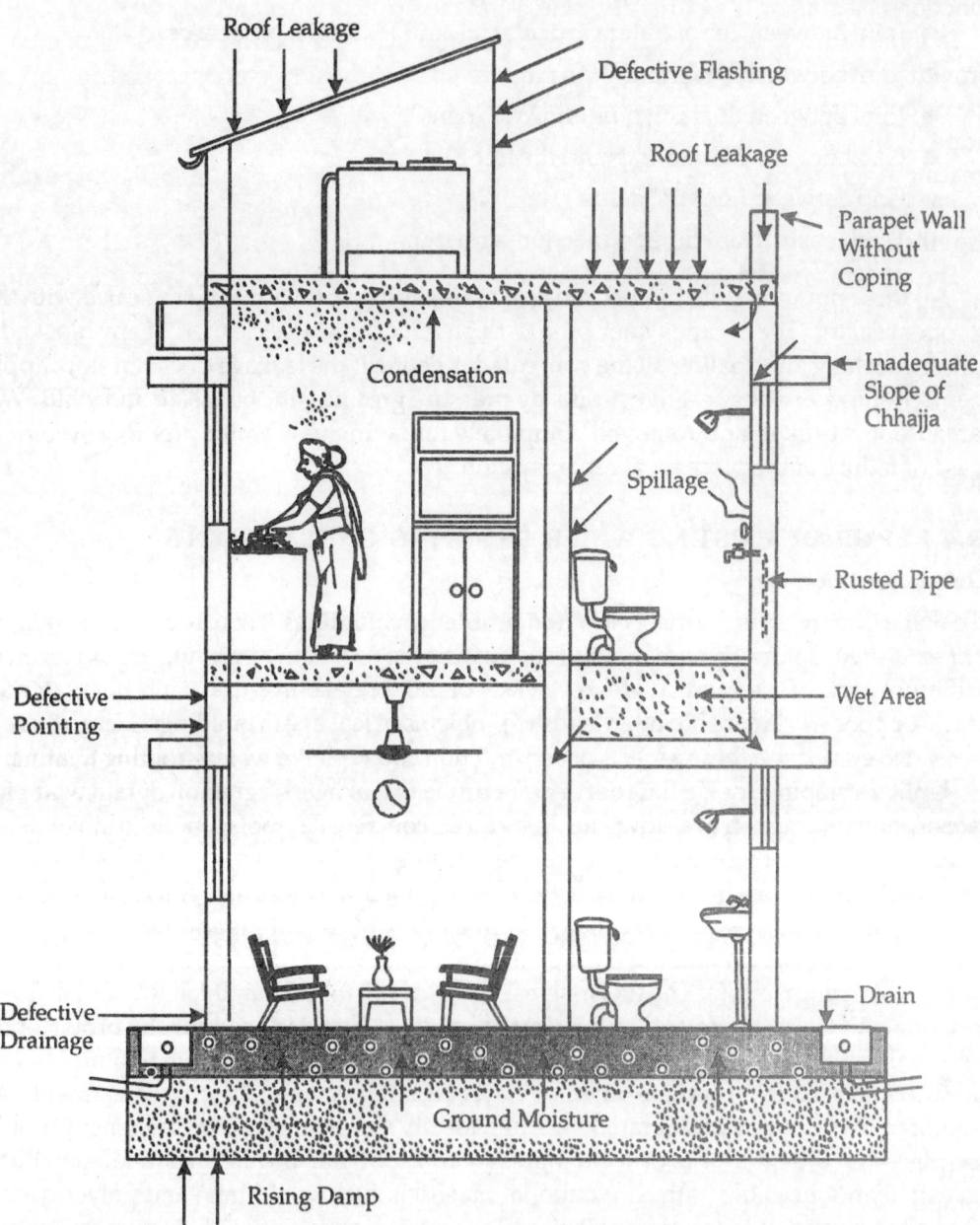

Fig. 9.1: Sources of Dampness in Building

The maintenance engineer has to consider the interaction of the water proofing system and thermal insulation and should devise suitable details so that the roof treatment functions satisfactorily against dampness and mechanical wear and tear

To eliminate leakage from roof, **stagnation** of water on roof should be avoided by providing **adequate slope**, limiting the area served by each rain water outlet and designing the outlet diameter to carry maximum expected discharge. It is preferable to provide a slope of 1 in 40 to 1 in 30. As it is not economical to provide the slope fully in the roof treatment, it is advisable to provide part of the slope in the structural slab itself. The rain water pipe should not be less than 100mm in diameter and one such pipe should be provided for each 40 sq.m of roof area in moderate rainfall zones.

The roof treatments, commonly used in the country can be broadly classified as **rigid**, **elastic** and **composite**. Lime concrete, brick coba and mud phuska with brick tile are rigid treatments commonly used for accessible roofs. However, they are liable to develop cracks due to temperature variations and need special care at the junction of roofs and parapet walls. These treatments are suitable for areas having low rain-fall. Bituminous felt, polyethylene sheets, glass fibre reinforced sheets are some examples of elastic treatments. They are easy to lay and provide satisfactory protection at junctions of walls and other projecting features. They do not provide thermal protection and have limited life span. Bituminous felt, being black in colour, absorbs lot of heat from solar radiation. **Composite treatment is a combination of elastic and rigid** treatments (such as felt laid on lime concrete, mudphuska laid on polyethylene). Such treatments can be designed to withstand any given conditions of rainfall in addition to providing adequate thermal insulation and protection.

9.4.2 Water proofing Systems

The type of water proofing systems should provide the required quality of water proofing. The systems should also include the critical components necessary to protect the structure from the damaging effects of water. The selection of water proofing system for repair and maintenance treatment has to pass through the several stages as mentioned below:

- Investigation and Diagnosis
- Selection and specifications of appropriate materials and systems
- Application, Quality Control and protection
- Post application inspection and its subsequent **maintenance**

Negligence at any stage can lead to failure of the system. Different membranes and materials are available to satisfy the need of water proofing of buildings. Drainage composites augment the system by providing an on-site drainage surface and protection course. Complimentary products including primers, surface conditioner, liquid membrane, mastic and adhesive protection board improves the system. **Following ideal characteristics of water proofing system** are required to be considered for making the system durable and effective:

- **Impermeability** to prevent passage of water
- Minimize water **absorption** by substrate
- Good **bonding** with substrate
- **Elasticity** to resist cracking and remain **flexible** under various climatic and environmental conditions
- Ease of **application** and minimize application errors
- **Compatibility** with substrate
- **Resistance** to ultraviolet rays
- **Resistance** to service **temperature** variations
- **Resistance to mechanical wear and tear**
- Long life and **durability** with resistance to chemicals present in the environment
- Adoptable to prefabricated drainage composites to protect the water proofing course
- Accept **flood tests**
- Economical

All building structures are also surrounded by protective envelopes. All such envelopes particularly below-grade areas and decks are sometimes subjected to hydrostatic pressure during their service life. Therefore, the use of a quality water proofing system is essential to preserve the water tightness of the protective building envelope.

9.4.3 Common Techniques of Laying Water Proofing

These systems cannot be repaired in patches and need to be replaced in totality. For the water proofing system generally factory made membranes are used. Common steps to carryout repair jobs of flat roof damp proofing are described as under :

- **Dismantle** all tiles, mud phuska, and bituminous coatings;
- **Remove** all dust and old coatings with wire brushes and wash with kerosene oil to make the surface smooth and clean;
- **Wash** substrate with clean water and dry it completely;
- **Prepare** the surface for appropriate slope and fill the cavities with water proof epoxy mortars;
- **Treat** the surface with appropriate **primer** or surface conditioner to accept the water proofing layer;
- **Treat** all inside and outside corners;
- **Apply** water proofing system of tarfelt or bituminous coats;
- Install additional liquid membrane to complete details;
- **Inspect** the water proofing system and repair cuts, fish mouths or tears;

- **Protect** the water proofing system from potential physical damage or direct ultra violet rays;
- Lay mud phuska and surface tiles to the required slope and specifications;
- Fill the surface tile joints with water proof cement mortars.

Various techniques are used to obstruct the passage of water/moisture through concrete and masonry to prevent leakage or seepage through building elements. The conventional techniques used in water proofing flat roof slabs are:

(a) Mud Phuska with brick tile toppings

(b) Lime concrete terracing

(c) Ferro cement toppings

(d) Brick coba

The roof slab should be repaired or replaced immediately in case of minor defects such as damage of gola, broken tiles and damaged pointing. However, major defects can be replaced by relaying the total system. The main treatment and relaying should be carried out carefully as described in subsequent paras.

(a) Mudphuska with Brick Tile Treatment

This treatment usually consists of the following steps:

- A coat of hot bitumen @ 1.7 kg per sq.m of roof area or polyethylene sheet laid over the substrate flat roof slab after surface preparation.
- 100 mm (average) thick mud phuska laid to slope with material consisting of puddled clay mixed with chopped straw (Bhusa) @ 8 to 10 kg. of straw per cubic metre of clay.
- Mud plaster laid to slope and consisting of puddled clay mixed with chopped straw @30 to 35 kg. of straw per cubic metre of earth clay.
- One or two coats of rendering by cow dung slurry (gobri) laid to slope and consisting of fine clay and cow dung in equal quantities, mixed with adequate quantity of water.
- One or two layers of burnt brick tiles laid to slope on a bed of mud mortar and jointed and pointed in cement mortar 1 : 3.
- Top surface pointing is done with highly plastic cement mortar prepared by adding water proofing admixtures to 1 : 3 cement; sand mortar.

Bitumen coating on the flat slab seals the pores of concrete surface and thus provides a barrier to the seepage of moisture through concrete. The coating should be done when the slab and the parapet walls are thoroughly dry. The surfaces should be firstly cleaned with wire brush and with cotton soaked in kerosene oil. Though residual type petroleum bitumen of grade 80/100 is commonly used, it has got a tendency to flow in hot summer days. Hence blown bitumen of grade 85/25 is preferred. Residual bitumen is heated to a temperature of 180° to 190° C while the blown bitumen is heated to 170° to 180° C and applied immediately with brushes by spreading quickly. It should be applied evenly over the entire roof surface and the vertical face of parapet wall upto the drip course, without leaving any blank patches. Air bubbles formed during the coating should be punctured and coated again with hot bitumen. Care should be taken to ensure that bitumen does not get hardened before applying, otherwise blank patches will be left on the painted surface.

Mud phuska is laid over the bitumen coated roof slab for thermal insulation and creating the roof drainage slope. Mudphuska acts as thermal insulation layer and absorbs ingressed of moisture to some extent from brick tiles and retain this moisture to protect RCC slabs. Clay suitable for brick making can be used for preparing mudphuska. It is mixed with straw and water and worked up with spades and allowed to mature. The mudphuska mortar should have only just enough moisture to retain the shape of the test ball. Mudphuska should be laid to the specified thickness as per required drainage slope. It should be compacted and allowed to dry. The cracks developed in the mudphuska are filled up by the application of mud plaster. Mudplaster is prepared like mudphuska with finely sieved clay and chopped straw. As the quantities of straw is more it is susceptible to less shrinkage. After the mudplaster driesout apply rendering (leeping) coats of cowdung slurry (gobri) to fill up cracks and pores in mudplaster.

The tiles are provided to withstand the driving effect of the rain and mechanical wear to serve as wearing surface for functional use. It also assists in draining off water by providing a hard and comparatively less absorbent smooth plane in slope. As the tiles are directly exposed to temperature variations, setting the tiles in cement mortar will accentuate the tendency to crack. They should be preferably bedded on a layer of mud, mortar, after the 'leeping' gets dry. Where rainfall is high, the tiles should be provided in two layers, so as to break joints between the two layers. In areas where rainfall is low, single layer of tiles is sufficient. It is advisable to use composite mortars of 1 : 1 : 6 or 1 : 1 :5 (cement : lime : sand) with crude oil 5% by the weight of cement added to it, for jointing the brick tiles . The joints may also be pointed with cement : sand or composite mortar.

(b) Lime Dhar Terrace

Lime concrete layer serves both as water proofing and thermal insulation layer. The slow setting property of lime permits prolonged beating of lime concrete for better compaction, sealing of pores and formation of a water tight skin on the top. The treatment has been very successful in areas, where the temperature variation and rainfall are low and where it is used over stone slabs. However, in areas where temperature variation is high, the thermal movement of RCC slabs may cause cracks in lime concrete, through which water may percolate. The water resistance of the treatment depends upon the type, quality of lime, grinding of lime and compaction achieved during laying. Surkhi may be used with fat lime for preparing the concrete. Strict control over proper bitumen coat or tarfelt layer, grading and mixing of the materials, workmanship and compaction are essential for good performance.

Fat lime, surkhi and graded brick ballast of size 25 mm and down are used for making lime concrete. The quality of lime should be tested, before using the same for lime concrete. One part of slaked fat lime and two parts of surkhi by volume should be mixed on a watertight platform. It should be sprinkled with the required quantity of water and should be grounded in a grinding mill or mechanical grinders. The lime concrete should be prepared by mixing the brick ballast and lime-surkhi mortar thoroughly in proportion of 2.5 : 1 parts by volume. Burnt brick aggregate should be thoroughly soaked in water for at least 6 hours, before using in concrete. Lime concrete is laid in single layer of about 100mm spread and compacted with wooden rammers to the required slope. To avoid surface cracking in lime concrete, fiberous materials like jute fibers, jaggery water and fenugreek powder are sprinkled on the top surface. The concrete is further consolidated by beating manually with wooden or bamboo 'thappies'. Special care is needed to consolidate the concrete properly at junctions with parapet wall. Beating is continued for 3 to 4 days until the mortar is almost set and wooden 'thapies' rebound from the surface. While beating, the surface is liberally sprinkled with **fenugreek powder** and a solution of **'gur' and boiled bael** fruit in the ratio of 1.75 kg of 'gur' to 1kg of bael fruit boiled in 60 litres of water.

For effective tamping of lime concrete, a machine has been developed at CBRI. The machine has three identical tamping units and two rollers fixed to an angle iron chassis on which an electrical motor or a petrol engine is mounted. The machine moves forward and backward on the two rollers. Using the machine, lime concrete can be thoroughly compacted. About 240m2 of roof terrace can be finished in two days time, using the machine.

In areas where the temperature variation is large or where good quality lime and surkhi are not available, the lime concrete should be further overlaid with one or two layers of flat tiles. In case flat tiles are not available, pressed tiles, Shahbad stone slab may be used. The joints should be grouted with 1:1:6 cement : lime :sand gauged mortar with 5 percent crude oil by weight of cement added to it. The joints can be pointed with cement : sand: or composite mortar and rubbed with iron float to have a hard polished finish. The disadvantage of lime terracing is its heavy dead weight on the substrate slab.

Lime concrete terracing should be repaired after an interval of 5 years by scrapping top surface and relaying the same with lime cream and composite mortar. This is done by thoroughly wetting the surface and compacting by beating with wooden thappies. The mortar is mixed thoroughly and solution sprinkled with fibres and fenugreek powder and gur-bael solution. The surface is finished smooth to the desired slope.

(c) Ferrocement for Water Proofing

Ferrocement is highly crack resistant , impervious and can withstand thermal changes very effectively. Ferrocement has been successfully used for providing water proofing treatment over R.C.C., reinforced brick, stone or even wooden slabs.

Ferrocement provides higher tensile and flexural strengths, better resistance to impact, fracture and failure. It also provides a crack free, tough, dependable surface, free from danger of leakage and corrosion. Since reinforcement in ferrocement is provided in the form of well distributed wiremesh layers, it can carry large strains without cracking. High specific surface and close distribution of reinforcement also results in an efficient and improved crack arrest mechanism. It can be safely adopted for water proofing treatment of structures constructed with reinforced concrete, reinforced brickwork and precast components.

Ferrocement can be used for water proofing of new as well as old flat and sloping roofs. In case of existing roofs which are not giving satisfactory performance, the roof treatment is removed to expose the roof slabs. Cracks, depression and honey combed surfaces are repaired using waterproof cement mortar before laying ferrocement

Ferro cement treatment is carried out after surface preparation and treatment.

Surface preparation requires the entire roof surface including parapet walls to be cleaned with wire brushes and washed with water. Surface cracks, if any, should be treated appropriately before undertaking surface treatment.

Surface treatment requires the porous and honeycombed surfaces of the roof slab to be grouted by **Cement slurry**. Cracks of more than 1mm should be chased out to 'V' shaped groove and filled with polymerised mortar. For efficient treatment the surface and parapet walls should be kept in saturated and surface dry condition.

Ferro cement roof laying comprises of first laying base course. The base course is laid on the treated surface dry roof slab. The base course consists of thin coat of cement slurry with admixture to fill cracks and pores in mudplaster. A bonding agent coat @ 1.3 Kg/ Sqm is painted by brush prior to cement slurry coat. The base course is of 6mm. thickness in 1:3 cement : sand mortar with maximum water cement ratio of 0.45 and mixed with non shrink **water proof additive** such as Conplast X-421IC @150 ml per bag of cement. The base course is made rough with hard coconut brush. For fixing wire mesh, wire nails are fixed on this base course at appropriate spacing with projection from the base course. Adequate curing should be done for 3 to 5 days.

Welded wire mesh 50mmX50 mm X10 gauge covered with GI 12 mmX12 mmX20 gauge chicken mesh on the top and bottom faces. Welded mesh should be stretched over the base course. The welded mesh should be tied together properly with the help of binding wire and nails at site.

Ferro cement mortar of 20mm thickness prepared by 1 : 3 cement-sand mortar with water cement ratio of 0.4 admixed with CONPLAST X 421IC or other additive is applied with trowels. This is finished with a coat of cement slurry. While applying the mortar, the wire mesh layers are lifted up using a hook so that a cover of about 4mm could be provided below the bottom layer of mesh. Hand trowelling is carried out over the surface for leveling and finishing with a wooden float. In case of very low temperature and high humidity, ferrocement mortar finish should be covered with wet gunny bags after twelve hours after finishing.. However, at higher temperature and low humidity, the surface should be covered with wet gunny bags after forty minutes of finishing the mortar. Twenty four hours after laying the treatment, the roof surface may be floated with water. After two weeks of water curing, the mortar is permitted to dry gradually.

9.5 MODERN REPAIR MATERIALS AND TECHNIQUES

Most of these techniques can be used for repair of existing water proofing systems or for complete replacement of damaged water proofing system. These modern techniques perform individually or in combination with other materials or techniques as a part of complete water proofing system.

9.5.1 Grouting

Grouting is a process whereby fluid-like material (either in **suspension** or **solution form**) is injected into subsurface soil, concrete or rock for one, or more of the following purposes.

- To reduce permeability
- To increase shear strength
- To reduce compressibility

Grouting is essentially used **to reduce the porosity or permeability** of building material as a first step towards water proofing. Highly fluid materials known as grouts are injected into the body of building element under pressure. Different types of grouts such as cement grouts, polymer modified cementitious grouts, epoxies and aqua-reactive polyurethane grouts are used for grouting. This is not an independent complete system in itself and needs further **treatment like coatings**.

The characteristics to be observed in the liquid grout are **flowability, expansion and bleeding**, while the important characteristics of the hardened grout are **strength, durability and density**. Grout properties are influenced by variations in proportions of materials and environmental condition such as:

- Type, age and brand of cement
- Type and quantity of the admixture
- Water cement ratio
- Air temperatures and humidity during mixing and pumping

Cement-based Grouts include neat cement, cement- admixture and cement-sand mortars. Cement grouting is the most widely used method in the construction industry for reducing the permeability and/or increasing the strength of structural elements. Cement-based grouting technique has yet not attained a degree of standardization especially with regards to its flow and strength characteristics. Ability of grouts to penetrate a formation is mainly a function of their fluidity. The change of fluidity in grout with time is balanced with stability for efficient functioning at the time of construction or repair.

Low viscosity **Polymer Sealing Liquids** are used in a similar way as cement grout for repair of cracks. It may be applied by brush application or by temporary ponding. When no further materials will penetrate the crack, the surplus materials are removed.

Polyurethane Injection chemical grouting (e.g. polyurethane formulations) technique is useful for non-structural water proofing purposes. Grout material is supplied in low viscosity formulation to enhance its penetrating capabilities. During remedial repairs injecting cracks can redirect water and moisture movement to stop or shift leakage in other areas of least resistance. Therefore, remedial water proofing treatment requires complete grout injection of cracks and application of water proofing system over the entire area under repair. **Polyurethane (PU) grouts** are divided in two major categories as **Hydrophobic**, and **Hydrophilic**.

Hydrophobic grout materials use water as a **reacting agent** and thus absorb very little water. The cured material is essentially free of water making it very resistant to post cure shrinkage.

Hydrophilic grouts can incorporate large amounts of water into their chemical structure creating a **gel** with variable water contents. The incorporated water, however, can evaporate in a dry environment causing the cured **hydrophilic material to shrink** or absorb more water causing the material to swell. The shrinking and swelling are some negative side effects of the hydrophilic gels.

9.5.2 Modified Bituminous Systems

Bitumen and Bituminous materials are still considered as very effective systems for water proofing. These are used either as hot applied or cold applied. The main advantages of these systems are:

- Low cost
- Cheap labour
- Easy application

However, the limitations are black colour, fast degradation under exposed conditions and UV attack and low resistance to high differential temperature. These shortcomings are overcome by modification of bitumen with elastomers like **SBS (Styrene Resin Styrene)**, **Polyurethane**, and plastomers like **Polypropylene** and **Polyethylene**. The addition of elastomers and plastomers in bitumen improves flexibility, elasticity and resistance to fracture and differential temperatures.

Woven fabrics of bitumen like hessian based felts, fibreglass mats are also available. These have better dimensional stability and resistance to fire These modern techniques perform individually or in combination with other techniques as a part of complete water proofing system. These materials have poor puncture resistance and mechanical strengths but the latest non-woven polyester mats have overcome these limitations. These materials have much better resistance to fungal attack.

9.5.3 Membrane Techniques

The research in polymers have developed single ply membranes like flexible PVC in the thermoplastic category and **EPDM** (Ethylene Propylene Diene Monomers) in the elastomeric category. The major advantages of single ply membranes are:

- Light weight
- Quick and neat job
- Attractive colours and designs
- Reduced fire hazards
- Safety of workers during installation.

9.5.4 Liquid Membrane Technique

These techniques of water proofing are based on liquid coatings of polymeric materials such as **silicates, silicones, coal tar, epoxies, polyurethane and acrylics**. The advantages of using these polymeric membranes for water proofing are easy and effective application including for regular preventive maintenance. Various liquid membranes are described in subsequent paragraphs.

Water based coatings comprise of different types of mono polymers or copolymers such as **SBR, SBS, PVA**, and **Acrylics**. These are used to resist the dampness and efflorescence in buildings. Water based coatings have superior performance due to breathability. They also have good compatibility with substrate and also adequate resistance to weathering.

Polymer Modified Cementitious Coating (PMCC) makes good water proofing systems due to adequate adhesion with substrate. The cementitious materials are combined with polymers to take advantages of **Impermeability, Film formation and Pore blockage** characteristics. Polymer Modified Cementitious coatings include epoxy, modified epoxy and polyurethanes.

Epoxy Coatings are mostly used for water proofing for internal applications. The advantages of the epoxy coatings are good adhesion with different substrate, high chemical resistance, good mechanical properties and impermeability. The disadvantages are limited resistance to UV rays and high cost. These can be used for regular maintenance of water proofing systems.

Modified epoxy coatings with other polymeric systems (such as polysulphides, coal tars and polyurethane) are used to overcome disadvantages of simple epoxy coatings for external applications.

Polyurethane coatings have high elasticity, excellent bonding with substrate, resistance to cracking and abrasion, resistance to weathering and UV rays. Polyurethane-based products are available in the form of one component or two component systems. One component systems are usually solvent based while two component systems may be solvent based or solvent free.

9.5.5 Cementitious Water proofing

Traditionally, masons have used mortar-parging and bituminous membranes to waterproof below-grade concrete block walls. Some cement-based coatings have also been used as alternatives to stop the passage of water under **hydrostatic pressure**. Block walls are still waterproofed in the traditional way but new coatings offer many benefits. First, these coatings **save time and labour** because they do not need to be covered by a second membrane. They can be **sprayed or brushed** in less time than it takes for mortar parging and application. Second, these require shorter cure time (about 7 days against 14 days for a parge coat) and allows back filling sooner. Third, these coatings can be both **effective and attractive** when applied to the interior side of an existing wall. These coatings can also be used above grade, though more commonly cement-based damp proofers are used.

9.6 PROTECTIVE COATINGS

9.6.1 Introduction

The exposed and critical surfaces are generally preserved for longer period by using protective coating. Protective coatings are usually needed in the following conditions:

- Buildings located in high rain prone region
- Buildings in polluted area
- White cement used in buildings

These coatings have short life and are required to be redone on a regular basis under preventive maintenance. Improper or inadequate preparation is probably the biggest reason for proper performance of protective coating. In case of improper preparation the protective coatings do not remain properly bonded to the substrate. In general the surface of the concrete or other material to which a coating will be applied should be clean and sound. Oil, grease and wax will inhibit bond and should be carefully removed from the surface. The best way of preparing the surface is by sandblasting. Sand blasting has the added advantage of removing laitance in addition to removing the foreign material on the surface.

A thorough etching with a 10 % muratic acid solution can be used on concrete slabs but great care must be taken to completely flush out all traces of the acid with clean water. If not, a deposit of chlorides will be left on the surface and these will act as bond breakers to some coatings especially the epoxies.

In some cases, it is impossible to remove the foreign matter from the surface specially drippings from automobiles which penetrate into the concrete. In these cases it may be necessary to chip out the contaminated area and apply a new concrete topping bonded to the substrate with protective coating.

Where the concrete surface is too rough for proper application of a thin coating, cement based coating can be provided prior to final protective finish coating.

Organic solvent based coating or epoxy resin coating cannot generally be applied to wet or damp concrete while water base material emulsions can be applied to wet or damp surfaces. The type of coatings and their application are described in subsequent sections.

9.6.2 Cementitious Coatings

Cement based coatings usually consist of a mixture of white portland cement, silica sand, pigment when required, accelerators and water repellents. The resulting powder is mixed with water to produce a mix of thick consistency. This mixture should be scrubbed well into the thoroughly wetted concrete surface.

The cement paints harden by hydration of the cement, the surface should be water cured after the paint is applied. Water curing can best be accomplished by using a sprayer which sprays a fine mist of water over the surface. These surface mistings should be carried out every three hours for two days after the application.

Cement based coatings are inherently porous and form a continuous film but are not efficient water barriers. In general use, these are applied more frequently on interior surfaces than on exterior surfaces. They have one advantage that they allow the surface to breathe and therefore are not subjected to vapour pressure build up behind the coating to break it loose.

Coloured cement paints should not generally be used because the concrete surface vary in porosity and the colour of the finished surface. Sometimes it is possible to coat with a second coat for complete uniformity. It is better to apply one or two coats of white cement paint and afterwards apply latex or alkyed type wall paint for decorative purposes.

9.6.3 Silicone Treatments

Silicone compounds are a very popular materials for damp proofing exterior walls above ground. Silicones form a good water barrier and prevent dampness in walls. These can be applied on the prepared surface either by brushing or spraying. It maintains the natural colour of the substrate surface. The initial cost is high but has longer durability.

9.6.4 Linseed Oil

Linseed oil shows a great deal of promise as a surface treatment for concrete pavement. By research linseed oil has been found effective in reducing cost of concrete maintenance especially for highways and bridges.

The Linseed oil compositions containing upto 97% of linseed oil get readily emulsify with water. The linseed oil compositions produce a low cost material that could be sprayed on concrete surface as a curing and antispalling agent. A 50-50 blend of linseed oil and mineral spirits is quite effective and is recommended in preference to the water emulsion blend.

The linseed oil compositions are prepared by mixing with water to form emulsions which when applied to concrete surface form a film. This film slows down the movement of the water in the fresh concrete as well as in cured concrete. Fast loss of water from fresh concrete is injurious in the proper hydration of the concrete and strength development, while absorption and movement of water in the cured concrete can cause spalling by freezing and thawing specially in subzero temperatures.

On new concrete, linseed oil compounds are most effective if applied after the completion of the initial curing period, usually 28 days, although these can be applied to new concrete after only 7 to 10 days of curing. Linseed oil, of course, can be applied to old concrete to inhibit further damage.

Two applications of linseed oil compounds should be sufficient to protect concrete from winter damage. The oil penetrates the porous surface to a depth of approximately 3mm and combines with the atmospheric **air to form a protective coating through which moisture and salt solutions can not penetrate**.

For maximum protection additional **applications of linseed** oil compounds should be made **each year for the first four years of the life of the concrete**.

One advantage of the linseed oil antispalling compounds is that these are penetrating sealers and not surface coatings. These do not wear away in traffic as fast as a coating would and therefore maintain their protection longer. In addition, the skid resistance of the concrete is unaffected.

9.6.5 Bituminous Coatings

Bituminous coating can be produced of either asphalt or tar. The coatings can vary from thin applications to built up coatings consisting of membranes of asphaltic impregnated felt with moppings of pitch or asphalt. Until recently the term water proofing was almost synonymous with bituminous materials, due to their universal use as water proofers.

Carefully controlled tests have shown that the diffusion of water through bituminous films is negligible and, for all practical purposes, the film is impermeable to water.

For the damp proofing of below grade concrete walls, a coat of sprayed or brushed cut back or emulsion is usually sufficient. If the concrete wall cracks or settles the integrity of the bituminous film will be damaged and leaking would occur at those points. It is possible to trowel on a slightly thicker coat so that it would have sufficient elasticity and flexibility to bridge over the smaller cracks.

The surface must be dry during its application. Emulsions, on the other hand , can be applied over damp concrete . In order to build up a thicker coating and give more flexibility to the film, the bituminous materials are often fortified with fillers such as asbestos fibre. These filler coatings are quite effective.

In exposed areas bituminous coatings are temperature sensitive and this is one of the problems with the material. At higher temperatures they flow somewhat readily and are liable to be tracked by vehicles or become rutted by the weight of the vehicular traffic. At lower temperatures, their ductility is reduced to such an extent that they crack easily.

Another disadvantage of the bituminous coatings is their deterioration by **oxidation** and by **ultraviolet light** when exposed to the actions of air and sunlight. These changes can cause the film to become brittle and consequently crack.

9.6.6 Surface Hardeners

Liquid floor hardeners are a popular and relatively inexpensive method of increasing the surface hardness of the concrete. In this application the surface of the concrete is hardened by the chemical reaction between the magnesium flousilicate in the hardener and the calcium hydroxide and calcium carbonate present in the concrete.

The chemical is available packaged both in liquid and crystal form. The liquid is more popular since it eliminates the problem of white residuals common to crystals. Two to three-coats application is recommended. The first coating usually consists of a solution of about 1.25mg of crystals per gallon of warm water. After thoroughly soaking the slab with the solution, the floor is left to dry. Puddles should be removed with a mop when the floor is completely dry. Second and third applications of the hardener can be made after the last coating has dried. The surface of the slab should be washed with water to remove any excess crystals which may have formed.

A very weak surface cannot be brought to the hardness of a good quality concrete floor without treatment. This treatment is suitable for interior floors such as warehouses, basements and floors that have heavy foot traffic only but it is **not suitable** for floors subjected to heavy industrial traffic.

Fluosilicate hardeners are useful for car showrooms and warehouses where a dust concrete surface is undesirable. The fluosilicates will provide a comparatively hard surface, which can be kept clean and maintained easily.

9.6.7 Epoxy Plastic System

The advent of epoxy compounds for commercial construction has brought about a radical change in surface protection of exposed concrete. Epoxy compounds can be used as paint type protective coatings to provide both water proof and vapour proof membranes. Where concrete slabs are badly deteriorated, and require skid resistance coatings both in the film and mortar types are installed in epoxy.

The biggest advantage of the epoxy coatings for floor protection is the fact that a epoxy coating is highly resistant to chemicals. It can be used to protect concrete in areas such as dairy barns, food packing plants, aircraft hangers and machine shops where excessive fluid spillage of fluid occurs. The epoxy protective coatings are particularly well suited to all these areas.

In order to satisfactorily install an epoxy resin protective coating, it is essential that the surface be properly cleaned and prepared. There will be little or no bond with the oily, dirty surface. The high bond to a weak surface, will result in chip off easily. Where several cleaning methods are available, sandblasting is considered the best. If this is not practical, some mechanical means of removing the dirt surface without polishing is recommended.

Wax or resin curing compounds should be used on the slab prior to application of epoxy topping. **Oil and silicone** treatments should not be used because these materials are **bond inhibitors**. If the epoxy coating is to be done over steel surfaces such as column baseplates, then the steel too should be sandblasted to a shiny brightness, removing all mill scale and oil or rust. All materials of epoxy should be thoroughly mixed so that the resulting mixture is homogenous.

Epoxy resin mortar coatings incorporating sand or aggregate is recommended where a highly chemical resistant floor with ability to withstand heavy traffic and high impact is required. For coatings of this type, the epoxy system should be 100 percent solid and should not contain solvents. In this epoxy mortar no more than $3\frac{1}{2}$ parts of the aggregate should be used for complete impermeability to liquids. One of the most satisfactory ways for concrete floor construction is to first install 6mm high Aluminium strips such as used for terrazzo. These strips are bonded to the concrete slab by means of the gel type epoxy adhesives. After mixing, the epoxy resin mortar is sprayed in alternate squares formed by the Aluminium strips and then compacted by using a large aluminum pipe roller. This roller should ride upon the aluminum strips. Two or three passes of the roller will thoroughly compact the mortar in the square with the added advantage that it does not pick up somewhat sticky mortar.

To help further to prevent any possible pick up of the epoxy mortar, the aggregate particle should have an angular shape. Crushed quartzite is one of the best aggregate for this purpose. It is also essential that the aggregate be bone dry. After the mortar starts to harden the surface can be steel troweled to the proper finish. After complete curing, this epoxy mortar provides resistance to heavy traffic, in addition to extremely high resistance to impact and chemical spillage of oils and greases.

There are also epoxy systems based on solid resins in a solvent solution. These systems provide the consistency of paint and have a very long pot life since **it is necessary before the system can cure.**

These coatings also provide very high chemical and abrasion resistance, and good flexibility. The coating is usually applied in minimum two coat application, each coat giving a thickness of 2 to 3mm. Since the solvent systems are somewhat slower in their reaction than the solid systems, the final cure usually takes 3 to 4 days.

One advantage of the solvent systems is the fact that they can be used with pigments to provide a variety of colours for decoration. The pigment will not reduce their chemical resistance or their other qualities. A solvent system can form top of a mortar system to add colour and to further increase its chemical resistance.

An epoxy coating system can also be reinforced with fibreglass and this is an excellent material for strengthening and protecting concrete floors which have been crazed or cracked.

This system consists of embedding fibre glass cloth in the epoxy resin mortar system. The neat epoxy system is rolled on the slab with a roller. The fibre glass cloth is laid over it and pressed in with a bladed knife or metal trowel. The second coat of the epoxy system is then applied over the top while the undercoat is still wet. This provides a very tough, chemical and abrasion resistant system, which acts like a layer of steel over the surface of the concrete. It should be used only on smooth floors, since there is no aggregate in the system to take up for the unevenness of the substrate.

It is possible to formulate coal tar epoxy resin systems to produce a black, heavy viscous material with a 1:1 ratio of components. This compound can be applied by automatic dispensing equipment to bridge decks. The coal tar acts as filler or diluent and reacts chemically with the system to form the cured material. The decks worn out or crack while concrete bridges, viaducts and overpasses can be protected from further damage with a single layer of the coal tar epoxy resin material. Aggregate can be spread while the film is still wet to provide skid resistance.

The coal tar epoxy system will protect the concrete substrate from the action of deicing salts, chemicals and water for a considerable period of time. This material is not particularly abrasion resistant but the introduction of aggregates into the film while it is still wet can provide skid resistance and also increases the abrasion resistance of the whole system. The coating does not have the high quality of the 100 percent solid epoxy mortar systems, the coal tar epoxy materials are preferred because of their much lower price and faster curing.

Epoxy systems should not be placed on exterior slabs on grade in climates where freezing and thawing occur. The epoxies form 100 percent vapour barrier. If placed on outdoor slabs, during warmer weather moisture from the subgrade and even from the concrete itself will be obstructed and will condense on the underside of the epoxy coating. Since good concrete is by nature watertight this water will remain accumulating at the surface of the concrete, below the epoxy coating, and cannot go back through the slab in to the ground. Upon freezing this water will expand and exert force that will result in spall of surface of the concrete taking off the epoxy coating with it.

The choice of the applied epoxy type is dependent upon the conditions and upon the requirements of the architect or engineer who is specifying the installation.

9.7 SUMMARY

Dampness problem is a very common phenomenon in all types of building structures and causes lot of defects. Water proofing should be done properly at the time of construction, otherwise it will always give trouble. Repair and maintenance of such water proofing system is very difficult and expensive and generally needs to be relaid. The possible points of leakage in roof and wet areas are near the joints in walls, roof, and sanitary systems. All joints need to be sealed properly. Water proofing systems such as Acrylic polymer modified cementitious coatings or liquid polymer membranes are laid during water proofing of wet areas.

A flat roof is more prone to leakage as it undergoes movements due to variation in temperature, moisture, loading or other causes. A provision of thermal insulation is necessary in roofs along with water proofing system,. Generally the water proofing treatments are **rigid, elastic** and **composite**. Lime concrete, mud phuska and brick-bat coba are rigid treatments and are suitable for low-rainfall areas. Bituminous felts, polyethylene sheets are elastic treatments and are easy to lay at junction of walls and other projecting areas. Composite treatment is combination of elastic and rigid systems such as mud phuska, **tarfelts,** and **multiplas.** Ideal characteristics such as **impermeability,** good **bonding, resistance to UV rays, wear** and **tear** and **economy** should be given due consideration while selecting a water proof system. In case of major defects, the conventional techniques like, mud phuska, lime concrete and ferrocement need to be relaid. The earlier damaged treatment is removed completely. Any defect such as cracks and honey combing in concrete should be repaired before relaying. Grouting with cement mortar grout, polymer sealing, and polyurethane injection are used to reduce porosity or permeability. However, they need further treatments of protective coatings. Modified bituminous systems have the advantage of low cost, cheap labour and easy application. Similarly membrane forming technique is also very useful and have advantages such as reduced weight, size, hazards and speed of work. Liquid membranes such as water based coatings, epoxy coatings, polymer modified coatings are used for water proofing.

Protective coatings are applied on exposed surfaces to protect these surfaces from moisture and pollutants. All surfaces should be prepared well by removing loose material by sandblasting. The various coatings are cementitious coatings, silicone treatment, (Linseed oil, bituminous coatings, surface hardeners and plastic coatings. These coatings are required to be used at regular intervals to provide protection to the substrate.

QUESTIONS

9.1 Describe importance of damp proofing building elements in not more than 100 words.

9.2 List sources of water causing dampness in building elements.

9.3 Explain briefly, in not more than 150 words, the technique of repair of water proofing systems in wet areas.

9.4 Describe importance of regular repair of water proofing of flat roofs.

9.5 List various stages involved in repair and maintenance of water proofing systems.

9.6 Describe ideal characteristics of any water proofing system in flat roofs.

9.7 Explain the following water proofing techniques for flat roofs:

(a) Mud phuska
(b) Lime concrete terracing
(c) Ferrocement
(d) Brick bat coba

9.8 Explain the patch repair techniques and protective treatment with coal tar proofing systems in flat roofs.

9.9 Name the modern repair materials for water proofing system in building elements.

9.10 Write short notes on:

(a) Cement based grouts
(b) Polyurethane injection grouts
(c) Modified bituminous systems
(d) Liquid membranes

9.11 Differentiate between water proofing with modified bituminous system and membrane techniques.

9.12 Explain the importance of protective coatings in repair and maintenance of buildings.

9.13 Write short notes on protective repairs:

(a) Silicone treatments
(b) Cementitious coatings
(c) Linseed Oil treatment
(d) Bituminous coatings
(e) Surface hardeners

9.14 Explain the use of epoxy plastic coating treatment of building elements.

Chapter 10

PROTECTION, REPAIR AND MAINTENANCE OF RCC ELEMENTS

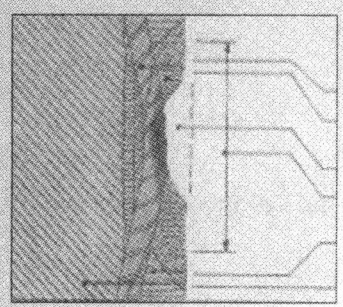

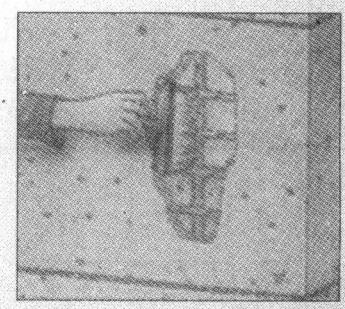
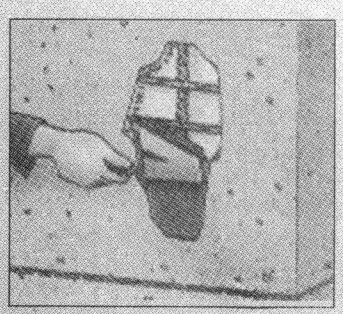
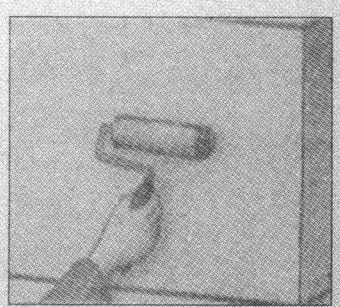

10

PROTECTION, REPAIR AND MAINTENANCE OF RCC ELEMENTS

LEARNING OBJECTIVES

After studying this chapter, the learner understands common repair and maintenance for RCC elements and will be able to:

- **Describe** necessary steps to prepare RCC elements for repair;
- **Explain** the material replacement technique for repairing RCC elements;
- **Describe** repair of surface defects of concrete;
- **Describe** removal of corrosion in existing reinforcement in RCC elements.

10.1 INTRODUCTION

Concrete decay is cancerous. It is commonly caused by 'carcinogenic' agents that corrode embedded reinforcing steel, deep inside the body of a structure, and deteriorate surrounding concrete. The problem of premature deterioration of concrete structures has assumed universal attention. Expenditure on maintenance, repair rehabilitation and protection is rising alarmingly, reaching 30-40 per cent of the total expenditure on construction.

There has been phenomenal advancement in the science and technology of repair, rehabilitation and protection. With the advancements in testing and monitoring methods and development of equipment, the engineers are able to understand and evaluate the deterioration and the condition of a structure in a fairly satisfactory manner.

For a successful repair and rehabilitation job, it is essential to determine the extent of damages and deficiencies, understand their causes and also identify their implications on the serviceability and safety of the structure.

For successful repair work following steps should be adopted:

  (i) Inspection and diagnosis

  (ii) Preparation techniques

 (iii) Primer coat for the steel reinforcement

 (iv) Primer coat on the prepared substrate

  (v) Repair mortar or micro-concrete

 (vi) Protective coating

The formulated repair product, whether applied by conventional means or by spraying, must play its role to full potential in the completed repair job. Satisfactory repairs require the appropriate material formulation and the application of right technique. There are basic requirements for achieving a durable patch repair work. These basic requirements hold true for virtually any repair job. For quality assurance of the repair job, following important considerations must be adopted:

- The repair material must be thoroughly bonded to the sound concrete of the element under repair

- The shrinkage of the repair material in the patch should be small enough not to jeopardize the bond

- The cracks in the patch work and its substrate should be filled with suitable materials

- The patch repair material and the old concrete must be compatible and should respond to changes in temperature, moisture and load similarly to avoid large differences in movement.

- The patch repair material should have sufficiently low permeability so that moisture will not migrate through it to the old substrate concrete

- The repair patch work should be resistant to weathering and occupancy condition at an early age whenever required.

In addition to these basic requirements, it is sometimes necessary to match the colour and texture of the patch work with the old concrete. If the old concrete has special exposure requirements such as the ability to resist chemical attack the patching material should also possess these qualities.

The repair material must be selected carefully. Ordinary hydraulic cement mortars have the advantages that coefficient of expansion is similar to that of the existing concrete, their appearance is similar, and are economical in cost. On the other hand such mortars have high water requirements, high cement contents and consequently appreciable drying shrinkage.

- For most repair jobs, labour cost constitutes the bulk of the expense so it is poor economy to opt for small savings in material cost at the risk of quality and appreciably shortening the life of repair.

General repair techniques for RCC structures need familiarity in their usage, applications, potentials and limitations. General methods of repair are almost standardized such as **Jacketing, grouting, trowelling** and **shotcreting**. Prior to repair of RCC elements, these need to be specially prepared for repair. Methodology of repair depends on the extent of damage of the particular structure. In case carbonation of concrete has taken place but the coating of required thickness can be placed. Preventive repair by protective coating should be immediately taken for structures which are in the initial stages of distress. The repair shall be comparatively cheaper if it is attended at an early stage of distress. The structure in which deterioration and distress has developed to a considerable extent affecting reinforcement, may need extensive repair and rehabilitation. The techniques explained in this chapter can be used for repair of RCC elements

10.2 PREVENTION OF CORROSION IN REINFORCEMENT

The construction industry is trying for a low cost, easy way to fight problem of corrosion in reinforcing steel in concrete. But a cheap quick fix miracle drug does not exist. However, for combating corrosion of reinforcement, many techniques do exist and are to be applied individually or in combination. These techniques are costly and may require change in construction practices but are worth the expense and effort because of the years they add to the life of a structure. RCC can be protected from corrosion in three ways:

- Seal the surface of the concrete to **prevent ingress of chlorides and moisture**
- Modify the concrete to reduce its permeability, thus retarding the flow of moisture and chlorides to reach the reinforcing steel.
- **Protect the reinforcing bars** to reduce the effects of chlorides when they do reach the steel.

More and more designers are specifying multiple levels of protection for structure that are subjected to the risk of corrosion. It is common to specify epoxy coated reinforcing steel and silica fume concrete for post tensioned structures.

Corrosion controlling steps in reinforcement are stated in subsequent sections.

(a) Good concrete practice

(b) Latex modified concrete

(c) Silica fume concrete

(d) Epoxy coated reinforcement

(e) Providing membranes and sealers

(f) Cathodic protection

(g) Electro osmosis

(h) Inorganic corrosion inhibitors

(i) Organic corrosion inhibitors

(a) Good Concrete Practice

Good concrete practices are required for all concrete jobs but some times these are not specified or the specifications are not enforced on the job site resulting in problems. The good concrete practices require following considerations.

- Specify and introduce **Chloride limit** as part of the good concrete. The code shows limits for various type of constructions and exposure conditions. The code commentary explains how to perform an evaluation of the total chloride content of the ingredients of the concrete.

- Maintain a **low water-cement ratio** in concrete. Concrete permeability decreases as the amount of water per cubic metre is reduced at a constant cement content,.

- Provide **adequate cover** to the reinforcing steel. This helps in increasing the time it takes for the chlorides to reach the steel. Any savings made by reducing cover are short lived.

- Provide **adequate curing** by not allowing fresh concrete to dry without curing for at least seven days of wet curing. Drying reduces hydration and increases permeability and shrinkage.

- Use of **water-reducing admixture** to give the concrete enough workability so that workers are able to **compact** concrete properly with ease.

- **Consolidate** the concrete thoroughly using suitable vibrators. The advantage gained from using a low water cement ratio will be lost if the concrete is not adequately consolidated to **high density**.

- Use post **tensioning to minimize cracking,** where appropriate.

- Include provision for **immediate repair of cracks** in the original specifications assuming that the concrete will crack. Repair or **seal the** cracks before the new structure is put into service.

Adoption of good concrete practices may not be enough for the required level of performance to face the severe **exposure conditions**. But good concrete practices are the most critical steps in controlling corrosion and other deterioration. Use of special concrete materials also help in limiting corrosion by **reducing the permeability**.

No slump concrete is concrete proportioned with high cement content and low water content to limit water cement ratio. Superplasticized dense concrete is more workable. Typical cement contents are about 450 kg per cubic metre with water cement ratios at or below 0.35. The high cement and low water content of low slump concrete lead to reduced permeability leading to reduction in corrosion.

Performance of this concrete is fair to good depending on the degree of consolidation and curing. Low slump concrete has **significantly low permeability** than conventional concrete.

The limited workability of the low slump concrete makes it difficult to place and consolidate. So only limited improvement in concrete performance can be achieved by reducing water content alone and plasticisers are necessary for better compaction.

(b) Use of Latex Modified Concrete

Latex modified concrete is prepared by adding liquid styrene butadiene latex to conventional concrete. A typical latex modified concrete mixture contains 15% latex solids by weight of cement, and has a water cement ratio of 0.35. The latex modifies the pore structure of the concrete and reduces its permeability. Rapid chloride permeability test on latex modified concrete shows fall in permeability to a very low range.

Latex modified concrete have been associated with some cracking problems and it should be placed in the evening or at night to reduce cracking. Relative cost of latex modified concrete is high but performance is good to excellent.

(c) Use of Silica Fume Concrete

Silica fume is an extremely effective pozzolanic material that reacts with calcium hydroxide produced in hydrated Portland cement paste to form additional cementitious material. As a result, the permeability of the concrete is significantly reduced. Typical silica fume concrete mixtures contain 410 kg of cement per cum, 8% to 10% silica fume by weight of cement, a water-powder (cement plus silica fume) ratio less than 0.40. Adequate plasticiser is provided to produce good workability with 50-200 mm slump. Performance of the concrete is good to excellent, when tested using the rapid chloride permeability test. Silica fume concrete is approximately equivalent to latex modified concrete in its chloride permeability. Plastic shrinkage cracking has been a recurrent problem in silica fume concrete. The silica fume concrete has a moderate cost.

(d) Epoxy Coated Reinforcing Bars

Pre-cleaned reinforcing steel bars are protected with a coating of powdered epoxy. The epoxy is fusion bonded in an assembly line process. The coating physically blocks chloride ions and the performance of these bars ranges from poor to excellent depending on effectiveness of coating.

Unless the bars are coated after bending, there is a potential for cracking and chipping of the epoxy coating during bending. Damage to the epoxy coating may also occur during field handling of the bars. The relative cost of epoxy coated steel is moderate.

(e) Use of Membranes and Sealers

Membranes and sealers help prevent chloride entry when applied to concrete surface. Urethanes, neoprene, or epoxies are usually used to built up in multiple layers of membranes. These multi layered membranes have the ability to bridge cracks in the concrete. Sealers range from linseed oil to sophisticated silanes and siloxanes. These liquid membranes and sealers perform excellently as claimed by manufacturers and researchers.

The performance of these materials vary depending upon the base of the sealer. Most sealers are not suited for sites where abrasion occurs. The effectiveness of all these materials decreases over time and they are required to be reapplied at a regular interval. Cost of these materials range from low to high. There is a continuous maintenance cost also and that should be included when comparing costs of different materials and techniques with reference to serviceable life.

(f) Cathodic Protection

Cathodic protection approach controls corrosion of steel embedded in concrete by applying direct current to the embedded steel by an external source. An electric current is applied to the concrete anode and the embedded steel. This action forces the steel in the concrete to become cathodic, which provides the protection. Cathodic protection is the only way to stop the on going corrosion in a concrete structure.

Cathodic protection is a complicated process that requires extensive pre installation engineering and extensive post installation monitoring. Its relative cost varies from high to moderate. Its performance can be termed as satisfactory.

Corrosion of steel in concrete is a electrochemical process. Cathodic protection is achieved by imposing a low voltage direct current from an anode system placed on the concrete surface through the concrete and on to the steel. The cathodic protection current opposes the current associated with the corrosion process. When sufficient current flow is achieved, the corrosion current will be suppressed. Areas of spalled or delaminated concrete are required to be repaired prior to the installation of an imposed current Cathodic protection system. There is need for periodic potential monitoring to ensure effectiveness of the system. Problems associated with corrosion at the boundaries of damaged areas are prevented by cathodic protection. Thus, cathodic protection has marked advantage over other procedures. The concept is simple but difficult to implement fully satisfactorily. If it is attended to when the damage is in initial stages, the cathodic protection is known to be cost effective. This is an emerging technique and has major potential to deal with chloride induced corrosion problem.

Cathodic protection is not recommended for carbonated concrete. This is because the carbonation increases the resistivity of the concrete making it more difficult to impose an electric current. The foregoing repair options are not mutually exclusive and could be used in combination.

Cathodic protection can be provided by conductive overlays, superficial anodes, conductive coating anode systems; wire and mesh anodes and combination. Cathodic protection needs following preparatory steps:

- Deterioration diagnosis
- Inspection mapping
- Repair of damaged portion

Deterioration diagnosis is carried out to assess the root causes and the extent of corrosion damage. Details of damage are found by **inspection mapping** involving following steps:

- Potential survey
- Delamination survey
- Loss of steel section determination
- Cover survey
- Reinforcement continuity survey
- Concrete resistivity survey
- Chloride content measurements
- Carbonation depth measurements
- Other aspect of concrete matrix

It is necessary to repair the damage in concrete structures before Cathodic Protection is installed. However, these techniques are more effective for steel structures such as pipe lines and oil platforms.

(g) Electro Osmosis

The solution to the carbonation problem depends upon the depth of the carbonation front. Electro Osmosis makes it possible to restore the initial alkalinity of the concrete. These techniques have advantage and disadvantages on respective counts. The chosen method would depend on the structure and its state of deterioration. "**Electro Osmosis**" is yet another effective method. Electro osmosis is considered for the re-alkalisation and de-salination. The introduction under pressure of an alkaline solution into the concrete pores gives a possible solution.

An externally applied alkaline gel covering the external conductor is drawn inside during the electro osmosis process. The gel is used to increase the pH of the pore water. The alkalinity increase of a carbonated concrete restores the passivation layer of the steel reinforcement. At the same time, the rust (Haematite) when present in moderate quantities—becomes "magnetite". This transformation goes with a volume reduction so that the pressure on the concrete diminishes.

The re-alkalisation takes places simultaneously by two phenomenas viz., an alkaline gel applied externally on the concrete is pumped inside during the process, and a cathodic reaction producing hydroxyl ions occurs around the reinforcement.

Without taking scattering into account, the principal motive force of the process is the potential difference between the reinforcement and the external conductor placed on the outside of the concrete surface. The reinforcement is connected to the negative pole (cathode) while the external conductor on the concrete surface is connected to the positive pole (anode). The pore liquid of the concrete encasement acts as the conductor. An

Chlorides   Oxygen   Water   Carbon Dioxide

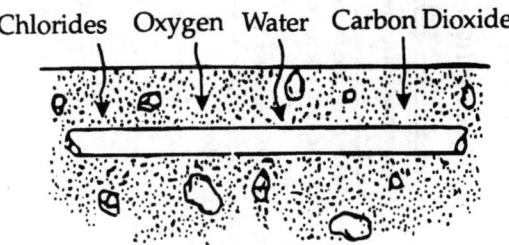

(a): Corrosive Environmental Agents

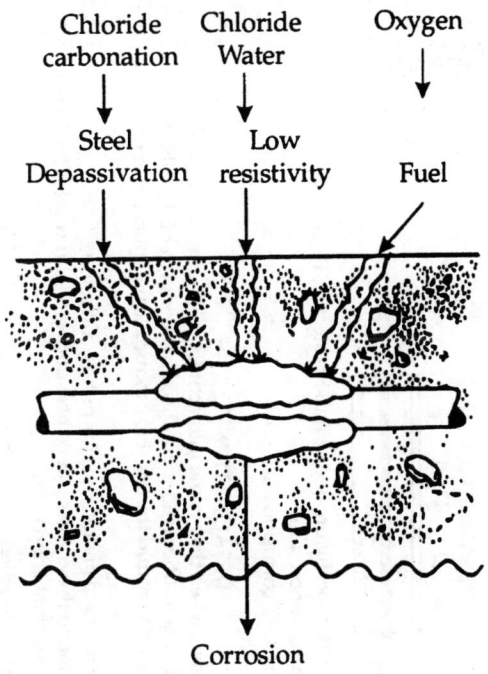

(b): Three Pronged Attack

Fig. 10.1: Corrosion-Damage Mechanism

DESALINATION - A

Reinforcement — Oxide

\# The passive oxide layer around reinforcing bar can be attacked by two forms of chloride contamination

\# Cast in chlorides used as an accelerator in the construction stage

\# Chlorides ingressed from an external source such as road salt

DESALINATION - B

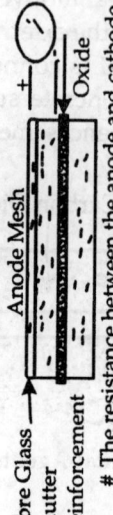

Reinforcement — Oxide

\# Chlorides locally break down the protective Fe : O layer in isolated anodic sites.

\# Macro Cell corrosion ensues

DESALINATION - C

\# Chlorides, when they come into contact with reinforcing steel to sufficient quantities lead to localised break down of the alkaline passive film around the bar this produces small pockets of intensive corrosion.

\# These are called anodic sites (or pitting corrosion)

\# Corrosion in these areas can be best understood if we think of a battery with a constant short circuit

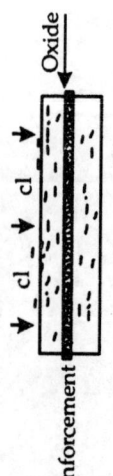

1. Steel Dissolves   2. Free electrons pass to chloride

Anode+   Cathode

4. OH ions transport back to the anode

5. Rust products are formed

3. Oxygen + Water + Electrons Hydroxyl ions OH

This diagram represents macro cell corrosion

HOW DO CHLORIDES PROMOTE CORROSION?

DESALINATION - D

Anode Mesh
oxide

Fibre Glass
Shutter
Reinforcement

\# An anode mesh is applied to the substrate encapsulate in a fibre glass tank

\# The tank is filled with water which acts as an electrolyte

\# A current density of 1 AMP per M is applied to the anode

DESALINATION - E

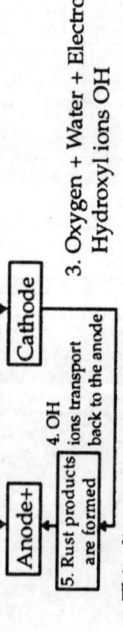

Anode Mesh
Oxide

Fibre Glass
Shutter
Reinforcement

\# The resistance between the anode and cathode drops

\# The chloride ions attracted towards positively charged anode

\# They pass into the water which is changed on regular basis

\# Hydroxyl ions are formed at the bar in the same manner as realkalisation

Fig. 10.2: Desalination

·electrolyte in the form of an alkaline gel made of fibres impregnated with an adhoc solution of Sodium Carbonate ($Na_2 CO_3$) is applied on the concrete surface. At the more porous places where the carbonation is the deepest, the first "points" of the re-alkalisation front can reach the reinforcement after a few hours.

The complete re-alkalisation of the places where the concrete density is higher takes more time. The time required for the complete treatment varies according to the basic construction and depends on the quality and strength of cover concrete around the reinforcement. The other factors affecting the treatment are such as the concrete porosity, the localisation of the carbonation and the dimensions of the external conductor. This period of treatment varies from a few days to a few weeks. At the end of the process, the pores will contain the alkaline solution.

The second phenomenon of the realkalisation process takes place around the reinforcement. A cathodic reaction occurs from the beginning of the process, the water and the oxygen around the reinforcement are transformed into (OH-) hydroxyl ions. The presence of these ions increases the pH of the pore water near to the reinforcement and raises its value to over 13. The re-alkalisation develops readily from the reinforcement to the concrete. This process is necessary in order to form a buffer of alkaline materials in the pore water behind the reinforcements. The process of corrosion, desalination and re-alkalisation are explained in Figure 10.1, 10.2 and 10.3.

Re-alkalisation - 'A'

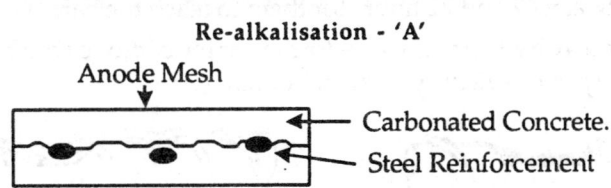

- Re-alkalisation is an electrochemical method to restore alkalinity that has been lost during the carbonation process.

- An anode mesh is suspended above the surface of a concrete substrate.

- Tests must be taken to ensure that there is continuity within the reinforcing bars inside the concrete and connections should be made at regular intervals.

Re-alkalisation - 'B'

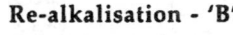

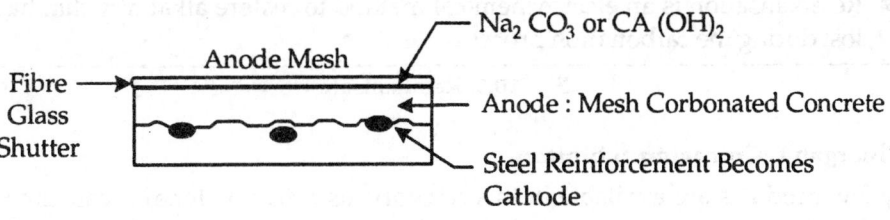

- A fibre glass tank is fixed to the concrete encapsulating the anode mesh and filled with a sodium carbonate or calcium hydroxide solution at a calculated molar solution.
- A current intensity of 1 Amp per sqm is applied to the anode.
- Sodium or calcium dissociate and are attracted towards the cathode.
- During this movement they carry their respective Na_2CO_3 or $CA(OH)_2$ particles through the pour water towards the reinforcement.

Re-alkalisation - 'C'

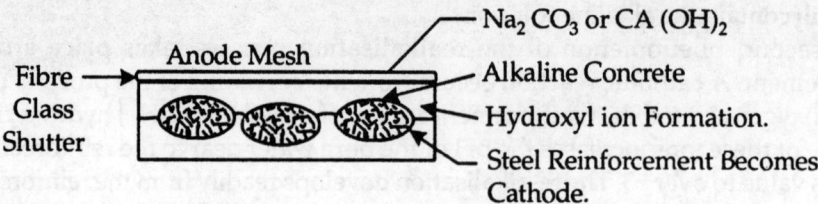

Fibre Glass Shutter — Anode Mesh — Na_2CO_3 or $CA(OH)_2$ — Alkaline Concrete — Hydroxyl ion Formation. — Steel Reinforcement Becomes Cathode.

- There is a drop in resistance between the anode and cathode soon after the system is powered.
- The dissociated metal ions are attracted towards the cathode.
- It takes between 42 and 72 hours for them to reach the bar.
- Simultaneously hydroxyl ions are formed at the cathode which further increases the alkalinity in the vicinity of the reinforcing bar.

Cathode Process:

$O_2 + 2H_2O + 4e \rightarrow 4OH$

Anode Process:

$Fe \rightarrow Fe^{++} + 2e$

- MCI 2020 Works by migration through hardened concrete occurs by liquid and vapor diffusion
- When MCI 2020 reaches reinforcing steel, it forms a molecular protective layer in both the anode and cathode areas. This effectively reduces the corrosion activity.

- Re-alkalisation is an electrochemical method to restore alkalinity that has been lost during the carbonation process.

Fig. 10.3: Re-alkalisation

(h) Inorganic Corrosion Inhibitors

Only few products are available in this category. Its active material is calcium nitrate. The admixture is added during batching of the original concrete construction. Calcium nitrate disrupts the corrosion process by enhancing the formation of the passivation layer on the surface of the reinforcing steel to act as corrosion inhibitors. The nitrate ions compete with chloride ions present in the concrete to react with the free iron ions of steel reinforcement. If the nitrate ions at the level of the steel is slightly greater than chloride ions then the reaction will be between nitrate and changes the iron into an oxide layer, which covers the passive layer on the steel. If the nitrate ions are less than chloride ions, the chloride ions will react with the iron ions to begin the corrosion process. During the chemical reaction between the nitrate and iron, the supply of nitrate ions is depleted.

The dosage of the calcium nitrate product must be determined based on the anticipated chloride loading of the structure over its expected design life. Actual dosage range from 10-30 litre per cum. Performance of this material varies from good to excellent.

Nitrate is also a component of some **accelerating admixtures** which accelerates setting of concrete even though used in the corrosion protection role. Retarders are frequently used with the calcium nitrate to balance the set of the concrete. When higher dosages of calcium nitrate are used, the retarder may be added at the job-site to reduce the problems associated with rapid loss of plasticity and rapid setting.

Over dosage also can be a problem. To select the appropriate amount of calcium nitrate, an engineer needs a correct assessment of the amount of chloride which the structure may likely to be exposed. The relative cost of nitrates is moderate.

(i) Organic Corrosion Inhibitors

The organic corrosion inhibitors combine other organic chemicals for successfully inhibiting corrosion in concrete reinforcement. The organic corrosion inhibitor forms a protective barrier on the reinforcing steel. This barrier prevents reaction between the iron and chloride ions. The material also reduces the permeability of the concrete to slow down the rate of chloride diffusion. The admixture is added during batching. Because there is no competing reaction between the organic corrosion inhibitor and the chloride, there is no need to estimate the chloride loading for the concrete structure. The dosage is about 6 litre per cum.

Laboratory tests show the organic corrosion inhibitor functions well even in cracked concrete. The protective barrier formed by organic inhibitor continues to function even when the chlorides have a straight path to the reinforcing steel after cracking. However, their long term performance is yet to be estimated. Concrete containing the organic corrosion inhibitor requires a higher dose of air entrainment. The relative cost of organic corrosion inhibitors is moderate.

The performance and the relative price of different approaches to control corrosion vary significantly based on the environment and material used. No miracle cure for corrosion is available, but effective techniques are to be selected on the basis of the element condition in respect of carbonation, chloride attack or moisture induced corrosion.

10.3 PREPARATION OF RCC FOR REPAIR

Preparing of RCC element for repair includes removal of all loose materials, cleaning of deteriorated area and removal of corrosion from reinforcement. Contaminated and loose concrete must be removed by hard steel brushing and washing thoroughly before undertaking any repair work.

Proper removal of loose or contaminated concrete, rebar cleaning and surface preparation are crucial for effective repair. Many repair failures are caused by improper performance of these operations. The following are the main steps to prepare reinforced concrete for the repair:

(a) Exposing and undercutting rebar
(b) Cleaning reinforcing steel
(c) Compensating reinforcement
(d) Edge and surface conditioning

(a) Exposing and Undercutting Rebar

The following procedure is used for repairs of horizontal, vertical, and over head surfaces. They are also applicable when removing concrete by hydro demolition, pneumatic, or hydraulic impact method. Various steps include:

- Remove loose or delaminated concrete above corroded reinforcing steel.
- Undercut all exposed corroded rebar after the initial removals.
- Undercutting exposes the blind side of the rebar for cleaning and will allow the repair material to fully incapsulate the rebar, securing the repair material structurally. Provide at least 15mm clearance between exposed rebars and the surrounding concrete or 15mm larger than the largest size of aggregate in the repair material which ever is greater.
- Continue removing concrete along corroded bars until corrosion free locations along the bar are reached and where the bars are well bonded to the surrounding concrete.
- Do not damage the rebar's bond to the surrounding concrete if non corroded reinforcing steel is exposed during the preparation process. Undercut the rebar if the bond between the rebar and surrounding concrete is broken.
- Secure the loose reinforcement in place by tying it to other secured bars or by other appropriate methods.

(b) Cleaning Reinforcing Steel

Remove all heavy corrosion and scale from the bar to promote maximum bond of the repair material. It can be done preferably by abrasive blasting method as already explained.

A tightly bonded with light rust on the surface of rebar is usually not detrimental to bond. Manufacturers recommendations should be followed for rebar preparation if protective coating is to be applied to the rebar.

(c) Compensating Reinforcement

If reinforcing steel has lost significant cross section, a structural engineer should be consulted. If repairs are required to the reinforcing steel, the bar can be replaced completely or supplementary bars can be placed over the affected section. Supplementary bars may be mechanically spliced to old bars or placed parallel to old bars approximately 15mm from existing bars. Lap lengths shall be determined in accordance with latest IS specifications and other design guidelines.

(d) Edge and Surface Conditioning

The surface conditioning steps are used for horizontal, vertical, and overhead surfaces. They are also applicable to concrete removal by hydrodemolition and electric, pneumatic, or hydraulic impact breakers.

After removing delaminated concrete and undercutting reinforcing steel, remove additional concrete as required to provide the minimum required thickness of the repair material. At edge location, provide right angle cuts to the concrete surface with either of the following methods:

- Saw cut 12mm or less as required. Avoid cutting reinforcing steel.
- Use power equipment, such as hydrodemolition or impact breakers. Avoid feather edges.
- Repair configuration should be kept as simple as possible preferably with squared corners.
- Remove bond inhibiting materials, such as dirt, concrete laitance, and loose aggregates by abrasive blasting or high-pressure water blasting with or without abrasive. Check, the concrete surface after cleaning to ensure that they are free from loose aggregates and delaminations.
- Cement and particulate slurry must be removed from the prepared surfaces before the slurry hardens.

10.4 REPAIR OF CORRODED RCC ELEMENTS

Repair of corroded reinforcement includes following main steps:

- Chip off all loose and cracked concrete in RCC element.
- Remove concrete in cover completely to expose corroded reinforcement.
- Remove all loose and deteriorated concrete by wire brush.
- Clean the reinforcement thoroughly to remove all rust of corroded steel by sand blasting or chemical wash.
- Wash and saturate all cavities in adjoining concrete with clean water and remove all free water before placing repair mortar.
- Fill all cracks with epoxy resins and apply bond coat just prior to placing patch repair mortar or concrete.

- Apply patch repair mortar or concrete or grout on the damp cavity surface.
- In case of loss of steel section, additional steel reinforcement may be tied to the existing bars prior to application of bond coat.
- Fill the cavity with repair mortar or concrete and consolidate thoroughly with rod and mallet or vibration.
- Alternatively fill prepacked aggregate concrete in the cavities after applying bond coat and finish with pressure grouting with repair mortar.
- Apply a protective coating on the finished surface of the repaired element.

The ordinary classical methods do not give satisfactory results and the problem reoccurs very fast due to the following two main reasons:

- Corrosion of steel is not removed totally and also the root cause is not eliminated adequately.
- Bonding between old surface and new repair material is not developed adequately,

There are some chemicals developed recently for inhibiting corrosion and these can also be used for effective repair of RCC elements. In cases where carbonation of concrete has taken place but corrosion of reinforcement has not yet started a protective coating of required thickness should be applied. Epoxy resin or polymer modified mortar or concrete provide adequate protection. Such coatings also stop penetration of chloride and other deleterious materials. Whenever the process of corrosion has set in, repair and restoration process depends on the extent of damage and condition of concrete around steel.

High **strength, resilient** materials which have high **resistance to attack** from chlorides and sulphates are normally used as repair materials. The polymer modified concrete (PMC) is one such material, which is commonly used for repair/rehabilitation.

The filling of crack is normally done by epoxy grouting. The epoxy grouting systems have high mechanical strength, adequate resilience and rapid hardening qualities. The epoxy systems are impermeable and immune to attack by sulphates and chlorides. These have high compressive as well as tensile strengths. Epoxy grouting systems can be injected into even hairline cracks and hence are very useful for repairing RCC elements effectively.

Polymer based paint coating is applied for enhancing the life of structure by preventing the future ingress of sulphates and chlorides. Thus, the entire structure is protected from corrosion and spalling problems. These coats must be applied periodically (normally at a interval of 5 years depending on the product).

Most of these chemicals/admixtures have been developed in other countries having quite different conditions, their use in India must be done with special care in respect of environmental and usage conditions. These chemicals must be tested first for suitability and manufacturer's literature should be studied well before using them in repair or construction job.

There are specific techniques of placing concrete in repair jobs depending on the nature of the element, conditions of placement, available resources (equipment, workman and materials). Some of the most common techniques of concrete placement are described briefly in the next section. Some of these techniques have also been specified under common techniques of repair.

10.5 CONCRETE PLACEMENT TECHNIQUES

Concrete repair often calls for special material placement techniques because the repair area may be only a small portion in existing member. Repair material can also be placed with the same technique as the original material. The selection of the appropriate placement technique depends on the needs of the job and the type of forms or equipment available. When selecting a placement method, consider one that allows for the **complete filling** of the void and is **most economical**. Ensure the use of repair materials with minimal shrinkage. Although some repair jobs are unique, following are the general **material placement techniques** that are most commonly used.

10.5.1 Pneumatically Applied (The Gunite Technique)

This technique is used in repair of vertical and overhead surfaces without congested reinforcement. Spraying can perhaps be as accurate as the pneumatic application of a finely graded cement mortar or concrete. The objective is to force the mix onto a surface so that it will adhere firmly and harden to maximum density. This is achieved by literally shooting a jet of the mix through a nozzle at considerable force. When a surface is to be built up to a desired thickness by spraying, it is essential to wash each layer thoroughly with water after it has hardened. For most repair jobs a **reinforcement mesh** should be placed over the surface, both to **increase the tensile strength** of the finished job and to provide the necessary degree of **support** for the mix. The sizes and extent of this reinforcement depends on the nature of the job, or more precisely on the thickness of cover needed. For repair work, the most commonly used material is a 100mm by 100mm mesh or 75mm by 150 mm mesh of 8 gauge wire. Galvanized, welded wire fabric should be used when constructing a sprayed concrete element. The wire in the fabric should be generally spaced not more than 100mm apart in each direction. Alternatively, galvanized expanded metal with openings not greater than 75 mm by 150 mm can be used.

The best technique for applying **sprayed concrete** is to hold the nozzle perpendicular to the surface to be sprayed and about 1 to 1.20m from the surface. To ensure that reinforcement is firmly embedded, it is advisable to tilt the nozzle slightly so that the concrete is forced behind the steel alternately from both sides. The concrete should be applied over large surfaces by moving the nozzle uniformly so that the spreading effect is limited to a small area. Moist curing of sprayed concrete is essential. The duration of the curing should be the same as for concrete placed conventionally.

The repair material is blown in to place by compressed air. There are two types of shotcrete mixes wet and dry. In the **wet mix** process, water is mixed with the dry materials before being shot onto the surface. In the **dry mix** process, the water is sprayed to the dry surface before application of dry material by the nozzle.

Advantages: Shotcrete usually has advantages of a low water-cement ratio and good compaction producing a repair material with high compressive and flexural strengths. The material can be applied quickly and almost in any shape or configuration. This placement technique can be performed without forms. **Shotcrete** also becomes cheaper than other methods, which require extensive formwork.

Disadvantages: The quality and economy of the final product is largely dependent on the skill of the nozzleman and the recollection of rebound material. Care must be taken to avoid formation of sand pockets within the mass because shotcrete usually does not contain coarse aggregate. It needs more cement paste than conventional concrete to properly coat the aggregate. Shotcrete results in excessive shrinkage because of too much cement.

10.5.2 Open Top Placement

Open top technique can be used for partial or full-depth repairs to horizontal surfaces. The repair material is poured directly into the top of a prepared surface of cavity or form. The repair material placed should be well vibrated and struck off so that the new material surface is flat and level with the surrounding concrete. This is the easiest and least expensive placement method.

10.5.3 Opening from the Top to Repair Bottom Face

Openings are made from the top near the affected bottom portion for pouring materials from the top. Pouring from the top be used for repairs to the bottom of relatively thin members such as slabs. Placement holes are drilled through the top of the slab into the void below. A form is installed under the prepared area to be repaired and the repair material is poured into the cavity through the drilled holes. The material can be placed through funnels to create a pressure head, which will help to fill the void.

The preferred method is to place needle vibrators into the repair cavity through the placement holes and also vibrate the form from below. Complete filling of the cavity depends on very good vibration of the repair material. The underside of the slab void should be prepared in such a way as to minimize or eliminate the entrapment of air as the new material fills the cavity. If creating these air pockets is unavoidable drill another hole through the slab into that area to vent the air out.

Advantages: In this method special equipment such as a concrete pump is not required to place the repair material because the placement is from top. Placement is done under gravity which helps to fill the cavity. Filling and plugging the placement holes anchor the repair material onto the original member.

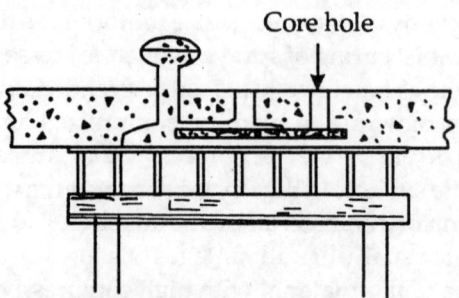

Fig. 10.4: Bottom Pours from the Top

Disadvantages: In this approach the placement holes can not always be drilled where needed because reinforcing steel or post-tensioning cables may be coming in the way. The cost of drilling the holes can also be high if contractors do not have their own equipment. In addition there is always a chance of trapping air between the new material and the top of the cavity. This technique is practical only for relatively thin members because thicker members require too much drilling and vibration of the repair material becomes very difficult.

10.5.4 Birds Mouth

Birds mouth technique (Fig. No. 10.5) can be used for repairing vertical cavities that cannot be filled from the top. The face of the repair cavity is formed just a few mm below the top. This section of the form is held tightly to the concrete surface. A slanted section or **"birds mouth"** is added from the top of the flat form projecting out from the finished surface at about a 45 degree angle and extending several mm above the top of the cavity. This forms a chute that allows the material to be poured into the cavity. It also allows material to be placed in the form higher than the top of the cavity to permit complete filling of the cavity. After the material has set, the extra material is removed.

It is very important that the throat created by the bird's mouth and the top of the cavity is large enough to allow a vibrator to be inserted into the repair area. Vibration is very important to ensure complete filling of the cavity and to remove air pockets.

Advantages: This placement technique does not require special equipment and often the repair material is moved from the mixer to the form by bucket. A pump can also be used to place the material into the form.

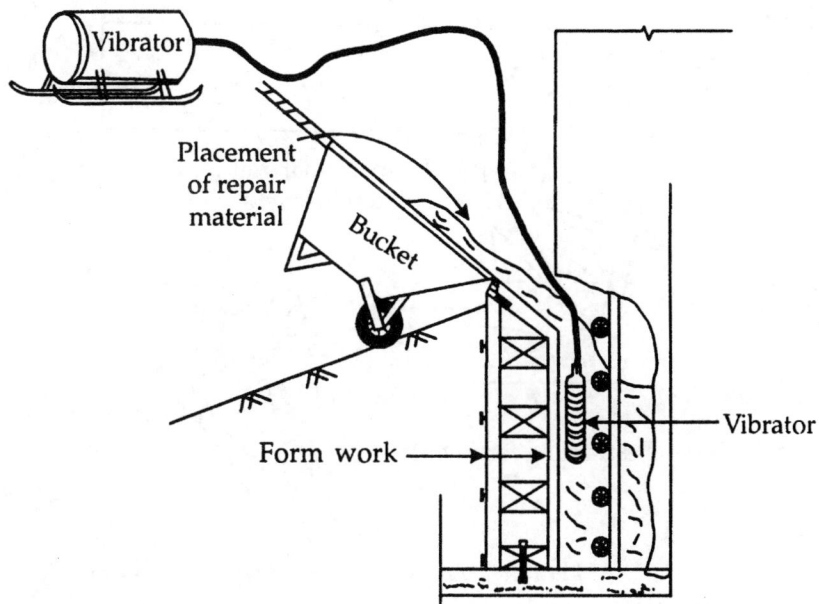

Fig. 10.5: Concrete Placement Method—Birds Mouth

Disadvantages: This method should be used only for placement that are not very deep because the aggregate may segregate from the mix as the material falls freely to the bottom of the form. The **bird's mouth** requires chipping and repair of the surface where the bird's mouth was removed.

10.5.5 Dry Packing

Dry packing technique (Fig. No. 10.6) can be used for repair of small confined areas such as the ribs of a waffle slab that have limited reinforcement or other obstructions. Dry packing approach uses a dry mortar that has enough moisture to form a cohesive ball when squeezed in hand but not enough moisture to make it plastic. A mallet and drew stick are used to **pack the material** into the place to form uniform compacted mass. A back form and a bottom form are used to provide confinement on ribs. After the material is placed and compacted the exposed surface is finished to match the existing concrete and adequate curing. Water misting over wet burlap is the preferred curing method.

Advantages: No specialized equipment is required and the repair mortars (sand and cement) are inexpensive. Dry packing produces a high quality repair material with minimal shrinkage, high early strength and good bond to the base material.

Disadvantages: Dry packing is some what labour intensive and requires skill in mixing the material to get the proper moisture content. If a few drops are added more, the mix will become plastic and when pressure is applied it will not pack tightly, but will only bulge and move from one area to another. Dry packing is not applicable on large surfaces or deep areas where closely spaced reinforcing steel prevents proper consolidation of the material in the cavity. Dry packing requires dry more skilled technician to place it.

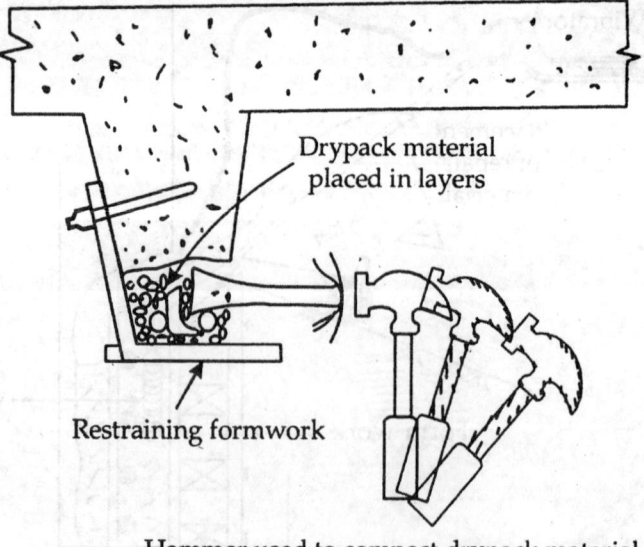

Drypack material placed in layers

Restraining formwork

Hammer used to compact drypack material

Fig. 10.6: Dry Packing

10.5.6 Form and Pump

Form and pump technique (Fig. No.10.7) can be used for overhead and vertical repair areas where other placement techniques are not appropriate due to the geometry of the member to be repaired. These include the under side of deep beam members with closely spaced reinforcing steel, and very large repair areas. The substrate concrete is properly prepared, and covered with forms. Often the form is completely closed and made watertight and is equipped with pumping ports, valves and air vents. Forms are made stronger than conventional formwork. to withstand the pumping pressure. The hose from the concrete pump is connected directly to the form and the repair material is pumped in. The form is completely closed so that it can be pressurised by the pump.

Advantages: The form and pump method allows faster placement of the repair material than most other methods. If the form is pressurized, it forces the repair material into tighter contact with the original concrete. This creates a better bond between the new and old concrete and allows for better covering of the reinforcing steel. It is less dependent on the skill of the worker than some of the other placement techniques.

Disadvantages: Formwork costs are higher due to stiffness and air tightness required in the form work. In addition this method involves the additional cost of the valves and vents and pumping equipment.

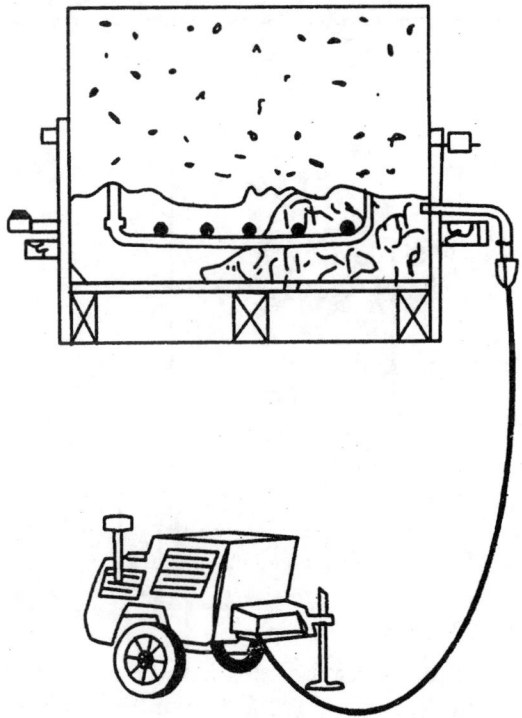

Fig. 10.7: Form and Pump

10.5.7 Preplaced Aggregate Concrete

Preplaced aggregate concrete technique (Fig. No.10.8) is used for repairs that require a very-low shrinkage of repair material to prevent settlement such as structural column repairs. After the cavity has been properly prepared and the form is installed, the cavity is completely **filled with a specially graded coarse aggregate**. Sand cement grout is then pumped in to the aggregate, completely filling the spaces between the aggregate. The grout is placed through ports in the form or through grouting needles that extend to the bottom of the form. As the grout is pumped through the needle, the needle is withdrawn gradually until the entire area has been grouted.

Advantage: The major advantage of this technique is the **reduction in shrinkage**. With the aggregates in point to point contact, there is little room for shrinkage to occur.

Disadvantages: Specially graded aggregates are required. It can be difficult to completely fill the cavity with the aggregate. If grouting needles are used, space is required above the repaired element to pull them out.

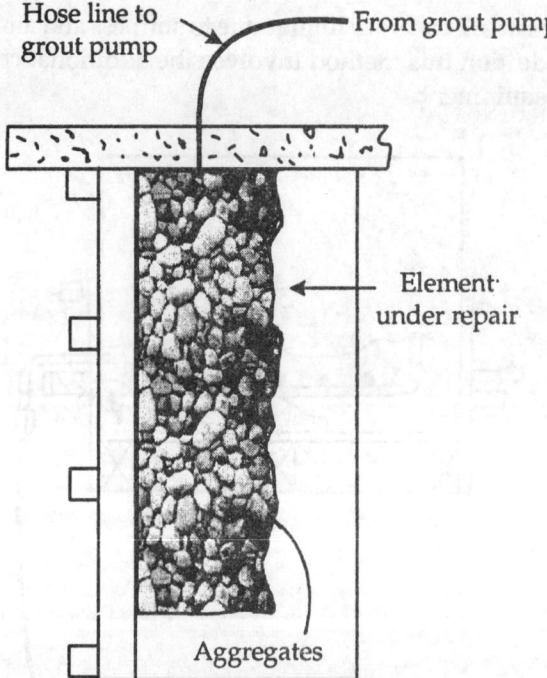

Fig. 10.8: Preplaced Aggregate Concrete

10.5.8 Trowel Applied Method

Trowel applied technique can be used for horizontal, vertical or overhead repair areas which are less than 25mm deep with little to no reinforcing steel. The repair material is mixed to the proper consistency. The worker uses a trowel to press the material in to the prepared cavity or apply it to the prepared surface. On horizontal surfaces such as decks and sidewalks a thin material is used to level the surface. On overhead surfaces a non-sag material must be used and it may be necessary to install the material in several coats to prevent sagging or falling.

Advantages: In trowel application method equipment requirements are minimal and workers can achieve different surface textures. This method is often used for small repair areas.

Disadvantages: The finish and the consolidation of the repair material depend on the ability and the skill of the worker. Because the repairs are thin and may need to be placed over head, special proprietary materials may be required for bonding. Repair must be carried out in several layers. Preparation is usually required between layers. Since no forms are used, special emphasis must be given to curing.

10.6 REPAIR OF SURFACE DEFECTS

Surface defects, such as **bug holes, form-tie holes,** and honeycomb, are unavoidable facts in concrete construction. Depending on project specifications, they may need to be repaired immediately before commissioning.

On many jobs, filling surface voids is a last task, so often it does not receive the attention it deserves. Engineers and contractors must be aware that these repairs are usually thin and are prone to failure. Failure of surface void repair often is caused by loss of moisture from the repair material. These thin repairs often fail because the mix water evaporates quickly from the surface and also gets absorbed by the existing concrete. This leaves little **water to hydrate the cement** to establish a strong bond. It is recommended to repair surface voids within **24 hours** after removing the forms so as to avoid drying of the base concrete. It is also advised to wet cure the repairs for at least 2 weeks to prevent evaporation and shrinkage.

Different techniques for repairing surface voids in form concrete are described in the subsequent sections. The repaired areas are usually darker than the surrounding concrete, so if colour matching is important, replace some of the grey cement. The right cement blend can be determined with white cement by installing test repairs on an inconspicuous area of the concrete.

10.6.1 Bug Holes

Bug holes are air bubbles trapped at the surface of cast in situ form concrete during placement and consolidation. They are commonly seen on vertical as cast surface and often do not require repair. But if repair is required, **sack rubbing** is an effective method. A good mix for sack rubbing contains **1 part** Portland cement, **1.5 to 2 parts** fine sand, and enough water to produce a consistency between that of thick paint and masonry mortar. Wet the existing concrete and apply the prepared grout to the entire surface with a rubber hand float, forcing the grout into all the voids. The next step is to rub the excess grout off the surface, and the timing of this step is crucial. If done too soon, some of the grout may be pulled from the bug holes; if done too late, the grout may be hard to remove from the surface.

A common way to remove excess grout is to rub the surface with burlap or sack. Closed cell polystyrene works better because removing the grout with burlap may create recess in the filled bug holes. After removing the excess grout curing should be done for adequate period.

10.6.2 Form Tie Holes

A common type of form tie uses plastic cones at the surface. The cones act as a spreader for the forms, aid in reducing grout leaks, and makes breaking back snap ties easier. But after removal, a typical cone leaves a hole in the surface about 25mm wide and 40mm deep. The most common way to fill **form tie holes** is to simply trowel mortar into them, but these repairs rarely hold up.

Form tie holes should be filled by dry packing as this technique has a history of producing durable repairs. Dry pack is a stiff sand cement mortar containing 1 part Portland cement, 2 to 3 parts sand, and enough water and suitable admixture to produce a mortar of adequate plasticity that will just stick together when moulded by hand into a ball. The low water content of dry pack results in minimal drying shrinkage, thereby improving durability. Ideally, the smooth inside surface of form-tie holes should be roughened to promote a strong bond. This can be done, by a 12mm drill by roughening the edges.

Dry pack can form an adequate bond with form tie holes if the base concrete is in a saturated surface- dry condition (the surface is saturated but contains no standing water). Using an epoxy bonding agent inside holes works even better because it prevents the surrounding concrete from sucking water out of the dry pack.

Pack the mortar in layers of about 10mm thick, overfilling the hole slightly. Then place the flat side of a piece of hardwood float against the hole and strike it several times with a hammer. If necessary, a few light strokes of the mortar with a rag may help the repair mortar to blend with the surrounding concrete.

10.6.3 Honeycomb and Larger Voids

Unlike bug holes and form-tie holes, which require specialized repair techniques, honeycomb and larger surface voids use procedures resembling those for typical concrete repairs. These techniques vary greatly. Some authorities do not require a minimum thickness, while others set limits, such as 25-50mm. The need for a bonding agent, whether an epoxy or cementitious grout also depends upon the repair requirements.

Use light chipping hammer weighing 5-6 kg or less to remove the required affected voided concrete. Indiscrete impact of hammers can fracture the remaining concrete surface also and therefore follow chipping with sandblasting or water blasting to remove fractured surface, if any.

A common way to place the repair material is to simply trowel it into the void. Better results can be achieved using a small pneumatic mortar gun to apply a grout similar to hand applied dry pack. A suitable mix contains 1 part Portland cement and 4 parts sand, adequate admixture and water cement ratio of about 0.35. If the repair is deeper than 25mm, apply the mortar in layers no thicker than **15-20mm** to avoid sagging and loss of bond. After placing each layer, wait about 30 minutes before placing the next layer. There is no need to scratch or otherwise prepare a preceding layer before placing the next, but do not let the in-place mortar dry. To complete the repair, overfill the void slightly and finish by trimming and trowelling. When finishing the repair, extreme care should be taken to avoid impairing the bond.

10.7 SUMMARY

For effective repair of RCC structures a systematic approach of inspection and diagnosis is necessary. Preparation techniques, application of primer coat, selection of repair materials and application of protective coatings are required to be properly supervised and carried out. Other important considerations necessary are proper bond to original surface, compatibility of repair materials, resistance of repair material to temperature variations and weathering.

There is no miracle low cost technique for repair of corrosion in reinforcement. However, for combating corrosion, techniques such as good concrete practices (**low water cement** ratio, adequate **cover**, proper **curing** and use of water reducing admixtures) should be adopted. Latex modified concrete and silica fume concrete can be used to reduce the penetration of moisture. Coating of reinforcing bars with epoxy or use of membranes and sealers for eliminating chloride penetration facilitates long life of RCC elements.

Cathodic protection controls corrosion of steel in concrete by applying an external source of direct current to embedded steel in concrete. It is not suitable for carbonated concrete. Electro-osmosis provides the solution to the carbonation problem.

Calcium Nitrate is an inorganic corrosion inhibitor and disrupts the corrosion process by enhancing the formation of passivation layer on the surface of reinforcing steel. Organic corrosion combines organic chemicals to prevent corrosion in concrete by forming a protective layer on the reinforcing steel.

Repair and prepared corrected area should be carried out systematically by removing all spalled concrete, cleaning the bars, applying protective coating and bonding material on the bars and concrete surface before applying the patch repair material.

Steps such as exposing and under cutting, repair, cleaning of reinforcing steel, compensation for significantly lost section, edge and surface cleaning are very important for preparing the area for repairs.

Appropriate techniques of concrete placement are important for effective repair work. Gunite technique is useful for repair of vertical and overhead surface without congested reinforcement. Open top technique can be used for partial or full depth repairs of horizontal surfaces. Pouring from top can be used for repairs to the bottom of relatively thin members such as slabs. Birds mouth technique can be used for repairing vertical cavities that can not be filled from the top. Dry pack technique by covering with form work can be used to repair small confined areas such as ribs of a waffle slab that have limited reinforcement. Form and pump technique is useful for overhead and vertical repairs where other placement techniques are not appropriate due to geometry of the member to be repaired. Replaced aggregate technique is used for repairs that require a very low shrinkage of repair material. Trowel applied technique is used for horizontal, vertical and overhead repairs which are less than 25mm deep with little to no reinforcing steel.

Surface defects such as bug holes, fc ·m-tie holes and honeycomb need immediate repairs. Bug holes are repaired by sack rubbing. Form-tie holes should be filled by dry packing. For repair of honeycombed concrete, chipping with hammer is followed by sand blasting or water blasting. After thorough cleaning, repair material is trowelled onto the surface. The need of bonding agent depends upon requirement of the repair job.

QUESTIONS

10.1 Describe importance of repairing RCC elements at various stages e.g before commissioning and during usage.

10.2 List six basic requirements of patch repair work in RCC elements.

10.3 Explain briefly the process of preventing corrosion in steel reinforcement in RCC elements.

10.4 Explain the importance of good concrete practices in reducing the repair and maintenance problems of RCC elements.

10.5 Explain the use of epoxy coated reinforcing bars in RCC elements.

10.6 Explain briefly cathodic protection to control corrosion in RCC elements.

10.7 Describe briefly the use of inorganic corrosion inhibitors in RCC elements.

10.8 Explain the principle of electro-osmosis in controlling the carbonation and corrosion in RCC elements.

10.9 Describe the main steps of preparations necessary for effective RCC repair.

10.10 Describe main steps in repair of corroded RCC elements.

10.11 List concrete placement techniques.

10.12 Explain briefly following techniques of concrete placement:

(a) Gunite technique
(b) Bird's mouth technique
(c) Dry pack technique
(d) Preplaced aggregate concrete
(e) Trowel applied technique.

10.13 Explain briefly the method of replacing corroded steel reinforcement.

10.14 Explain repair techniques of surface defects such as:

(a) Bugholes
(b) Formtie holes
(c) Honey comb.

Chapter 11

REPAIR & MAINTENANCE OF FOUNDATIONS, BASEMENTS AND DPC

11

REPAIR AND MAINTENANCE OF FOUNDATIONS, BASEMENTS AND DPC

LEARNING OBJECTIVES

After studying this chapter, the learner understands common repair and maintenance of foundations, Basements and DPC and will be able to:

- **Explain** stabilising the foundation soil;
- **Explain** the technique of underpinning for repair of foundations;
- **Explain** the repair of raft foundation slab;
- **Explain** various techniques for repair of basements with respect to:
  - Structural problems
  - Dampness problems
- **Explain** the repair of DPC against rising dampness.

11.1 INTRODUCTION

Foundations are all structural elements below plinth and these elements distribute the load of building ground and soil under it. Foundations play an important role in service life of the building structure under various conditions. Repair of foundation becomes necessary when the super structure is subjected to distress due to inadequate foundation and soil settlement. This may be due to poor bearing capacity of soil, inadequate depth or width of foundations. Sometimes inappropriate quality and type of foundation may result in distress in building super structure.

When soil settlement or expansion causes distress in one or two storey building, deepening or rebuilding the foundation will usually solve the problem. When foundations are rebuilt, permanent shoring jacks are necessary to hold the building in its original position. First put exterior ribbon bracing to prevent lateral movement of the sill plate. Existing foundation is removed and repaired section by section by placing permanent shoring jack every **2 m or at each load bearing point**, whichever comes first.

The base of the jack rests on a block of strong concrete so the steel jack does not make contact with wet mud ground. Proper placement of the jacks is crucial to ensure that the load don't move.

The jacks remain in place even after the foundation concrete is fully poured. Reinforcement in the foundation is placed alongside the jacks, and a high quality low slump concrete is used to ensure contact between the new concrete and the sill plate.

While deepening foundation, permanent shoring jacks support the structure as the area around and beneath the existing foundation is excavated. When the needed repairs are carried out, the shoring jacks still remain in place within the new foundation system till the curing is complete.

Close working with a structural engineer is necessary for providing structural details needed for good workmanship and the success of repair. This is essential from the safety of tenants and occupants of building under repair and other adjoining buildings in neighbourhood.

11.2 STABILISING FOUNDATIONS

A small percentage of foundations fail or get distressed, and the deficiency usually puts much more load than the foundation capacity. Sometimes repairing a foundation requires beyond fixing cracks or adding steel reinforcement. In some cases the source of the failure is not the concrete or steel but relates to the under lying or adjacent soil condition. Problems range from improperly compacted soil fill, drop in the water table, corrosion, consolidation, or effect of neighboring new construction. The soil stabilization increases the foundation bearing capacity with steel pipes that are pushed or screwed into the ground beneath the footing or floor slab. Jacking against the pillers raise foundation back to its original level. In 2-3 days a structure can be raised enough to close cracks and carry out other cosmetic distress repair.

The repair specifications are developed through a team effort involving the repair contractor, who has the practical experience and the consulting engineer, who does the investigation and design calculations. Repairs should be designed and carried out to last as long as the remaining life of the structure being repaired. Some techniques used successfully to stabilise foundations are:

(a) Hydraulic piles

(b) Construction of grout columns

(c) Compaction grouting and steel piers

(a) Hydraulic Piles: If the soft soil causing the settlement is extended to about 6m below the bottom of the existing footing, the most practical and economical means of underpinning is hydraulically driven steel piles.

To install the piles, earth is excavated to expose the footing bottom. A portion of the spread footing is removed to locate the pile as close as possible to the wall. Next a structural steel head is installed beneath the footing to transfer the load of the building to the pile. Steel pile sections 100mm outside diameter, 1m long are pushed into the soil with a hydraulic ram. The sections are coupled with an inner sleeve coupler. The hydraulic pressure for each pile section is measured to record soil strength and determine pile capacity. **Numbers, size** and **spacing** of these piles depend on the **load transfer** necessary for the building.

(b) Construction of Grout Columns: Compaction grouting involves injecting a stiff cementitious grout into soil. The grout displaces and compacts the surrounding soil to increase strength and stiffness. Compaction grouting with conventional cement is best suited for relatively loose materials such as earth fills and loose sands. Compaction grouting techniques can be used to create piles that can resist tension and compression loads.

Construction of grout pile begins with drilling or driving an injection casing to the specified bearing layer of bedrock or other sound material. Grout is then injected along with gradual withdrawal of the casing. Injection pressures and grout quantity should be monitored closely as required for the intended soil compaction and pile cross section.

(c) Compaction Grouting and Steel Piles: Soft organic silts and clays are not normally densified with compaction grouting alone, but steel piles are also necessary to play supporting role. The grout should have a specified minimum compressive strength of **10 MPa**. The grout pile cross section should be **0.35 sqm** and grout **pumping pressure** should be limited to **4 MPa**. After several grout piles are installed, the grouted sand layer provides adequate bearing capacity. After the grout piles are installed additional compaction grouting of the weaker soil layer may be undertaken using the steel piles. Once the pile gets raised, additional grouting restore positive support of the pile by the grout columns.

11.3 UNDERPINNING

Underpinning

Underpinning may be defined as placing of new permanent support under existing foundation. Underpinning can be used to perform following functions:

- Protect existing buildings from damage during the process of deep excavations for other structures close to their foundation.
- Prevent settlement due to inadequate or improper type of foundation.
- Repair foundations for the damage caused due to change in conditions such as decay of timber piles and lowering of ground water table.

Underpinning of heavy structures requires great skill and care. The operations are required to be carried out in adverse conditions. Such jobs are required to be completed with minimum inconvenience to occupants and without damage to structure being underpinned or adjoining ones.

The building in proper condition to receive underpinning installed, care must be taken that during underpinning the earth beneath the footings must not be lost or loosened. The loss of ground moves soil from outside into the excavation and hence forms voids outside the sheeted line of underpinning work.

Wedge up and test the underpinning to bring the load to it or else the load will be carried subsequently to underpinning by settlement of structure. Moderate loads are handled with needle and shores, wedges may be driven to relieve them of their loads. For heavier loads, the underpinning work should be installed to regulate the ensuing settlement which should be uniform. This settlement is considered troublesome if it is 25mm or more.

11.3.1 Foundation Support

No additional support is required for small excavations or in case of buildings with sufficient safety. All other buildings require additional support. The support can be provided using the following methods:

(a) Shoring
(b) Needling

(a) Shoring

Shoring is used when loading is not excessive. Wooden shores are placed in an inclined position against the wall in suitable niches and supported at the bottom by wooden foot blocks. As far as possible, make shores vertical to minimize lateral thrust on the wall. Shores are reinforced by channels or I beams bolted to it. Provide ties and bracing for lateral support. For carrying part of the wall load from head of the shore to ground, shores are combined with needle called 'springing needle'.

(b) Needling

Temporary beams called needles are installed, to support the structure during underpinning. For light loads, I-beams resting on suitable block (wooden cribs) and wedging are used. Heavy loads and long spans are supported by built up sections and hydraulic jacks. For uniform bearing and to avoid crushing of masonry at edges while wedging, bearing pads or wooden cushion blocks are put above the beams.

The lateral slopping of needles must be prevented (especially in case of I beams). For this they are used in pairs with wooden fillers or concrete at edges. A combination of needle and shore beam can also be adopted. The needle carries weight of wall from head of the shore, which transfers the load to ground.

In many cases needles and shores are not needed to strengthen before excavating beneath them. Their use is somewhat limited due to necessity of space inside and outside building walls.

11.3.2 Underpinning Methods

(i) Pit Underpinning

Pit excavations are made preferably above water level and the pits are filled with masonry or concrete. After hardening of pit concrete wedges with steel plates and I-beams are installed under the foundation. Pits are excavated away from the building to act as effective cut off between building and deep excavations in vicinity for protecting the building against foundation failure.

(ii) Piles Underpinning

Pile underpinning is used for protecting structure against nearby excavation as well as arresting its own settlement. Pile under pinning is suited when foundations carry heavy loads or where the foundation is required to reach a hard strata (difficult by using piers) or where ground water is present. It provides a quick method giving prompt support for structures endangered by quicksand or other difficult conditions.

When the unit pressure on foundation is more, wedging is done by "pretesting". Steel sheets (25 cm to 40 cm wide and 1 cm thick) are driven in excavated ground in short sections and then concreted. These are driven by means of falling or pneumatic hammers in accordance with available head room. When head room is limited hydraulic rams are used. Care must be taken to check large upward thrust offered by piles.

As the earth is excavated, resistance to driving of pile disappears. Excavation done by augers and orange peel buckets, jetting and suction in a diaphragm pump. Keep the pile full of water. The last case suction should not get too near the bottom to avoid loss of ground.

After driving upto the required depth, excavated pile is concreted. It is tested for a load by reapplication of the load using hydraulic rams. A settlement of 3 to 6mm represent a load of 80 tonne (for 35 cm dia pile) in coarse sand or gravel and even greater values for fine soils. The soil also gets compressed due to forming of a concrete bulb.

(iii) Caisson Underpinning

Water in foundations make pit underpinning impracticable when existing piles are to be removed. Caisson underpinning may be used in such conditions. Caissons are driven under compressed air to prevent loss of ground. In this method a person can be inside the working chamber. Artificial sinking of footings may be done by jacking in order to compact soil underneath for arresting settlement. Combination of various techniques can also be adopted for underpinning.

(iv) Pier Underpinning

Two different types of pier underpinning systems are available. **One system** uses pier section of **galvanised or epoxy coated steel pipes** for piers. The piers have pointed bearing and are driven to bearing strata with a hydraulic drive unit. Another system uses screw piles with steel shafts. The load section, with one or more helixes attached provides the needed bearing capacity. The piers are screwed into the ground with a hydraulic drive unit and extensions with or without helixes. Extensions are added during driving until bearing strata is reached. Both systems can be installed from inside or outside the buildings. Both hydraulic or screwed driving techniques produce no harmful vibrations. For light buildings steel underpinning piers are considered attractive to concrete piers. Steel piers are hydraulically pushed or screwed into the soil and attached to brackets installed beneath and anchored to footings.

The basic steps in underpinning involved in either the pipe or screw type pier system consist of driving one or more steel piers to the rock base or a suitable soil bearing layer. These are connected to the foundation through a metal head assembly. Hydraulic jacks, attached to the embedded steel piers, are used to raise the foundation back to its original elevation.

Although the steel piers can be pushed or screwed into the soil from above, they must be connected to the foundation. A small excavation, typically a 900mm x900mm square hole extending about 300mm below the footing, exposes the bottom of the footing at each location where a pier is driven. Because the holes are small, there is little damage to landscaping.

In the pipe pier process, an **L-shaped bracket** that transfers the load from the footing to the pier is first attached to the footing (Fig. No.11.1). Then the pier is driven. For the screw pier process, the pier is driven first and then connected to t-shaped lifting bracket and attached to the footing.

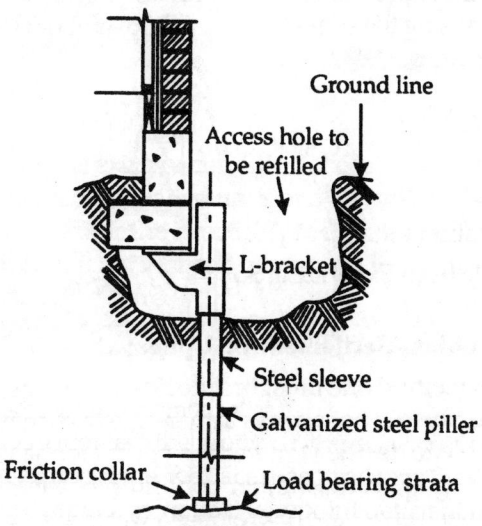

Fig 11.1: Foundation Underpinning

To prepare for transferring load to the foundation, workers use a chipping hammer to smooth the outside face and bottom of the footing and create a nearly right- angle corner. This allows the **L-shaped pier bracket** to be placed directly under the foundation of wall. The bracket is seated under and bolted to the side of the footing.

Non-shrink grout is placed on the vertical and horizontal faces of the steel L- bracket to provide continuous bearing support. For foundations requiring more support an additional steel beam can be placed under the foundation with the L-bracket seated against the steel beam. The L- bracket transfer the load from the foundation to the in-place steel pier. It is also used as a jacking platform to raise the foundation. All steel elements are protected with epoxy resins and other chemicals against corrosion.

11.4 REPAIR OF RAFT SLAB FOUNDATIONS

Bulging floors, cracked walls, are the symptoms of foundation distress. This may represent anything from a simple nuisance to a catastrophe, depending on how severe the problem becomes. Assess the causes of the damage before attempting repairs, methods of repair and to prevent the problem from recurring.

Water is the main culprit in causing most of the expansive soil problems. There is either too much water causing the soil to swell, or not enough water causing the soil to shrink. If all the soil beneath foundation swells uniformly or shrinks uniformly it is unlikely to cause a problem. But when only part of the foundation soil heaves or settles, differential movement causes cracks and other damage in floor, wall and foundation slab.

Most differential movement is caused by variations in soil moisture. After construction, soil beneath part of the foundation becomes wetter or drier than the rest of the soil. Soil moisture content at the time of construction may be abnormally high or low. If the moisture content is low in soil when a slab on grade is poured in foundation, soil at outer edges regains moisture directly from rain or plant watering. Water does not move rapidly through confined impermeable expansive clay soils so as to return to normal moisture content. This slow movement of moisture under the central portion of the slab results in swelling of soil at the slab edges, causing the edges to cup up. Repair of raft slab foundation involves the following steps and stages:

(a) Diagnosis of the problem

(b) Repair methods

(c) Preventing further damage

(d) Correcting foundation settlements

(e) Preventing trouble from Expanding Soil

(a) Diagnosing the Problem-Settlement or Upheaval

Before choosing a repair method one must first decide whether settlement or heaving is causing the distress. The cause is not always apparent. If a corner of the building is low, for instance, we are unable to decide whether the perimeter has settled or the center of the building has heaved up. Recent weather condition of the place gives some clues. Upheaval is more likely after heavy rains or prolonged wet spells. Look for poor site grading that traps water near the building.

Examine mouldings and skirting/dado inside the building. If most movement occurs at the corner because of settlement, there are usually big gaps in the skirting. If the center has heaved up however, dado in the corners won't be much distorted.

Mortar joints in bricks also give clues about the source of problems. Observe site along a mortar joint at the presumed low corner. If the centre has heaved, the joints will be reasonably straight up to the high point. If the corner has settled there will be a kink in the mortar joint near the corner.

For masonry buildings, also examine windows near the low area. If settlement is the problem, brick veneer towards low corner will have pulled away from the window. If the building center has heaved up there won't be much separation at the window.

(b) Repair Methods

The two most common methods of repair are mechanical jacking and slab jacking. Mechanical jacks can exert very high local pressures. A careless or incompetent operator can damage the foundation. However, the high lifting force is sometimes needed under concentrated loads or to overcome mechanical binding.

In a slab jacking operation, grout is pumped beneath a slab or beam to produce a lifting force that restores the member to its original elevation. The grout is a mixture of water, cement, and sand and or flyash. Lime also may be added to the grout. Sometimes mechanical jacks are used along with slab jacking. The repair method used depends on the type of distress being treated. The types of distresses are:

(i) Edge Settlement

(ii) Interior slab Heaving

(iii) Edge Upheaval

(i) Edge Settlement

If the slab has settled at the edges, dry soil at the perimeter can sometimes be watered enough to correct the problem by partially reducing differential moisture (Fig. 11.2). If watering does not adequately raise the perimeter, slab jacking will probably work if edge settlement does not exceed about 50 mm.

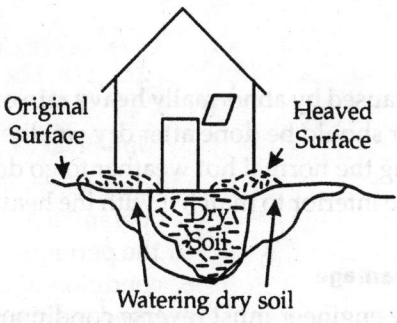

Fig 11.2: Swelling at Edges

(ii) Interior Slab Heaving

Upheaval of a interior slab is the most difficult problem to correct (Fig 11.3). If the interior slab has heaved, the only solution is to break out the floor slab, excavate the soil to the proper elevation, prepare the base, and repair the floor slab. This usually is **not feasible specially when the floor is structural raft slab** having reinforcement. It also costs too much and completely **disrupts the usage** of the building.

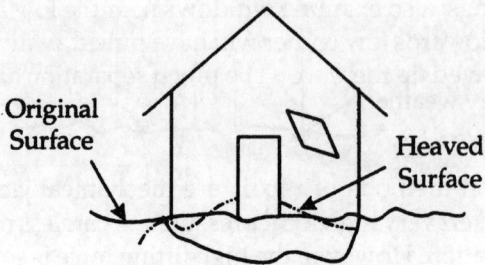

Original Surface

Heaved Surface

Fig 11.3: Heaving of Middle Slab

If total heave of the interior slab is less than **200 to 250 mm** it may be possible to partially correct the problem by combining mechanical jacking and slab edge jacking. The perimeter beam is raised enough to match the heaved crown. The beam is underpinned using spread footings and cast in place or precast piers. Finally grout is pumped underneath the slab edge to fill voids left by the raising of slab edges.

Raising the perimeter of a heaved slab is a delicate operation because the original level of the walls is being changed. Lifting the beams may damage the walls. The center of the slab may also rise when the perimeter beams are jacked leaving the hump as bad as it was before jacking. Also during slab jacking, grout may push into the heaved area, aggravating rather than correcting the problem. Usually the slab deflects at the center as the perimeter rises and much of the difference in level can be corrected.

It is not advisable to level a slab foundation by mechanical means alone. Rather than filling voids and stabilizing the soil, mechanical jacking creates voids that the slab must bridge. Foundation slabs are not usually designed as structural bridging members and may fail if left unsupported.

(iii) Edge Upheaval

Edge upheaval is usually caused by abnormally **heavy rain** or snow fall (Fig. 11.4) due to wet soil near edges. Repair should be done after dry weather for effectiveness. The soil may dry sufficiently during the normal hot weather to go down by itself. If it does not, carry out slab jacking at the interior to match it with the heaved perimeter beam.

(c) Preventing Further Damage

To test repair, the property engineer must reverse conditions that caused the problem. Soil swelling can usually be stopped by cutting off the moisture supply. Plumbing leaks are a major cause of soil wetting and swelling. The most common plumbing problems in slab on grade are caused by faulty or omitted shower pans or incorrectly installed water closet, drains, waste pipes and joints. Sometimes locating and repairing plumbing leaks can solve a recurring problem of heaving.

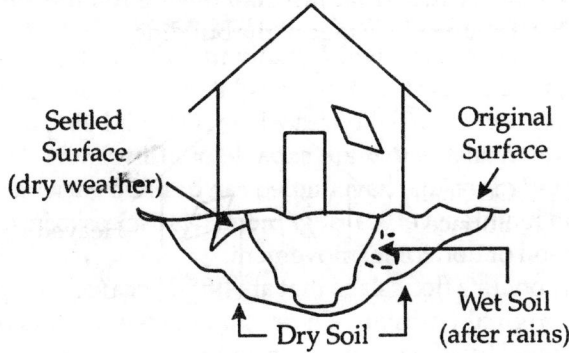

Settled Surface (dry weather)

Original Surface

Wet Soil (after rains)

Dry Soil

Fig 11.4: Soil Wetting at Edges

Poor drainage of subsoil near the foundation is another common cause of soil heaving. The owner may have to regrade the lot to make sure of draining away rainwater from the building foundation. The BIS recommends ground slopes of not less than **1 in 20** away from a building. Down spouts or gutter should carry water away from the foundation and there should be no place where water ponds near the building during rains.

Occasionally subsurface aquifers cause problems. During wet weather these carry water to swelling clays beneath the foundation. An interceptor drain from the water supply line must be used to divert water away from the foundation.

Regular lawn watering during dry periods helps to prevent edge settlement caused by soil shrinkage. The watering also prevent development of large shrinkage cracks that provide a conduit for water to penetrate beneath the foundation.

Swelling clays can be treated with a chemical stabilizing agent as part of the repair procedure. The agent alters clay by decreasing its moisture absorbing capacity. This is a promising method for solving swelling problems but field records for foundation performance on stabilised clays span only 4 to 5 years.

Preventing a problem is always desirable than trying to solve one. Correct a soil problem by **eliminating** or **reducing the root causes** making sure to prevent recurrence.

(d) Correcting Foundation Settlements

The most common method of stopping or retarding foundation settlement is to install steel or concrete piers to rest on suitable bearing ground. Depending upon the type of foundation and the weight of the structure carried the support piers may be spaced from 2 to 3m on centers. The size of the piers is determined by the specific type of bearing ground. Mud jacking of the settled foundation should never be done without proof that the grout or mortar used will not cause additional damage from expanding soil. These piers may be installed from the inside or outside of the foundation, depending up the depth of the piers. The need for heavy equipment makes it impossible to install deep concrete piers from the inside.

If the foundations cannot be lifted as much as it has supposedly settled then it may be possible that damage from expanding soil has been mistakenly diagnosed as settlement. A foundation that gives the impression of settlement because of damage from expanding soil cannot be raised without seriously damaging the walls, because it has not moved from its original position. Lifting will actually involve tilting the foundation into an unnatural position and because building foundations are not designed for this type of stress, large cracks may develop in the area where lifting is being done.

Hydraulic jacks are available that are capable of lifting multistorey buildings. With such equipment, foundations that have settled can be lifted back into place. Care should be taken to check and maintain vertical alignment. It is useless to install piers under walls affected by heaving soil or horizontal movement.

Foundations and concrete floor slabs that are lifted because of expanding soil cannot be lowered to their original positions. There is one exception that occasionally a short stem wall can be lowered by slicing soil out from the underside of the footing. However, this is not practical except in isolated cases. Further if steel columns extend through the basement floor slab so that the slab has distorted the first floor, and if wooden partitions have contributed to this distortion, the floor can be levelled by removing the partitions and cutting off the steel columns the same amount that the floor has been raised. In order to avoid additional lifting, the floor slab should be cut loose at any point where it is not free of the columns, foundation walls, fire-place masonry and plumbing stacks. It is necessary to isolate the basement floor slab from any element that connects it with the floor framing above.

(e) Preventing Trouble From Expanding Soils

It is easier to prevent trouble from expanding soils than commonly thought. Expansion of the soil beneath the floor slabs can be eliminated by grading for good drainage and by watering the yard consistently until normal weather conditions prevail. The yard should be watered to appropriate moisture level to keep the surface sealed against deep penetration of water through the shrinkage cracks. These requirements also apply to the prevention of foundation settlement.

Properly installed steel or concrete piers will control foundation settlement. If a floor or foundation continues to move after piers have been installed, then it can be concluded that the foundation was **treated for the wrong problem**. The real problem of upward movement of the foundation and the basement floor slab is a result of expanding soil. Since these kinds of foundation movements exhibit similar signs, it may be necessary to establish bench marks for monthly checks of component levels.

11.5 REPAIR OF BASEMENTS

11.5.1 Techniques for Repairing Cracked or Bulging Walls

For repairing of basement walls one or combination of the following methods may be used depending on the type of problem:

(a) Reduce Loading

(b) Add Butteress or Pilaster

(c) Install Earth Anchors or Tie Backs

(d) Add Rebar and Grout

(e) Inject Epoxy

(f) Apply Surface Bonding

(g) Installing Helical Anchors

(h) Vertical Steel Tube Reinforcement

(i) Stud Wall

(j) Exterior Augmentation

(k) Cantilever Retaining Wall

(l) Buttress Block and Grid

(a) Reduce Loading

Excessive lateral earth pressures are frequently caused by saturated clay backfill. The most effective way of reducing this pressure is to remove the clay backfill and replace it with compacted, free **draining granular** fill.

Reducing the amount of backfill also reduce the lateral earth pressures. In fact reducing the backfill height from 2.5m to 2.0m reduces the pressure on the wall by more than 50%. Surrounding steps, side walks, driveways, and garages often make reduction of back fill impractical. **Wet soil exerts much greater pressure** on the wall than dry soil. Always slope the ground away from the walls and place gutter down so that rainwater runs away from the building walls.

To transfer earth loads from the basement wall to the basement slab, **buttresses** must be anchored to the slab. If the slab is not strong enough, buttresses must be extended through the slab to the top of the footing.

(b) Add Buttresses or Pilasters

Back walls laid in running bond are sharing about 30% loads horizontally $\left(\dfrac{1-\sin\theta}{1+\sin\theta}\right)$ rather than only vertically. To make the basement wall to transfer loads horizontally, intersecting walls, buttresses (Fig. 11.5), or pilasters can be placed at spacing of **less than 4 m**. The exact size and spacing should be determined by a structural engineer, based on the length and thickness of the wall, height of the unbalanced fill and type of backfill.

Buttresses or pilasters can be constructed from concrete masonry blocks filled with rebar and grout.

(c) Install Earth Anchors or Tie Backs

Earth anchors or tiebacks can be used to strengthen the wall against the earth. One way of doing this is to drill a hole through the wall from inside the basement and to dig a pit in the outside ground several metre away from the wall (Fig. 11.6). Then working from inside

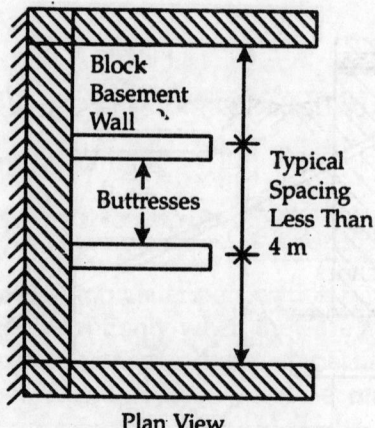

Plan View

Fig 11.5: Adding Buttress

the basement, push the tieback rod horizontally into the ground until it appears in the pit. Anchor the tie rod on inside of the basement wall by a faceplate and nuts. Connect the tie rod on the outside, to a steel or concrete deadman that is set in the pit and buried.

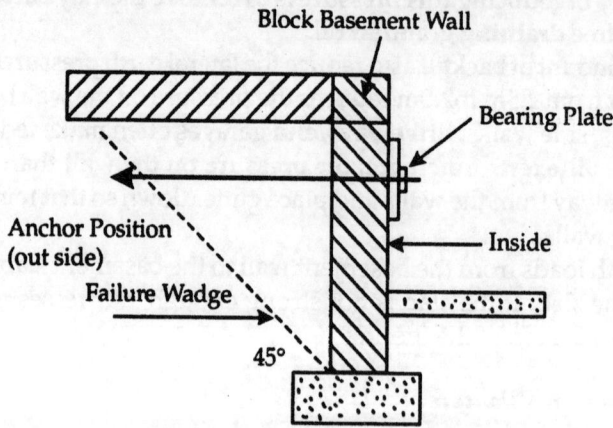

Fig. 11.6: Installing Anchor

Instead of a concrete deadman a special tieback rod that works like a toggle bolt can also be used. The rod is inserted through the wall into the ground, then pulled back toward the basement to expand a set of wings that hold the rod in place. Alternatively the rod can be grouted into the drilled hole.

For earth anchors to be effective the tieback rod must be long enough to anchor into the stable soil mass beyond the failure wedge as shown in Fig 11.7. The rod is anchored by the weight of the soil and the deadman.

Tieback rod anchors only the block to which the rod is attached. Earth pressures acting on the surrounding wall can thus create a punching shear. To avoid this, the tieback rod can be anchored to a vertical wooden or steel beam placed against the interior face of the basement walls with a reinforced grout column on inside of the wall.

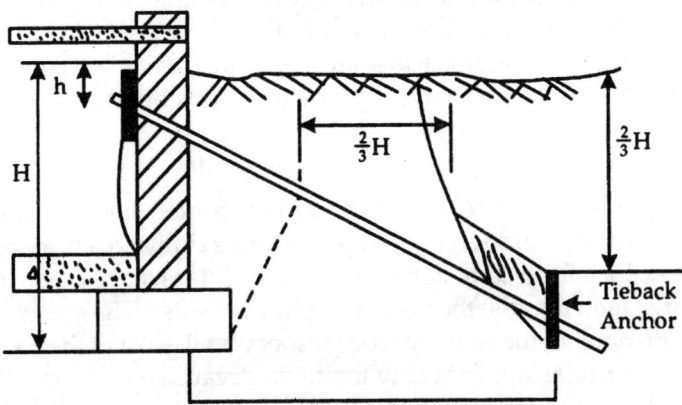

Fig. 11.7: Tieback Anchor

Depending on the wall height and the lateral load, one or more soil anchors may be required. For multiple anchors, a horizontal steel beam placed against the inside wall face can connect the rods and support the wall.

Though tieback rods require little excavation and do not disturb plantings near the building, but they may not be practical where there are underground obstructions. Also when the wall is drilled to accommodate the anchor, the opening allows water leakage in the face of the wall if not treated properly.

(d) Add Rebar and Grout

Un-reinforced block wall can be strengthened by installing vertical rebar in designated cores and filling the cores with grout. This is done from the outside by removing selected blocks in the top course or by cutting an opening in the outer face shell of the top block course. Grout can be placed by shovel or more easily by pump as shown in Fig 11.8. A vibrator can be used to consolidate the grout so that it fills the core without leaving voids.

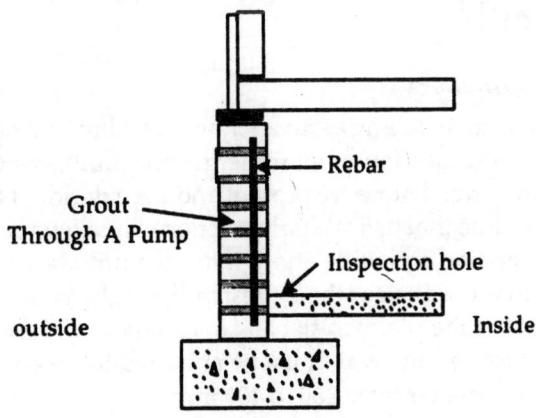

Fig.11.8: Add Rebar and Grout

To make sure that the grout reaches the bottom of the wall inside the basement, an opening can be cut into the bottom block. However, opening should be sealed quickly if the grouting is successful. Typical spacings are 2 to 2.5 m to centre to centre. Exact spacings for a given repair should be determined by structural considerations.

(e) Inject Epoxy

If the basement wall is made of solid block or is fully grouted, injecting epoxy into cracks can restore the wall's original strength. The procedure is similar to epoxy impregnation of cracks in concrete. Install injection ports and seal the surface of the crack. Inject the epoxy through the ports, then remove the ports and plug the holes. The epoxy can be colored, but it usually won't match the colour of the masonry wall. Cost of the epoxy, equipment and epoxy repair sepecialist is probably too high. Because epoxy also fills the cores of hollow block, this method is not practical for repairing cracks in hollow block masonry walls. It does not strengthen the wall beyond its original strength so it should be used only when cracks are caused by a one-time overload.

(f) Apply Surface Bonding

Surface bonding is another technique that can be used when cracks are caused by one time overload. Bonding material can be applied easily to the inside of a cracked block basement wall after surface preparation. This also increases the tensile strength of the wall face. A glass fibre reinforced cement mortar can be used with two coats of epoxy reinforced with glass fibre mesh.

Surface bonding mortar should not be mixed at site. Dry ready mixed surface bonding mortars of good quality are generally available. Mix the mortar according to the manufacturer's directions and apply it to a clean prepared wall. Surface bonding does not adhere well to a painted wall. The pores of the block must be open to achieve a good mechanical bond prior to application of such bonding mortar.

Surface bonding application is quick, simple, and relatively inexpensive, but like all other basement repair methods, it should be suitably designed and repair carried out under competent engineer/worker.

(g) Installing Helical Anchors

This approach is similar to installing earth anchors or tie backs. Thoroughly investigate the possible existence and location of all underground utilities situated at or near the work area before excavating. If none are present and the exterior of the wall is accessible, dig a hole next to the wall so that helical starter section can be lowered into it. Dig the hole 200mm below the bulge line and wide enough to accommodate the helix. Then drill a 80mm diameter hole in the wall near the largest bulge in the wall. Align the core hole at an angle matching that of the planned tieback installation. If the exterior of the wall is inaccessible the hole through the wall must be enlarged to accept the largest diameter helix. Helixes range in diameter from 150 to 400 mm.

A cone along cable system is used during the installation of the first section to provide a staring force in the direction of installation. With the helical starter section outside the wall, mount a steel helical extension to the starter section and attach the shaft to the torque motor. Additional helical intermediate sections and extension sections are then installed until adequate torque is achieved. The last extender section is then removed so that the end of the previous extender section is outside the wall. A continuous thread lag stud is concreted with basement wall to an adapter that is bolted to the helical anchor shaft.

The installed system can be load tested for pull out capacity with load testing equipment offered by manufacturers. A double hydraulic ram is used to apply the load against a reaction plate mounted over the end of the anchor and pushed against the face of the soil or wall, depending on the local conditions.

The hydraulic system is then actuated imparting a restoration force against the wall. The wall is moved slowly and monitored for any evidence of new structural cracking in wall or floor.

After the wall is aligned, remove the shoring columns and secure the floor joists to the top of the wall framing. Restoration efforts should be conducted under the supervision of a registered professional engineer.

When locking a plate against the wall the longest side of the plate is usually positioned in the direction of the longest span of the wall. The plate washer and hexnuts are tightened against the plate and any excess length of the lagstud is cut and removed. Once the tieback rod is locked to the wall, all equipment including the shoring columns, should be restored by compacting the removed earth in dug holes for helical anchors in 300mm loose lifts and replanting any shrubery or hedge near the ground.

Plug and Patch the holes in the concrete wall with a fast setting repair mortar. Prepare the hole for patching by cleaning out all dirt and debris from the area. Apply a bonding agent to the existing concrete before installing the repair mortar.

(h) Vertical Steel Tube Reinforcement

Place vertical steel tubes inside the basement to prevent further deformation and stabilize the basement walls which are bending vertically or horizontally but are not yet badly deformed. This method is relatively inexpensive and requires no exterior excavation (Fig. 11.9).

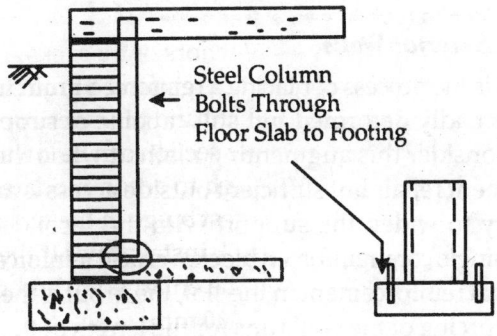

Steel Column
Bolts Through
Floor Slab to Footing

Fig. 11.9: Adding Vertical Steel Tube Reinforcement

Many people find the steel columns unsightly. But the rectangular tubes can be panelled over or painted the same colour as the basement wall.

Sometimes plumbing systems must be removed first to accommodate the steel columns, which are bolted both to the floor joists above and below the floor. Any remaining space should then be filled in with non-shrink grout. This space should be filled completely to prevent the wall from bending further and to distribute the wall pressure evenly over the full height of the column.

(i) Stud Wall

Alternately a stud wall (Fig. 11.10) that resists lateral loads and part of the vertical load can be installed within the basement. Though inexpensive and relatively easy to construct, stud wall reinforcement has a significant disadvantage of storage capacities and restrict its use.

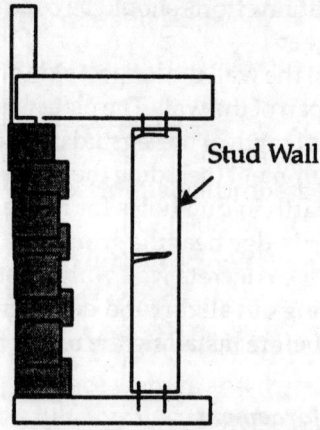

Stud Wall

Fig. 11.10: Stud Wall

The studs act as vertical beams resisting horizontal earth pressures on basement walls. Pressure treated boards located at third points enable the studs to transfer loads efficiently to the basement slab and the first floor framing.

(j) Augmentation by Exterior Walls

Exterior augmentation is the process of placing a reinforced **grout filled block wall** outside existing wall which is badly deformed but still capable of supporting the structure as shown in Fig. 11.11. Consider this augmenting method when there is no interior access available for the basement repair but sufficient outside access is available for excavation.

It may be necessary to widen the support with the second wall, which should be constructed of special single core-reinforced block. Vertical reinforcement is placed through the block and horizontal reinforcement in the slot, the grout is then poured into the cores of the block. This reinforcing of the wall runs for the length of the existing wall. The new wall is similar to a reinforced concrete retaining wall.

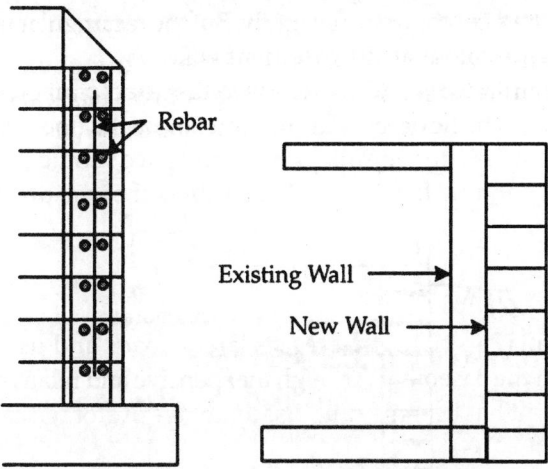

Fig. 11.11: Exterior Augmentation

(k) Cantilever Retaining Wall

A second wall can be built inside or outside the basement. A reinforced concrete retaining wall as shown in Fig. 11.12 inside the existing wall will resist lateral loads, while the existing wall supports the vertical loads. This technique can be designed to resist substantial lateral pressure caused by high water table and wet clay soils.

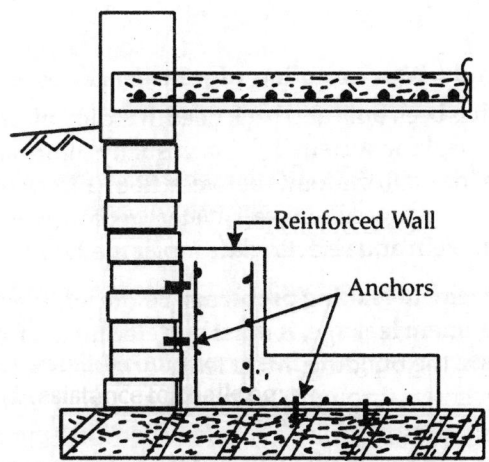

Fig. 11.12: Cantilever Retraining Wall

Because the concrete retaining wall does not rely on the floor framing above for lateral support, the top of the wall can be set at any height necessary to support the old wall. This is particularly important benefit when obstacles such as pipe and ductwork are present.

(l) Buttress Block and Grid

This method is designed for heavier loads as compared to the stud wall but lighter loads as compared to the cantilever retaining wall. The buttress block and grid method shown in Fig. 11.13 is a simple and flexible solution for distressed walls.

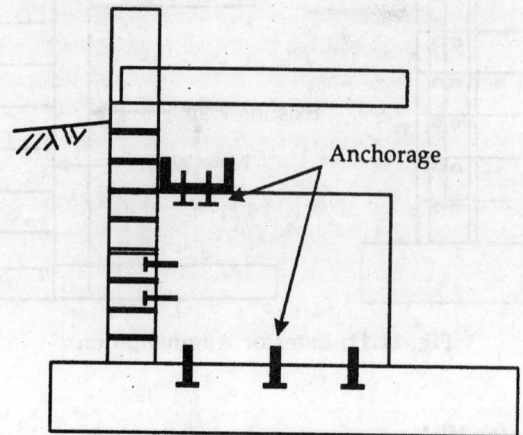

Fig. 11.13: Buttress Block

Large masses of concrete are poured at a designed spacing. Place buttresses and grids (horizontal beams resisting horizontal loads) to brace the existing wall.

In most situations, the grids should be made of structural steel. But when the loads are light and the buttress spacings relatively close, the grid could be made of heavy pressure treated timber beams.

11.5.2 Post Construction Waterproofing of a Leaky Basement

(a) Once a basement has been built and back filled, it is difficult and also more expensive to keep it dry from leaking water. Problems usually centre around water that leaks through cracks in the wall, the joints between floor and wall, defective form work, holes or undrained window sills in wall. Water proofing of leaky basement involves diagnosing the problem and selecting the repair method.

(a) **Diagnosing basement leaking problems:** Before we select an appropriate repair method for basement leakage, it must be determined, from where the water is actually entering the building. Most leaks originate below grade, but in some instances water enters through a leaky roof and travels down through the walls to end up in the basement.

In some cases wet basement walls and floors may not be caused by seepage at all, but results from condensation of excessive interior moisture. A simple way to test for condensation is to check the cold water pipes in the basement. If they are sweating, then condensation is definitely contributing to the dampness. Close basement window and doors to dehumidify the basement to eliminate condensation and determine if condensation is the only problem. If the basement is not leaking at the time of examination, one need to make it leak to determine the location of water entry. A simple way to do this is to turn on the water hose that is placed upside down adjacent to the outside of the basement wall foundation. If leakage occurs in basement walls, then it can be concluded that the dampness problem originates below ground and not from above.

(b) **Choosing a repair method:** After determining the cause of the leak and finding the point of water entry, a repair method can be selected. Some leakage problems may require a combination of repair techniques.

Repair methods for leaking basements fall into two general categories:

- **Prevention** of water infiltration into the crack
- **Diversion** of water

Prevention of water infiltration into the crack

It is usually better method although it may be too costly or too difficult to access the leak location. In such cases, installing a new drainage system or repairing an existing one can facilitate prevention of water infiltration.

Diversion

Prevention of water infiltration alone may not prevent water from entering the structure completely and is not always a long-term solution. Water infiltration through cracks can be prevented from the outside or the inside of the wall, depending on the source. To make repairs to outside of the basement wall, wide trenches must be dug on the outside to work in and extend it to the bottom of the footing. If the ground adjacent to the affected wall is inaccessible, then cracks can be repaired from the inside. The epoxy injection procedure for repairing cracks from inside the wall is recommended. No matter how carefully we analyse and plan for a repair, we may have to change our approach after the repair site is opened up and some unforeseen facts are revealed.

One method for eliminating seepage involves digging down alongside the crack on the outside and pouring in **bentonite** to fill and seal the crack. The bentonite fill should not be disturbed when it gets wet and swells up. It should be left in place for two to three months. After that the excess bentonite can be removed within 25 mm of the crack and fill in with black cinder. Bentonite does not harm plants or animals. Sometimes the extension of the rain water spouts facilitates elimination of leakage through walls.

With the **bentonite**, method a crack need not be sealed from the inside. But if the crack appears unsightly, it may have to be filled. In case of wide cracks bentonite will seep through from the outside and the cracks need to be sealed with hydraulic cement mortar or epoxy resin.

Foundation cracks in locations where it is impossible to get access to the outside of a wall such as those beneath a thick slab, have to be handled differently. The common way is to seal cracks from the inside with mortar or epoxy, however, it may not last long. The seal gives way soon and seepage starts again. The only sure way to stop such a leak from the inside is with a bleeder system. It is called a bleeder system because an opening is left in the chipped-out-crack. The opening extends down to the top of the footing and bleeds all seepage into the inside drain. If there is no inside drain, the bleeder must be installed to carry water to the nearest foot drain on the floor.

To build a bleeder an opening is first cut through the basement floor. The opening should be 25 mm wide and centered on the crack. It extends 250 mm away from the wall to give access to the inside drain tile. Then the crack is chipped out in a V shape from top to bottom 40-50 mm deep.

After cleaning and dampening the crack, a short length of smooth rope is placed into the point of the V groove. Next hydraulic **cement mortar is packed tightly into the crack over the rope**. While there is still enough rope not covered by mortar to permit pull in it, cementing is stopped and the mortar is allowed to set. The rope is pulled downward. This leaves the crack filled with cement mortar and flush on the basement side, but allows hollow drain on the inside created by rope pulling.

Filling mortar and pulling the rope is continued until the full crack is filled from the top to slightly below the floor level. The crack is left open at bottom for drainage. The area of the floor that was chipped out is filled with stone to within 25 mm of the top of the floor. A patch of hydraulic cement mortar is placed and finished over this.

To complete the job, epoxy resin is used to cover the entire crack, from floor level and up to 25 mm on each edge. This prevents cracking of the mortar and bleeding of water through the mortar.

Sometimes water that runs down the joint between a sidewalk or driveway and the foundation can be stopped without the crack filling. If the slab/wall joint is 6 mm or wider, a line of bentonite can be spread along the joint by pushing against the wall. As the bentonite level drops into the crack, more may be added. Reclaim the excess bentonite after filling the crack with bentonite.

The most sophisticated way to repair leaks in existing basements is with pressure - grouting equipment. Bentonite, mixed with water to form a slurry, is pumped through a pipe driven up against the foundation at 300 mm c/c around the perimeter of the foundation and at several levels from the footing up to the level grade. The grouting pipe should be driven in at an angle, rather than straight down. There must be enough pressure on the grout to make the bentonite mix with soil next to foundation and form a watertight seal over the entire wall.

Pressure grouting, however, will not remedy a leaking concrete block foundation or a basement floor with water under the floor. Water coming through the joint between the basement floor and the wall is a right case for constructing an inside drain and sump. Water is forced through the floor and wall joint as water pressure builds up under floor. It is possible to install epoxy cover on the inside of the basement to cut off the water but this method is often unsuccessful. An inside drain tile system is the only sure way of stopping floor and wall joint leakage.

A wet basement problem from overflowing window walls can be avoided by installing an inside drain tile system around the perimeter of the basement. A 25 mm hole is drilled through the basement wall about 50 mm below the sill of a problem window wall. A 25 mm by 250 mm section is cut out of the basement floor directly below the hole, giving access to the inside drain tile, as was described for making a bleeder. A 25 mm plastic elbow is pushed into the hole and a 25 mm plastic pipe is run from the elbow down the basement wall into the cut-out section of the basement floor. The pipe permits excess water in the window wall to come through the hole and flow into the inside drain tile. Admittedly, the plastic pipe shows on the basement wall, but it keeps the window sill wall from overflowing. The floor cut-out is finished as previously described with suitable tiles.

To sum up, it is the best approach to waterproof a basement during original construction. If that has not been done properly, then the repair should be based on the problem analysis. Bentonite on the outside cures leaking cracks and a drain tile with a sump pump may cure leakage problems on the inside.

11.5.3 New Basements

Basements in all types and sizes of buildings-whether residential or commercial are becoming increasingly popular due to shrinking spaces and rising cost of real estate. basements are being put to varied uses but one thing that is unfortunately true in most cases is the fact that more often than not they start leaking sooner or later. Flooding is also not uncommon. The reasons are not far to seek. A proper systematic approach towards the construction techniques, specifications coupled with judicious choice of materials can go a long way in avoiding such problems. First identify the sources of water which is affecting the basement. These are the ground water table (GWT), water seeping in from the ground surface, leaking from defective water supply and sanitary fitments, not only in the same building but also in the neighbor's building. The GWT that we are referring to is very seldom the GWT from which we draw out water through hand pumps and shallow wells. What afflicts the basements generally is the water that gets added to the soil artifically due to a combination of the causes listed above. One need not be an expert to realize that most of these sources are such that we do not have much control over their elimination.

There is a general misconception in the minds of many builders and building owners that in most places the GWT is very low, so special measures are not needed to waterproof the basements built in such areas. This is really not true. The result is obvious. So the remedy lies in making the basement structure as waterlight as possible.

Waterproofing of basements is a specialised task. Each situation requires careful study of the site conditions to evolve workable and cost effective specifications. There cannot be any one single foolproof remedy to cater for every situation. However there are some general guidelines, which have been discussed here. Inspect the site to check the location of municipal sewer and water lines. Pay special attention to the portions of the basement in proximity of these lines. Unless these are inescapable, try and avoid construction of toilets in the basements. If they are a must, they need special attention. Use very high quality pipes both for water supply as well as for sanitary fitting. It is suggested to use 'Post-chlorinated Poly Vinyl Chloride (CPVC) pipes and fitting for plumbing. 'Astral' CVPC (Flowguard) is one standard name in this category. For sanitary pipes one should use brands like 'Kissan', 'Supreme', 'Prince', etc. Care should, however, be taken to use the respective jointing cement/glue to ensure leak free joints. Coming to the specifications for construction, all construction in the basement should preferably be of RCC using M-20 grade (i.e. 1 : 1½ : 3) of concrete. Integral waterproofing compounds (IWPC) like IW302, Santulan or 'Polycrete' must be added to the concrete. It is also advisable to use polyester fiber like 'Fair Fiber', 'Mr Fiber' or 'Recron 3s' in the concrete mix to improve its water tightness and reduction in shrinkage cracks. Polymer modified bond coats must be applied between succesive lifts of concreting to make the junctions water tight. Electrical conduit pipes have also been found to act as major routes of water inlet if these are not properly sealed at their entry points into the basements. Such entry points must be sealed with special sealing compounds used for electrical cable junction boxes. Adequate ventilation of basements needs special attention for the atmospheric humidity.

A	=	Base Slab	B =	Outer Protective Wall
B	=	Horizontal Water-proofing System	D =	Vertical Water-proofing System
C	=	Brick Flat or Cement Concrete	F =	Inner Protective Walls
G	=	RCC structural wall or floor thickness and reinforcement will be designed according to the depth and maximum water pressure.		

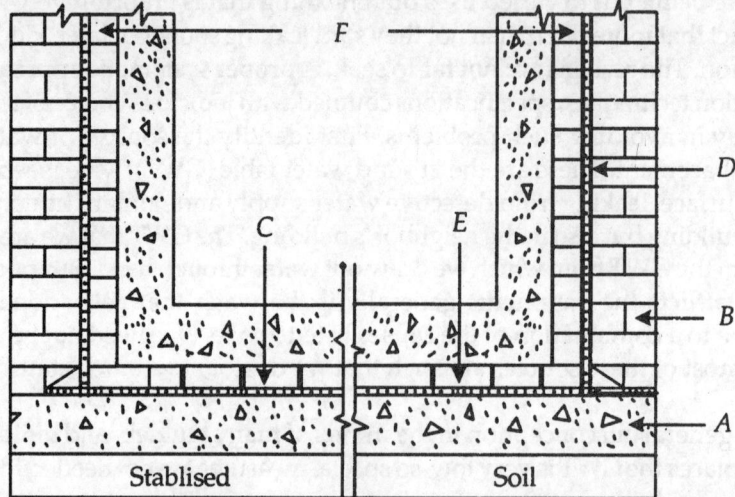

Fig. 11.14: Water Proofing Treatment for Basement in New Building Where Space is Restricted

Actual Sequence and Technique of Construction: First lay a lean mat of plain cement concrete 1 : 4 : 8 for the base. Then provide one-inch thick plaster in cement sand mortar 1 : 4 (with IWPC). Over this provide two coats of water proofing treatment with 'Hydrotite', 'Brush Bond', 'Conpro WP-2' or 'EP Tar-202'. Now provide the RCC layer as per the design. If the GWT is high or is likely to rise above the base level, it would be prudent to provide two coats of water proofing treatement over the RCC base also. The external sides of the vertical walls below ground level should be given two coats of coal tar epoxy like 'EP Tar-202' or 'Nitocote ET 550'. However, the internal surfaces should be given two coats of 'Hydrotite' or 'Brush Bond'. There is a general tendency among builders to use 'Kota' stone lining for the waterproofing of the walls as well as the base. This is a cumbersome, time consuming, costly and above all an ineffective method of waterproofing because the protection offered by them is not seamless as there is no effective method of **sealing the numerous joints**. More over, these stones are not impervious. They will give way to the passage of moisture sooner or later. In case coating on the external faces is not feasible due to constraints of working space one could resort to provision of half brick wall externally in place of the normal shuttering and grouting the gap between the concrete and this wall with non shrink polymer modified cementitous grouts like 'Centricrete' or 'Conbextra GP-1' progressively as the work proceeds. Placing black polythene sheets externally or grouting with neat cement alone would not serve the purpose. Polythene sheets will make the water flow faster due to the slippery surface and there is no way to seal the joints between the overlapping sheets. Neat cement would eventually shrink and crack thereby allowing water to pass through.

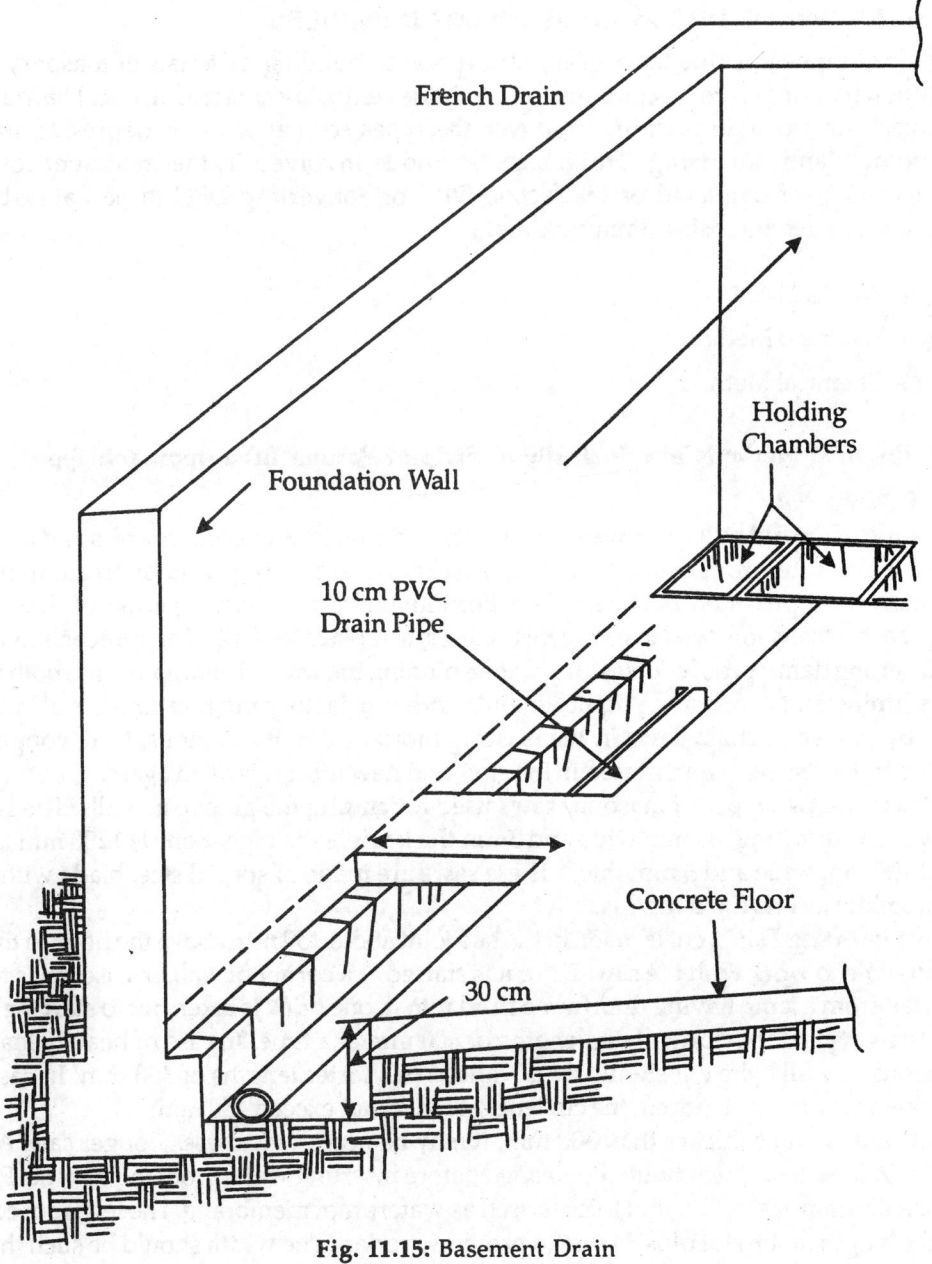

Fig. 11.15: Basement Drain

If we take the measures listed above we would be able to avoid the problems of damp and leaky basement to a large extent. Though there are methods now available to treat/ rectify the leaking basement effectively but all of them are expensive and put the owners to a lot of inconvenience. So the remedy lies in the age-old adage; **'Prevention is better than cure'.**

11.6 REPAIR OF DPC AGAINST RISING DAMPNESS

Rising dampness is due to capillary absorption in building materials of masonry. The solution therefore is to discontinue the capillaries either by creating physical barriers or convert the wettable surfaces to unwettabe types so that water is depressed in the capillaries and not rising. Three basic methods involved in the treatment for the replacement of damaged or ineffective DPC or converting DPC to non-absorbtive capillaries arresting rising dampness are:

(a) Physical Method

(b) Electrical Method

(c) Chemical Method

(a) Physical Methods are Basically of Strip or Porous Tube Approach Type
Strip Approach

Many of the old buildings were constructed without damp proof courses and as such their walls suffer from rising dampness. In some other buildings, although damp proof courses were provided but these have become ineffective over a period of time. The approach usually followed to rectify this defect is to replace the DPC. The approach consists of inserting damp proof courses after under-pinning the walls. This approach is both time consuming and expensive. A quicker and more satisfactory method consists of cutting slot by power or chain saws in the existing mortar joint. Bituminous, lead, copper or polyethylene sheets, are inserted in the slots and new mortar is filled again.

There are two types of masonry saws used for cutting the groove in walls. The small saw is 350 mm long, 80 mm wide and 3mm thick while the bigger one is 1200 mm long, and 100 mm wide and 3 mm thick. These saws are made of special steel blade with 4x4 mm stellite inserts fixed in slots.

For inserting D.P.C, cut is made in the bed joint about 150 mm above the floor so that it is possible to work with the saw. The cut is started at a corner of wall. In case it has to be started from a jamb having doorframe fixed into it, one brick is taken out to start the cut. The sawing is done normally in lengths of 600 mm at a time. In case of heavily loaded sections of walls, the cut should be advanced in shorter lengths of 300 mm. In case of brickwork with mud mortar, the cut length should not exceed 450 mm.

In case of walls thicker than 300 mm, it may be necessary to use a longer saw. After sawing, the cut surface should be cleaned before inserting waterproof membrane. Fibre based bituminous felt 2.5 mm thick is used as waterproof membrane. The felt is cut equal to the length of the slot plus 80 mm to provide overlap. The width should be such that it

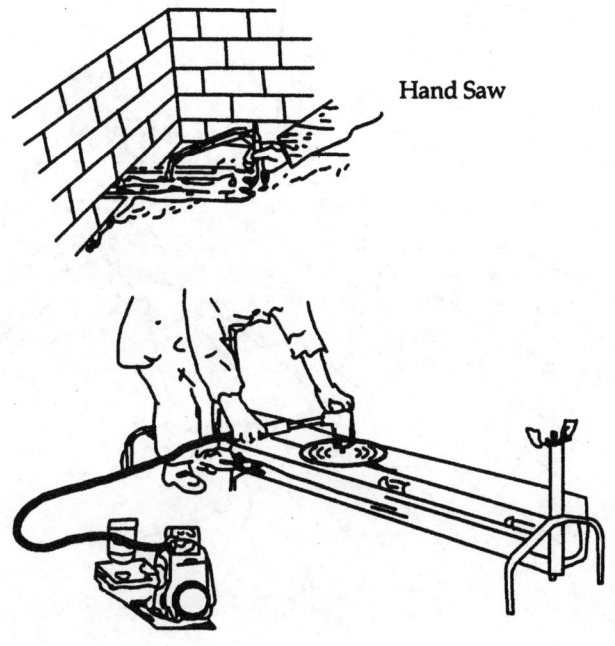

Hand Saw

(a) Cleaning with Grinding Disc

(b) Dust Collector

(c)

Holes

(d)

Fig. 11.16: Inserting Damp Proof Course

projects out by about 5 mm on each side of the wall. Bitumen confirming to IS 702-1961 for blown type or IS: 73-1961 for residual type is used. Bitumen is applied at 1.2 kg per m2 on both sides after heating the bitumen to the required temperature. The bituminous felt is next inserted in the slot with thin steel flats. Since the slot cut by the saw is about 4mm while bitumen felt is only 2.5 mm thick, the felt can be easily inserted. At the location of the over laps the cut is thickened by sawing for a second time. The gap over the bituminous felt is filled with **1 : 3 cement: sand grout** with water proof admixture. The laying of bituminous felt in the slot is continued in this manner along the entire wall length. The wall surface is made good by plastering or pointing.

Porous Tube Method

Porous tube method comprises of drilling holes in the existing masonry and embedding porous ceramic tubes in the wall externally or internally. The gaps around porous tubes is filled with porous mortar. This facilitates the inside water to find an access to these tubes and then get evapourated or drained out as the case may be.

(b) Electro Osmosis Method

The basic principle is reversing the flow of electrical charge. It is in fact the difference in the electrical charge between the ground and the wall that enables the water to rise. Electrodes connected by copper strip are installed in the wall and in turn connected to ground via junction box. The minute electric charges on the wall are then discharged to the ground. This facilitates prevention of rising dampness above ground.

(c) Chemical Method

Holes are drilled in the masonry or wall materials and water repellant, gel forming chemicals are introduced. These chemicals convert the capillaries to non-wettable type to provide chemical DPC. These chemical damp proof courses form barrier against rising dampness. **Latex siliconnate** is the most commonly used chemical DPC repair material.

Installation of Latex-Siliconate

Chemical D.P.C. method obviates the necessity of cutting the wall. In this method, holes are drilled in the mortar joints just above the floor at suitable intervals. A specially formulated chemical composition in the form of liquid is injected into the holes. The solution spreads into the masonry through the holes or pores and forms a continuous horizontal damp proof layer in the wall above the floor.

The damp proofing composition is a mixture of pore blocking and water repellent compounds. It is prepared at site from natural rubber latex and sodium methyl siliconate. For preparing 10 kg of composition the following quantities are required:

- Sodium Methyl Siliconate   =   1.0 kg
  (30% concentration)
- Rubber Latex   =   1.5 kg
  (60% concentration)
- Water   =   7.5 kg

The rubber latex is diluted with nearly half the water to be used and the remaining water is added to siliconate solution. The two solutions are intermixed by pouring siliconate into the rubber latex in small quantities at a time with constant stirring. The composition should be used within 8 hours after preparation.

For injecting the damp proof composition, holes of about 10 mm diameter **are drilled at 80 to 100 mm** centers in horizontal mortar joint, just above the floor using carbide - tipped masonry drills, to reach within 20 to 40 mm from the other face of the wall. In case, the wall thickness is more than 350 mm, holes have to be drilled from both the faces. Pressure rubber tubing about 30 mm long and of 12mm external diameter is fixed into the mouth of hole such that it projects out of the hole by about 5mm. For applying the composition, a tapered nozzle is pressed into the rubber piece. The composition is applied under a pressure of 0.5 to 0.7 N/mm^2 using compressed air from an airtight vessel. The rubber tube is removed after every injection and used in the next hole. The liquid feed into the hole is continued till the masonry around the hole gets saturated. In the absence of a pressure set up, the alternative is to feed the liquid into the hole under gravity. Depending upon the type and section of the wall, the operation will take **2 to 8 hours**. After the treatment is over, the **holes are filled** with semi-dry cement sand mortar.

The effectiveness of this treatment depends on the extent to which damp proofing solution has **spread horizontally in to mortar joint** and have lost water by evaporation. It is preferable to carry out the treatment in summer. Artificial drying by blow lamps can also be resorted to, where necessary. Though the treatment is very effective in walls built with bricks or stones in cement or lime mortar, it is not very useful in walls built with mud mortars.

The chemical approach is the simplest as well as the most practical method for **repairing existing DPC**. The greatest advantage of this method is that it is **universal for all building materials** and it does not disturb the structure comparatively. The only limitation of this chemical infusion method is that proper precautions should be taken for effective and **full saturation**. This method is not used for walls over 50 cm thick. Chemical method (Fig. 11.15) involves the following main steps:

- Drilling of holes
- Water test
- Introduction of the chemicals
- Finishing treatment

The horizontal holes should be drilled by rotary drills at lowest possible points. The inclination should be close to 30°. Holes can be drilled either from inside or outside wherever accessible and from both the sides if walls are more than 35 cm thick. The depth of the hole should be about 85 to 90 percent of the wall thickness.

Water should be filled to check if there are any cracks or voids to save the chemical. If water is drained out very fast then the hole should be filled with cement mortar and redrilled after the mortar has hardened.

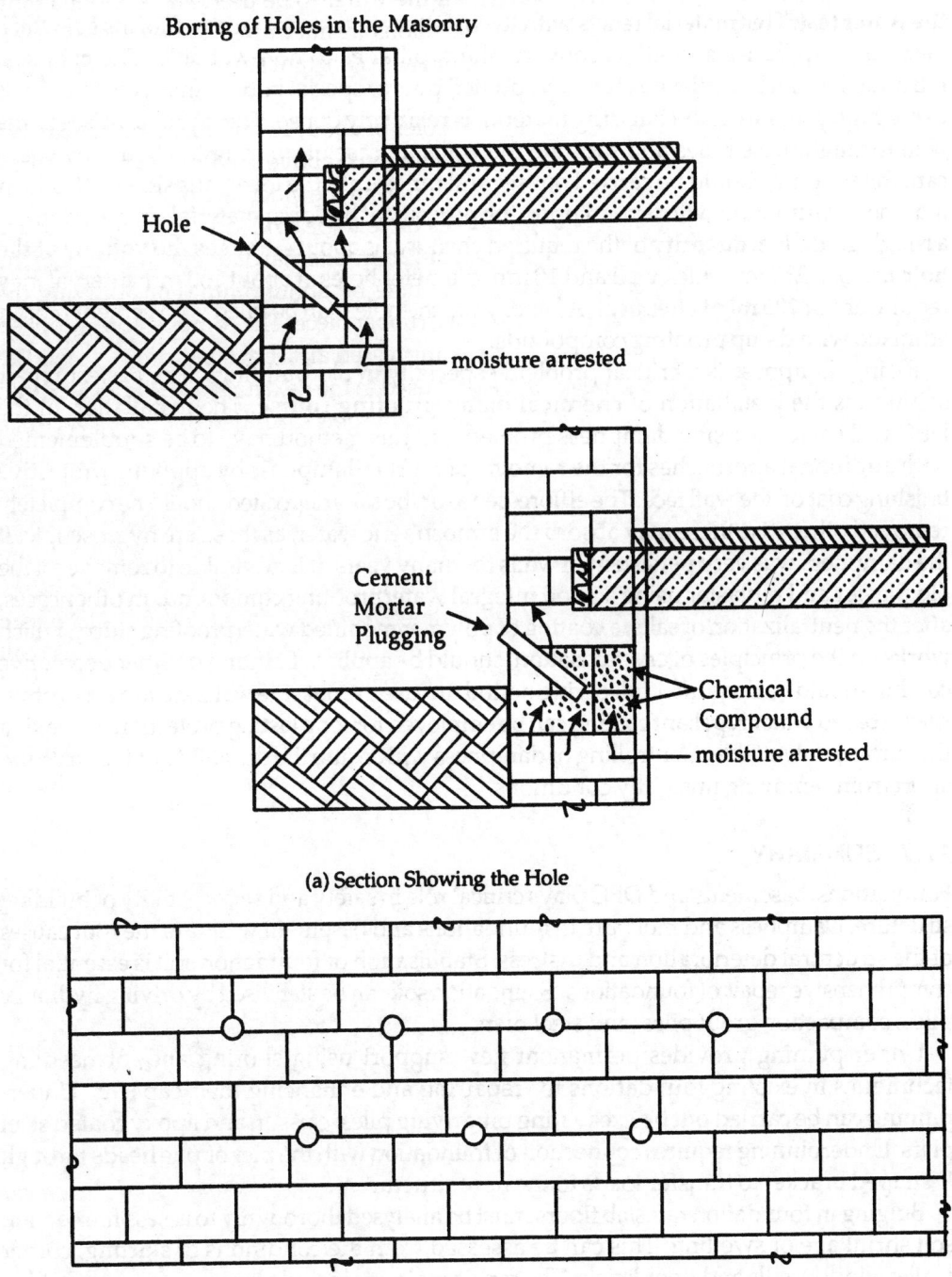

Boring of Holes in the Masonry

Hole

moisture arrested

Cement
Mortar
Plugging

Chemical
Compound

moisture arrested

(a) Section Showing the Hole

(b) Elevation Showing the Holes

Fig. 11.17: Provision of Chemical D.P.C.

The infusion of damp proofing material should not be introduced before six **hours after the water test**. This material reacts with the moisture in the capillary and forms gel which block the capillaries as well as converts the capillaries to non-wettable. The chemical introduction can be achieved either by spouted cans or special bottles. Inject the chemicals using high pressure if the building material is relatively dense. The liquid diffuses in the wall through these holes and saturate the walls. In case of pouring the liquid by spout can, the pouring should be carried out several times until the level falls slowly. It is clear that the consumption will depend upon the porosity of building materials in question. As a rough guideline, **quantity** of the required chemical is about **4-5 times the volume of the hole** *i.e.* for a 230 mm thick wall and 10 mm diameter holes at about 100 mm interval may require about **200 ml of chemical**. After drying, the holes can be filled with cement mortar admixed with damp proofing compounds.

Rising dampness is a critical problem especially in old buildings. The easiest repair solution is the installation of **chemical damp proofing course**. The installation of the DPC will solve the rising dampness problem but this method have to be supplemented with additional approaches for the removal of mortar dampness by applying protective finishing coat on the wall face. The efflorescence or the salts deposited should be completely removed otherwise these may absorb the atmospheric water, as these are hygroscopic. If the dampness was very persistent in walls for many years, it is advisable to remove all the old plaster and replaster using a good integral waterproofing compound. In other cases, after the neutralization of salts, a coating of polymer modified waterproofing slurry, which works on the principles of crystallisation, should be applied. Painting or other decorative coating should be done after full curing period. In case the out side walls are too absorptive, silicon based water repellant coating can be employed for long lasting protection by making the surface unwettable. Controlling of dampness will ensure the durability of the structure apart from removing unsightly conditions.

11.7 SUMMARY

Foundations, basements and DPC plays critical role in safety and serviceability of building structure. Dampness and moisture in foundations and basement walls are the root causes of the structural deterioration and distress. Stabilisation of foundation soil is essential for comprehensive repair of foundations. Foundation soil can be stabilised by driving hydraulic piles, compaction grout piles, and steel piers.

Under pinning provides permanent new support using shoring and/or needling techniques in existing foundations for repairing and enhancing their capacity. Under-pinning can be carried out by excavating pit driving piles, caisson and epoxy coated steel piers. Underpinning requires connection of foundation with the pier or pile heads through L-shaped bracket to transfer loads to the new structure.

Bulging in foundation raft slab floors must be analysed thoroughly to assess foundation soil shrinkage or swelling. This can be assessed from the conditions of skirting, corner walls, window sills and floor levels. The repair methods include local mechanical jacking or slab jacking by pressure grout or combination of these techniques. It is difficult to remove distress completely but it is better to reduce the distress by repair efforts. Problems from expanding soils can be minimised by proper drainage of foundation soil and maintaining its optimum moisture level to eliminate soil expansion or shrinkage.

Cracks or bulgings of basement walls can be corrected by reducing lateral loads, providing lateral supports and anchors. The restoration and repair of foundations and basement walls consist of Butteress, tie anchors, stud walls, cantilever retaining walls, placing additional reinforcement and grouts, surface bonding and exterior augmentation. Tie backs are installed by drilling holes through wall and ground and anchoring to the constructed dead man. Cracks in basement walls can be filled by injecting epoxy resins and appropriately treating the surface. Stud walls are also constructed upto certain height to share part of lateral loads. Exterior augmentation can also be achieved by placing reinforced grout filled block wall on the outside of the existing wall.

Water proofing of leaky basements involves diagnosing the problem and selecting the right method and material. Effective repair method also ensures diversion of water to avoid infiltration in the wall crack. Sealing with bentonite slurry for creation of water proof film barrier is also sometimes essential. Appropriate drainage system can be created by installing bleeders along the crack and system of floor drains.

Capillary absorption in building materials in foundation masonry causes rising dampness in walls if effective DPC is not provided. Repair of existing DPC against rising dampness is carried out by physically replacing damp proofing system or chemical treatment. Physical method of repairing and replacing DPC consists of making saw cuts in stages of 400-600 mm lengths just above DPC. Glass reinforced plymeric sheets or bituminous coated tarfelt sheets are introduced in sawcuts after cleaning and coating in stages of 400-600 mm depending on the type of walls. Electrosmosis method may also be used to reverse the rising dampness. Chemical mixture of rubber latex and sodium methy/ siliconate is injected in water saturated holes drilled just above plinth level at suitable intervals. After drying this chemical forms unwettable material barrier at the plinth. The holes are plugged with cement mortars admixed with damp proofing materials and then finished. Chemical method is the most practical and economical approach for repairing existing DPC.

QUESTIONS

11.1   Describe stabilising of soils for repair of foundation

11.2   Write short notes on:

(a) Hydraulic piles

(b) Compaction grout piles

(c) Shoring

(d) Needling.

11.3   Explain the purpose of underpinning.

11.4   Explain briefly the following techniques of underpinning:

(a) Piles underpinning

(b) Pier underpinning.

11.5   Explain repair technique for interior raft slab heaving.
11.6   Explain repair techniques for foundation settlements.
11.7   List 10 important techniques of repairing bulging basement walls.
11.8   Explain the repair techniques for basements by:

    (a) Installing tiebacks
    (b) Adding Rebars
    (c) Applying surface bonding
    (d) Stud wall construction
    (e) Buttress block and grid.

11.9   Explain briefly the main steps in repairing leaky basements.
11.10 Explain physical method of strip approach to repair DPC.
11.11 Explain briefly chemical method to repair existing DPC.

Chapter 12

REPAIR OF FINISHES

REPAIR OF FINISHES

LEARNING OBJECTIVES

After studying this chapter, the learner understands repair of building finishes and will be able to:

- **Explain** repair of external wall finishes
- **Explain** repair of internal wall finishes
- **Describe** decorative coating repair
- **Describe** repair of industrial floor finishes

12.1 INTRODUCTION

Building finishes are provided for protecting the exposed surfaces and to have a good aesthetic appearance. Generally the selection of finishing material is based on factors such as appearance, economy, thermal insulation, location, building bye-laws and maintenance norms. Main elements for application of finishes are walls (both external and internal), flooring, ceiling, columns, terrace, doors and windows. Broad classification of building finishes can be done as follows:

	Type	Examples
(a)	Brush applied finishes,	White wash, colour wash, dry distemper, oil bound distempers, emulsion paints, cement paints, Enamel paints and other decorative paints
(b)	Panelling and partition	Aluminium, plastics, applied timber, plywood, tile and decorative stones
(c)	Renderings	Plastering, Pointing, Rough texture finishes and peagrit of different colours
(d)	Wall Claddings	Stone, Bricks, Tiling and plastics
(e)	Flooring	Cement concrete, Hardonate concrete, Terrazo Tile, Stone, Timber, PVC and Industrial flooring
(f)	False ceiling/ceiling	ACC Sheets, Thermocol, Glass wool, Aluminium, Timber, Plastics, Gypsumtile (POP), Painting and plywood
(g)	Roof terrace	Brick tiles, coba, lime concrete, Tarfelt, Multiplastics, Stone tiles, Bituminous coating, cement concrete and PVC tile

If various finishes in buildings are not maintained, these may develop defects and may not serve the desired function. Repair and remedial measures are required to remove the defects to make the element suitable to the expected functions. Repair of various type of finishes is discussed in subsequent paras.

12.2 REPAIR OF WALL FINISHES

Wall finishes in buildings are applied both on external and internal faces. External finishes are applied to protect the walls from the environmental forces attacking the building structure. Sometimes external faces of walls are given natural finish of stone or brick. Such walls are protected by pointing on the external faces with waterproof cementitious mortars. Unplastered external wall surfaces require more frequent repair and maintenance. External wall surfaces are repaired and maintained by repairing mortar joints at certain intervals depending on the weather, occupancy condition, occupancy volume and importance of the structure.

12.2.1 Repair of Mortar Joints

Mortar joints are repaired against leakage of moisture through walls by using following methods:

(a) Tuck Pointing

(b) Covering the entire wall surface with cementitious grout

(c) Surface grouting joints only and coating with clear waterproof film

(a) Tuck pointing repair involves the following steps:

- Racking and cutting all mortar joints to a depth of atleast 6 mm;
- Sprinkling water on the whole surface an hour prior to removal of debris and wetting the masonry;
- Preparing special waterproof mortar by adding water proof compound and Acrylic Polymer into ordinary cement sand mortar of 1:3 proportion;
- Filling the specially prepared mortar in to the cleaned and moistened joints and finish the joints forminng either concave joints or struck flush;
- Curing the repaired surface for a minimum period of 7 days.

(b) Surface grouting the entire wall involves the following steps:

- Cleaning the surface of joints by scraping and wetting the joints;
- Allowing the surface to dry;
- Preparing mortar grout taking 1 part cement, 3 parts fine sand and 1/4 part hydrated lime and adding adequate water to make fluid consistency;
- Applying the prepared grout fluid into the joints with the help of stiff nylon brushes;
- Painting the entire wall surface with a diluted coating of the same grout;
- After treating all joints this may result in change of appearance and protection of the wall appreciably.

(c) Surface grouting joints only and coating involves the following steps:

- Preparing the joints by racking and cleaning them;
- Cleaning and wetting the wall;
- Applying prepared grout or proprietary mortar grout to the joints of wall with small nylon brushes;
- Applying clear transparent waterproof coatings on the entire surface.

12.2.2 Masonry Repair by Grout Injection

Grout injection has gained popularity in recent years as an effective method for repairing or strengthening masonry walls. This technique involves low pressure injection of fine cement based grout into cracks, voids, collar joints, or cavities within masonry.

(a) Situations for applications

Grout injection has several applications to repair and stop the spread of cracks to increase the life of buildings particularly historic structures and those in seismic areas.

Left unchecked, cracks can sometimes threaten the safety of a building structure and hence it is necessary to repair them. Grout injected into a crack not only blocks the moisture path, but also restores material continuity across the crack. Cracks as narrow as 0.18mm can be injected with micro fine cement slurry, cracks 0.50mm or more in width are commonly repaired by grout injection.

Masonry wall systems are some times designed and built without adequate drainage spaces behind the exterior walls and without the proper flashing and weep holes. Water can enter these walls and eventually reach the inside surface through voids in header courses and bed joints, partially filled collar joints, and through cracks.

Most of these older structures were constructed using high quality face bricks or stones on the facade and lower quality common bricks or rubble for interior walls. Construction in junctions of such walls are generally poor, and the internal joints are also not filled properly. With passage of time these joints and header courses get deteriorated.

Grout injected into thermal cracks and cavities strengthens masonry by bonding together all elements of the wall. This also increases structural integrity and overall stability, by improving resistance to seismic forces. Parapets, veneers, bearing walls, and shear walls can be strengthened using the same technique.

Grout injection is especially useful for restoring or stabilizing historic structures because the technique does not alter the building appearance. Special grouts are formulated with modern building materials. This increases masonry durability and overall service life by bonding wall elements together and reducing moisture infiltration.

(b) Grout mixes

Grout mixes are formulated for each individual application. Grout mix design includes selection of type and proportions of Portland cement, fly ash, lime, and admixtures. To fill larger voids, fine sand aggregate should also be added to the grout mix.

The mixture must be prepared carefully to ensure fluidity of the grout that can penetrate into fine cracks and fissures and also resists segregation and shrinkage. Some projects require multiple grout mixes to achieve the desired results.

(c) Evaluation of building condition

Before grout injection, assess the exterior condition of each wall area to be repaired. Note any surface cracks, missing mortar, spalled units or other visible damage that could have an impact on grout confinement or the injection process. Use a fibre optic instrument to examine the internal wall condition at randomly sealed location. Evaluate the ability of the wall ties to resist injection pressure.

(d) Surface repairs

Seal (using mortar, epoxy, or another material capable of resisting injection pressure) the discovered cracks and voids to contain the grout. Give special attention to masonry around windows, doors and the wall openings. Replace masonry units that show serious damage, such as extensive cracking or spalling. Rake and repoint deficient mortar joints not likely to withstand injection pressures.

(e) Injection procedures

Each application is unique, and procedure must be modified to fulfill each repair project objectives. Correctly repaired walls through injection process can strengthen masonry while preserving the building appearance. Observe the general guidelines during injection as described below:

Drill the smallest possible diameter injection holes into mortar joints to prevent chipping and damaging masonry units. Grout is injected through holes drilled to intercept cracks and internal cavities. Hole diameters as small as 25mm have been used for injecting grout in masonry built with thin mortar joints. A properly formulated sanded grout can be injected through holes up to **75 mm** in diameter. For injecting very fine cracks, holes are spaced as close as **100 mm centre to centre**. Use a maximum spacing of up to 800mm centre to centre for injecting open collar joints and large void spaces.

Water flushing should be done eight hours before grout injection begins to remove dust and clean internal surfaces by flushing all injection holes with water. Regulate water pressure keep the pressure less than 1 N/mm² (1 MPa). Starting at the top of the repair area, use as little water as necessary to flush dust and debris downward. Continue working down the wall until the second row from the bottom, and the water flowing from the bottom holes is clear.

To prevent damage while strengthening already fragile masonry, use a grout pump capable of injecting grout with the pressure of about 1MPa. For collar joint injection, begin grouting holes at the base of the wall and proceed to the top, moving across the wall horizontally and then upward. Plug holes after grout has flown out from the next row of holes.

Use low lift heights of about 600mm or less to reach every portion of wall below. A particular level is reached when grout flows from one or more holes at the designated lift height. Allow the grout to stiffen before continuing with the next lift.

To restore the masonry to its original appearance follow injections by simply pointing the holes with mortar. Quality control is maintained through standard flow tests and observation of grout flowing from adjacent injection holes. After the grout has hardened, verify grout coverage using nondestructive testing.

Wall cleaning should be done immediately after completing the injection for any grout spills from the masonry surface using a water hose. After completing the injection process, use water and a stiff non metallic brush to remove any surplus remaining materials.

12.2.3 Surface Mask Grouting to Repair Mortar Joints

When joints are weathered but not badly deteriorated, this technique can stop leaks more economically than repointing. Surface or mask grouting consists of applying a thin coat of cement based grout to the mortar joints and the mortar/brick interface. A mixture of 4-part cement, 1 part lime and 5 parts fine silica sand or premixed proprietary surface grouting mortars can be used. Before planning such work on a building of historical significance consider the repair of the joints which change the appearance to some extent. Surface grouts contain only very fine aggregate, and it is not apparent in a finished job. Mask grouting should be done only when the outside air temperature is at least 10 °C. The recommended procedures of mask grouting is as follows:

(i)   Fill small holes and repair seriously deteriorated mortar joints. Prepare the surface for grout application by removing dust, dirt, efflorescence, paint or previous water repellent treatments.

(ii) Cover the brick faces with strips of masking tape as wide as the height of the brick. Use a treated hardened blade held tight against the wall to cut the strips of tape.

(iii) Place dry grout material in a clean container and gradually add clean water stirring until the mixture has the consistency of thick batter. Let it set for 10 to 15 minutes, then mix thoroughly as much material as you can apply within 2 hours. The material can be retempered as needed within that time, but avoid retempering of pigmented grout to maintain consistent colour.

(iv) Dampen the mortar joints but leave no free water. Then following the joint lines, apply the grout with a brush. Make short diagonal strokes to seal the brick mortar interface. Alternatively, the grout can be applied with a sponge float, using circular motion.

(v)   Carefully strip off the masking tape before the grout sets.

The process is laborious, but can be both effective and economical under the right circumstances. Surface grouting is not a panacea for every leaky brick wall but it can be an effective repair method for some of them. Mask grouting works when the joints are slightly eroded but basically in good shape. Fill holes and repair deteriorated joint mortar before applying mortar or grout.

12.2.4 Efflorescence Removal

Efflorescence appears in various forms on masonry surfaces. It, sometimes, is crystalline or crusty and at other times it is powdery or fluffy. In its worst form, it appears as scum or as unattractive green or brown stains. Although efflorescence is not directly harmful to masonry, but it indicates other problems, such as excessive moisture entering the wall cavities and presence of injurious salts.

Efflorescence results when moisture comes in contact with soluble salts within masonry units or mortar. The moisture dissolves the salts. The transporting forces such as gravity/hydrostatic pressure, evaporation or a similar mechanism transports the salt solution to the surface, where it appears as efflorescence.

Efflorescence can be minimized by correct selection of materials, careful design details, and good construction practices. Before removing efflorescence, determine the source of the problem and take steps to correct them. This will help to prevent efflorescence from reappearing a short time latter.

Observe for all possible points where water may be entering a wall, or where joints have been left without mortar or installed incorrectly. Check for cracks in masonry, voids in mortar joints, deterioration in sealant joints, flashing that do not extend through the wall, and weep holes that are blocked or absorbent walls especially vulnerable to moisture penetration. Check parapet copings for open seams, joints or holes.

Identification of the salt causing the efflorescence helps in selecting a cleaning agent that effectively dissolves the efflorescence, but has minimal effect on the masonry. Many proprietary cleaners are available to remove specific types of efflorescence. If the wrong cleaning solution is used, it can cause soluble salts to become insoluble, creating an even greater problem. Improper cleaning of masonry surfaces can also significantly change its colour, eat away cementitious materials, leave aggregate exposed, and cause additional efflorescence to occur.

Hydrochioric acid with 33% of water is used to remove the efflorescence on exterior walls. Potable or canal water can also be used .Apply waterproofers to interior walls to initially reverse the evaporation gradient to get rid of moisture from external wall.

Applying a sealer or other coating as a treatment for efflorescence is **not recommended**. Water entering a coated wall through the sealer or another path will still dissolve salts in the masonry. When the salt solution travels toward the surface of the wall, the water passes through the coating as vapour, but the salts remain behind as concentrate. These salts crystallize and create pressures that can spall and crack brick.

This type of efflorescence occurring on surface is known as **crypt** to efflorescence.

To minimize the effects of cleaning on a wall, always start with the gentlest method possible and progress towards harsher measures as needed. The most gentle method usually is dry brushing. If that is not effective, try using a brush dipped in a bucket of water, or use a gentle water spray. Proprietary masonry cleaners, acids, and other chemicals should be used only when simple techniques prove unsuccessful.

Testing should include determining pressure intensities of water suited to each method. Test various chemical concentrations. Test patches should be approximately 2.0 square metre in size and should be on an inconspicuous part of the structure. Make sure to choose an area that accurately represents the type and amount of the total efflorescence on the building.

A building may have several types of masonry materials and surface textures, several types and degrees of efflorescence may be present in the same building. In these cases, each area should be tested separately as the same method may not work on all areas. Test areas should be evaluated only after complete drying. It may usually take about one week.

Once a cleaning method has been chosen, the test areas serve as benchmark for determining the number of applications of the cleaning and duration of treatment to be done on the surface and required water pressures for effecting cleaning and rinsing.

Use caution when working with any cleaning solution, especially acid solutions. Many of the acid solutions are dangerous to mix and use. The use of these solutions should be studied correctly from manufacturer's literature and suppliers demonstration .

Use appropriate safety gloves and aprons when working with chemical and acid solutions. Basic safety requirements includes goggles or a face shield and rubber gloves. Some acid solutions also may require the use of rubberized pant, jacket / or apron.

Wet the entire wall surface with clean water before applying any acid solutions. This allows the cleaning solution to remain on the surface of the masonry by reducing absorption by masonry units and mortar. Once cleaning begins, work areas are delineated by structural or architectural features. This helps prevent mottling and uneven colour variation that can result when cleaning is interrupted.

Protect landscaping, windows, doors, and frames from contact with cleaning agents. Acid solutions can corrode metal door and window frames and stain adjacent masonry materials, such as limestone and cast stone. Alkaline solutions, such as wash water, may cause pitting of aluminum window frames.

To prevent or eliminate efflorescence, remove any one of the three points of the efflorescence triangle.

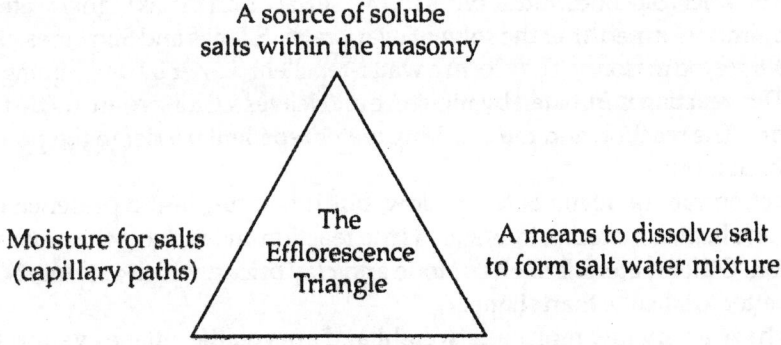

Fig. 12.1: The Efflorescence Triangle

The following special precautions may help keep efflorescence from appearing on new construction:

- Keep masonry materials dry by storing them off the ground and under a protective covering. Do not allow sand to be contaminated by dirt or plant matter.
- Cover unfinished walls during work stoppages, inclement weather, and especially at the end of the day to prevent mortar joint washout and to keep masonry dry.
- Protect the base of walls from rain splashed mud and mortar splatter by using straw, sand, or saw dust extending about 1.0 m from the wall, and by placing plastic sheeting vertically 600 to 800mm up the wall surface.
- Turn scaffold boards near the wall on edge at the end of the day to prevent rainwater from splashing dirt onto the wall.
- Avoid rubbing or pressing mortar particles into brick or block faces when brushing excess mortar from the wall after tooling mortar joints.

12.2.5 Protective Coatings

When moisture gets into masonry walls it can cause many problems such as efflorescence, freeze thaw damage, steel corrosion, mildew, and loss of insulation value. A clear water repellent coating when selected and applied with care, helps to reduce moisture penetration in masonry, thereby minimizing the above said problems. There are five main categories of water repellents available in the Indian market. These are:

- Acrylics
- Silicones
- Silanes
- Siloxanes
- Blends of silane and siloxanes

Each of these has unique properties and suitability to different substrates. No product works equally well on all masonry materials e.g. Silicones perform better on concrete block than on brick. Silicones and acrylics form a surface film that is highly water repellent. The film is simply formed after the solvent evaporates. Silanes and Siloxanes, on the other hand, penetrate the masonry. They form a water-repellent barrier by undergoing a chemical reaction. This reaction is initiated by moisture and leaves silicone resin inside the pores of the masonry. The reaction and the resulting water repellent barrier in the pores does not occur without moisture.

The reaction rate of silane is fairly slow, but it is faster in the presence of alkalies. Concrete, an alkaline material, thus speeds the reaction and reduces the amount of silane evaporation. Other materials such as stone and clay bricks, which are not alkaline, may respond better to silicone than silane.

Silanes have a very low molecular weight and are volatile. Silane evaporates rapidly and on hot, windy days, silane may evaporate rapidly and thus making the treatment ineffective. To compensate for this evaporation, silanes must have a high solids content (20% to 40%). Other water-repellent products have a lower solids content usually (5% to 20%).

For effective protective coating on the wall surface, it is necessary to adopt the following steps:

(a) Selection basis
(b) Application procedure
(c) Type of water repellent
(d) Salient characteristics

(a) Selection

Before selecting water repellent coating, we should analyze all the conditions and necessity for the coat. We should know the condition of the substrate and correct requirements specifications of the repair job. When evaluating transparent water repellent, we should review the manufacturers literature thoroughly for checking the following specific properties.

- Water repellent
- Water vapour transmission
- Surface gloss
- Weathering and ultraviolet stability

- Resistance to efflorescence
- Water permeability

The material selected for coating must ensure the properties reported in the literature for the substrate under consideration. Before making a final selection, carry out an actual trial with a small sample. Choose an inconspicuous location so that an inappropriate selection may not affect the overall appearance and performance of the wall. Some acrylics provide a glossy sheen or may slightly darken the wall. We need to evaluate the permeability of the selected water repellent on the actual leaky patch on masonry walls to ensure proper selection of coating material.

(b) Application procedure

Water repellent work effectively only if they are properly applied. The masonry surface must be thoroughly prepared by cleaning and drying. Only silanes and siloxanes can be applied to damp surfaces, but even they cannot penetrate masonry whose pores are already full of moisture. So wait at least 48 hours after heavy rain before applying water repellent. Do not apply the protective coatings immediately on new construction. Allow construction water to evaporate. Ideally wait for application of repellent at least 30 days after the completion of a new wall.

It is preferable to apply water repellent when the air temperature is between 10 °C to 30 °C. Complete all caulking, joint sealing or repair work at least 72 hours before applying the sealer. Some water repellent can affect the curing of these materials and even cause them to lose adhesion. For application on large jobs use of low-pressure airless sprayer or pesticide sprayer is recommended. Apply the sealer from top to downward. Make sure to eliminate any heavy runs or drips.

Apply dosage at the recommended rates as per manufacturer specifications. As a general rule, coverage rate for bricks may vary from 1.5 to 5.0 square metre per litre depending on porosity of the substrate.

(c) Type of water repellents

Two different types of silicone materials are available for use in water repellent formulations. One is soluble in water and the other in mineral spirits of hydrocarbon solvents. The water-soluble materials are preferred from environmental pollution considerations. They are non-flammable and may be applied both on dry or damp surfaces.

Solvent type repellents are preferred for brick, cement and cinder block masonry. The other type of mineral spirit solvent can be applied successfully only on dry surfaces in case of below freezing temperature.

For effective performance of silicone repellents, the formulation must contain sufficient silicone concentration. The industry recommends not less than 5% silicone for the solvent based materials and not less than 2 percent for the water soluble repellent.

Silicones function by forming lining in the capillary pores and thus for effective performance, the surface should be clean to permit silicones full access to the surface pores.

(d) Salient characteristics

Clear water repellent can reduce water leakage and problems caused by water leakage. Water repellents are certainly not a substitute for proper detailing and good workmanship in original construction.

These have the following limitations:

- Water repellents do not last forever. Its life depends on the abrasive environment. A clear water repellent needs preventive maintenance and requires to be reapplied.

- Water repellents do not stop water penetration through below grade. These can minimize efflorescence, but only when they are applied to a dry wall. If white deposits are still appearing on the surface, that means moisture is in the wall. Silanes and Siloxanes resist efflorescence better than other sealers because their high vapour transmission allows moisture in the wall to escape.

- Water repellent does not bridge cracks. Some water repellents seal cracks that are less than 250 microns width only.

12.3 DECORATIVE COATINGS

The protective coatings are used for decorative purposes and also act as protective coatings. The application of these decorative coatings slow down or retard the process of deterioration. The understanding of application, capabilities and limitations of these coatings facilitate in design of repair system to improve the service life of various surfaces which are coated. Different type of these coatings are:

(i) Cement Paints
(ii) Oilfree Urethanes
(iii) Acrylics
(iv) Oil modified Urethanes
(v) Catalyzed Epoxies
(vi) Latex Emulsions
(vii) Alkyd
(viii) Two compound Urethanes
(ix) Epoxy Esters
(x) Polyesters
(xi) Vinyls
(xii) Phenolics
(xiii) Chlorinated Rubber

(i) Water based portland cement paints

These are supplied in powder form to be mixed with water before using. A cement wash can result in properties similar to those of the concrete surface itself. These are not true film forming systems but are highly filled. These are not coatings in real sense but are mentioned here because of their common use in stucco work and new construction for decorative finish and texture coat.

Capabilities

- These provide a decorative colour and a dense texture for concrete surfaces commonly used for stucco work;
- These are easily prepared on site;
- These can be applied even on the damp concrete.

Limitations

- Portland cement paints require skilled application
- Cement paints are prone to water marks and are not recommended for damp services
- It is difficult to obtain a uniform colour on flat surfaces
- Cement paints readily catch dust and dirt
- Cement paints cannot be applied over other coatings but only to concrete.

(ii) Oil free urethanes (moisture cure)

Moisture cure urethanes generally form a film as the solvent evaporates and reacts with moisture in air. These coatings are often similar to epoxy esters but give a much harder film. Certain moisture curing urethane form films that result in the catalyzed urethanes systems. Some urethanes are ultraviolet resistant. Urethanes generally contain solvents, but water borne formulations are now available in the market. In general, urethane systems are high- performance coatings and its specific properties vary from supplier to supplier due to the versatility of chemistry involved.

Capabilities

- The urethane adhere well to concrete due to their capability of wetting substrate;
- These resist a wide variety of chemicals and are not affected by the alkalinity of concrete;
- Urethanes are hard to resist abrasion and are good solvents;
- These are easily applied on a dry substrate;
- Certain aliphatic type of urethanes possess excellent exterior weathering properties.

Limitations

- Most of these urethanes fail on exterior exposure. Only the aliphatic are suitable for exterior exposures;
- Some moisture cure urethanes embrittle with age;
- These are limited to pre formulated colours;
- These must be applied to dry surfaces;

- These are difficult to recoat once fully cured unless lightly abraded or weathered;
- These cannot be applied in high build coats;

(iii) Acrylics

Acrylics are among the most widely used coatings of the generic classes. There are many formulations available, which have excellent physical properties and are easy to apply. Solvent-based acrylic coatings can be formulated as low solids penetrating sealers and as high solids permitting finish coats. These resins vary widely in their physical properties. Coatings of low cost and limited performance are available in this generic class. The resistance to concrete alkalinity and excellent weathering characteristics make them attractive as coatings for architectural features and concrete protection.

Capabilities

- Acrylics are easy to apply and tint;
- Solvent acrylics have excellent wetting properties in concrete and therefore function well as sealers and colour stains. These are also film forming;
- Acrylic coatings are easily recoated and have a fair degree of chemical resistance;
- These posses excellent outdoor weathering properties;
- These can be used to formulate coatings of wide range of specific physical properties.

Limitations

- Acrylic coatings may lack certain chemical resistance properties, depending on the formulation. Verify with the manufacturer that the specific acrylic is satisfactory for the intended service;
- These lack the abrasion and wear resistance;
- Use of solid acrylics should be limited for water sealing and primer applications as opposed to barrier applications.

(iv) Oil-modified urethanes

These coatings are prepared from combining alkyd type oils and isocyanate. These coatings are also called urethane resin and tends to produce a harder film with better chemical and abrasion resistance. The modifying urethane resins used in these coatings are usually low performance types to save on cost. Therefore, these oil modified urethanes are suitable for interior systems but readily cracks and becomes yellow when exposed to sunlight. These also tend to embrittle with age. These are widely used in architectural applications as wood finishes. Pre pigmented versions are also available. Their use on concrete is highly questionable, since urethane content is generally too low to overcome the soaponification problems of the alkyd oils.

(v) Catalyzed epoxies

An epoxy resin is usually based on bisphenol alcohol and diluent. It reacts with a curing agent hardener to produce the epoxy coating film. Finish properties vary widely depending on the resins diluents, and hardeners used. Epoxies are tough adhesives and are durable. These are commonly used in concrete floors and on walls because they are easily cleaned. These resist traffic abrasion and chemical attack. Water borne solvent epoxies with 100% solids are widely used to repair floor surface. These may be combined with aggregate of decorative quartz to produce trowel or squeeze applied decorative floors for patching and repair systems. Catalyzed epoxies are often used as primers or base coats for two component urethanes.

Capabilities

- Catalyzed epoxies resist a wide variety of chemicals including alkalis and many acids;
- They adhere well to concrete and have excellent wear resistance;
- Epoxies are water resistant and can be used in damp proofing and immersion service;
- Their hard tough surface resist mildew fungus growth and withstands repeated scrubbings;
- Epoxies can form high build films by repeated applications before curing of previous layer.

Limitations

- Some epoxies are subject to application problems in high humidity;
- These must be usually applied at temperatures above 50 °C and are sensitive to cold concrete surfaces;
- Epoxies cannot be tinted onsite and they tend to embrittle with age;
- These are difficult to recoat once cured unless abraded;
- Their surface appearance is subject to scuff damage;
- High solids epoxies tend to entrap air in the system and substrate;. These should be applied only by skilled applicators familiar with short pot life systems;
- Epoxies often require recoat many times.

(vi) Latex emulsions

Probably the most common types of coating applied to concrete are latex water-based coatings. As water evaporates from the film, the latex particles coalesce. Latex binders are synthetic materials usually acrylic polyvinylacetate or butadiene styrene and they can vary in hardness, flexibility and gloss. These coatings still contain volatile compounds or coalesent agents. These are available at hardware stores as water based paints which, typically do not meet HIPAC standards, but these are listed because of their availability, low cost and decorative value. There are new latex resins available which perform better as latex and elastomeric paints.

Capabilities

- Latex paints are easy to apply and clean up;
- These retain good colour on exterior exposures in non-hostile environments;
- These do not have the odour problems associated with solvent based coatings and are normally VOC compliant;
- These dry quickly and are easily recoated with other water based coatings;
- Latexes have a higher vapour permeability value per mm than most other coatings. These can breathe;
- These are easily tinted;
- These are not affected by concrete alkali attack;
- These can be formulated as high build elastomeric and textured coatings.

Limitations

- Latexes generally lack in high performance properties, particularly chemical and abrasion resistance;
- These are subject to mildew and fungus attack;
- These face curing problems in cold temperatures especially on very damp surfaces;
- These often contain volatile emulsifiers which are subject to VOC regulations;
- These are generally strong surface wetters and therefore face adhesion problems when pointed with a strong film forming type coating.

(vii) Alkyds

Alkyds are the familiar **oil solvent based paints** in which the vehicle is a modified oil resin resulting from both evaporation of solvent and oxidation of the oil. These can be formulated to posses varying degrees of gloss, drying time and elasticity. These are primarily decorative paints, used for colour and gloss retention. These are often more economical to a high performance system and gain other property advantages in the process.

Water reducible alkyds are available in the market. Aikyds are widely used in architectural applications. Alkyds usually do not meet HIPAC criteria and they cannot be applied directly to concrete since they are subjected to saponification reactions. The coatings react with alkalis present in the concrete to form a soap making the coating to delaminate or result in non-adherence to the concrete.

Capabilities

- Alkyds are easy to apply and have a fair degree of coat retention in mild environments;

- These can be tinted on site;
- Alkyds are easily recoated;
- These have good impact resistance.

Limitations

- Alkyds have poor alkali resistance and should not be applied to concrete without an appropriate primer or sealer coat;
- These can't be used in immersion service;
- These deteriorate rapidly in chemical environments and are not easily recoated with other solvent borne coatings because of poor solvent resistance;
- Although widely used in exterior applications these are affected by ultraviolet attack. These eventually become brittle and are subject to ailigatoring;
- The maximum solids content of alkyds is somewhat limited and these are becoming subject to environmental limitations due to solvent emission restrictions;
- These dry slowly.

(viii) Two Component Urethanes

The two package urethanes are available in two broad types. The aromatic type should be limited to interior application because of chalking and yellowing when exposed to sunlight. Aliphatics, however, posses the attributes of excellent colour and gloss stability on external exposure. Both type may be applied at much lower temperature than epoxies and are not subjected to long-term embrittlement. These are most commonly used on concrete as floor coatings and antigraffic coatings. These also are used as decorative, chemical and abrasion resistant toppings for epoxy base coats.

Capabilities

- Catalyzed urethanes have excellent chemical resistance to a broad range of reagents over a wide range of service temperatures;
- These resist solvents and abrasion well, yet, these are flexible and resist impact;
- The urethane reaction is not affected by low temperatures and can totally cure at lower temperatures;
- The aliphatics have excellent outdoor weathering properties including colour and gloss retention and resistance to chalking;
- Urethane resist fungus and bacteria growth even when subjected to high humidity and temperature;
- These form a tile like surface that withstand harsh detergent and solvent scrubbing.

Limitations

- Urethanes are sensitive to surface moisture and humidity and must be applied to dry substrates;

- These cannot be tinted onsite;
- These are difficult to recoat once fully cured unless abraded or weathered;
- Urethanes can't be used in water immersion service;
- These can't be applied as high build systems without special application equipment and techniques.

(ix) Epoxy Esters

Epoxy Esters are produced by oil modified expoxies. These oil-modified epoxies produce a coating that dries by oxidation of oil. These can't be used in exterior exposures because these chalk when exposed to sunlight. These are good to use as interior decorative barrier systems often in flooring applications.

Capabilities

- The epoxy esters have an intermediate degree of chemical resistance.
- These are easy to apply.
- Their films are moderately hard, tough, and abrasion resistant.

Limitations

- Epoxy esters readily chalk and fade on exterior surfaces;
- Since epoxy esters are oil modified coatings, it should be ensured that the system chosen is formulated to be resistant to saponification;
- These may not be VOC compliant.

(x) Polyesters

Of the three major catalyzed systems the polyester systems are the least commonly used. These are primarily protective coatings rather than cosmetic finishes although these are available in different colours. These have good chemical resistance hence these are often used as vessel linings.

Capabilities

- Polyesters have high degree of resistance to a wide range of aggressive chemicals;
- These can be used in immersion service and damp proofing applications;
- These can withstand temperatures upto 400 °C;
- Polyesters resist impact well.

Limitations

- Polyesters are difficult to pigment;
- These coatings have more critical effect on concrete whose alkalinity requires to be neutralized;

- These must be applied at temperatures above 50 °C;
- Some polyesters are subject to saponifications problems.

(xi) Vinyls

The vinyl type polymeric coatings are plasticized copolymers of vinyl chloride and vinyl acetate dissolved in strong solvents. These dry rapidly by solvent evaporation thus spray application is preferred. Surface preparation is critical for application of Vinyl's. Traditional Vinyls have been low in solids and some are water borne

Capabilities

- Vinyl's resist a wide range of chemicals;
- These have fair colour retention and can be used in exterior surface, although they should be lined to noncritical decorative applications;
- Vinyls resist water and are often used in damp proofing applications and on submerged surfaces;
- These have excellent recoatability except when topcoated with other paints containing strong solvents;
- These have excellent flexibility;
- These resist moisture vapour transmission;
- These resist food and lead byproducts

Limitations

- Vinyls have poor solvent resistance. Extreme care must be taken when over coated with other paints or strong cleaning agents and where solvent splashing may occur;
- Because the solvent evaporates very fast and application of many vinyls are difficult by brush or roller;
- These have poor wetting characteristics which may facilitate in barrier applications;
- Vinyals should not be used in continuous service at temperature above 150 °C;
- These chalk and embrittle on outdoor exposure;
- These have poor resistance to organic acids;
- These are usually not VOC compliant.

(xii) Phenolics

These coatings are made from oil modified phenolic resins. These dry as the solvent evaporates and the oil oxidizes. Therefore to apply these coatings to concrete, a barrier coat may be required. Their properties are similar to those of chlorinated rubbers. These are suitable for use in damp proofing and in submerged service. Use them with caution in chemical environments

Capabilities

- Phenolics have good water resistance;
- These have good resistance in damp chemical environments and to chemical vapours;
- These have good general chemical resistance.

Limitations

- Phenolics are generally softened by strong solvents;
- These discolor with age;
- These become hard and difficult to recoat with age.

(xiii) Chlorinated Rubber

Chlorinated rubber coatings dry by solvent evaporation and have good chemical and water resistance. These are widely used as utility coatings in damp service areas subject to vapour attack

Capabilities

- Chlorinated rubber coatings are easy to apply;
- These can be used for underwater damp proofing application because they resist water and air salts;
- These resist acids and alkalis;
- These are easily recoated;
- These have good moist chemical and fume resistance.

Limitations

- Since chlorinated rubbers chalk on exterior exposure these should be used only in case of concrete at interior decorative application;
- These have good resistance to solvents such as animal and vegetable oils and fats;
- These have limited resistance to heat, they deteriorate when exposed to temperatures above 150 °C for prolonged periods;
- Since these are low in solids, chlorinated rubber coatings are not VOC compliant.

12.4 REPAIR OF INTERNAL WALL FINISHES

For internal walls in buildings plastering, painting and wall paper are some common finishes. These finishes are used for protection and decorative pleasing appearance.

such as good mixing of materials and proper application of plaster coats should be undertaken. The main causes of defects in plaster work are : use of very rich mortar, pure cement coat in top layer, weak substrate backing structure, poor bond with base material, weak rendering under coat, entrapped moisture in between two coats, unsuitable material and impact due to accidents.

For repairs, it is necessary to hack away the defective plaster right upto the wall surface and to brush away all loose mortar, dirt and dust. Mortar joints must be raked, cleaned and washed with water. The porous and loose old plaster must be sealed with cement or paint in order to prevent moisture absorption. Corners must be repaired by using expanded metal or wire-mesh to reinforce the corner. For this, cut the plaster on both sides of corner to expose 100-150 mm of wall surfaces. Fix the wire mesh at the edge of the corner for the entire height of the repair and then finish with suitable non shrink plaster mortar admixed with bonding agent.

Efflorescence is removed by rubbing brushes on the affected surface. The affected surface is washed with diluted solution of one part hydrochloric acid or sulphuric acid and five parts clean water. In case of reappearance, the process is repeated till all the salts causing efflorescence are leached out. Sometimes ready made solutions/chemicals can be used to remove efflorescence.

12.4.2 Repair of Paint

Paint is applied for protection and decorative appearance. The inside appearance of building is an important aesthetic need. Provision of protective coating on to the substrate is an essential requirement for enhancing serviceable life of the substrate. The failure of paint work is due to:

- Poor workmanship in preparing surfaces;
- Use of incorrect type of paint;
- Lack of adhesion of point to surface due to presence of moisture and humid environment;

Common defects in paint work are cracking of paint, flaking off paint, efflorescence, blistering, cissing, loss of gloss, grinning gelling and sagging. Common defects and remedial measures are given in Table 12.1.

Repainting

All paints have their life. Right time for repainting is when the paint is just worn out but not completely worn out and is still bonded to the substrate surface. The old paint has to be scrapped out to expose the original surface when the old film is allowed to deteriorate to the extent of excessive cracking, peeling and curling.

Repainting too often should also be avoided. As a thumb rule, white-wash with lime should be repeated after a year, dry distemper after two years, oil bound distemper or plastic emulsion after three-four years and good enamel or cement paint after about five years.

Checklist for Painting and Repainting Work

Following points should be given due consideration while doing a painting job:

TABLE 12.1: COMMON PAINTING DEFECTS AND PRECAUTIONS

Common Defects	Remedies/Precautions
Patchiness or uneven finish	Apply extra coat of primer
Cracking/chipping of paint film	Avoid using excess putty or a thick coat of paint.
Flaking off of paint film	Cover putty completely with the primer coat with no gaps before applying paint
Efflorescence (formation of white powdery deposit on walls after painting)	(a) Give enough time gap between plastering and painting (about six months) to dry moisture inside of plaster.
	(b) Use a paint forming porous film e.g. Emulsion and distempers to allow inside moisture to escape.
Blistering, or swelling of the paint	(c) Allow the surface to dry fully before applying paint film especially between two coats of paint
	(d) Avoid painting under direct sunlight
Cissing, or tiny craters	Clean the surface thoroughly with soap solution and clean water
Non-drying, very slow drying of paint film	Protect from the moist atmospheric conditions to the extent possible. Scrub oily/greasy surface with a rag soaked in white spirit and wash with soap/water
Loss of gloss	Clean the surface thoroughly and ensure proper steps in surface preparation and drying
Brushmarks	Ensure the right viscosity and a good brush for paint application

(i) Check that all masonry surface voids, pores and cracks are filled with filler. Check surface preparation

(ii) Check surface preparation of wood surfaces. Ensure that all nail holes, cracks and other defects are properly filled and smoothened.

(iii) Check and ensure surface preparation for steel elements.

(iv) Check and ensure use of primer prior to painting of new surfaces.

(v) Check and ensure adherence and soundness of old paint in repainting jobs.

(vi) Check and ensure suitability of paint quality and make.

(vii) Check and ensure uniformity of thickness for application of each coat of paint or repaint.

(viii) Observe painted surface minutely for paint drops, dabs, brush marks, waves and for variation in colour, texture and finish.

(ix) Check those places that are difficult to reach such as edges, bolts, holes, corners and crevices. Ensure that corners, cracks welds, bolts and nuts are properly painted.

(x) Check and ensure complete cleaning of the paint which falls on floor, walls and glass panes.

Common Defects Noticed in Painting Works

(i) Surface not prepared before painting;

(ii) Nail holes left unfilled in wood work and plaster;

(iii) Variation in paint shades wherever large areas are painted, especially the external surfaces;

(iv) Brush marks seen;

(v) Paint adhesion poor leading to flaking;

(vi) Substandard quality of paints used;

(vii) Proper primers not used while painting steel and wooden surfaces;

(viii) Painting left out in:

> Sides, top and bottom edges of door and window shutters

> Fan hooks

> Rebates in joinery and unexposed surfaces of beads in glazing

TABLE 12.2: PROTECTIVE COMMERCIAL COATINGS FOR WOOD AND STEEL

Coating Name	Nature of Compound	Uses
Satya Rubber	Single pack Epoxy based rubber paint. No primer required.	Protective coating Antirust, Antichemicals, Anti abrasion, 100% waterproof and effective sealant.
Techcost	Polyurethane rubber based and two part coating	Protective coating on wood and steel, protects the surface from heat, water, weathering and insect attack.
Bitumen Paint	Bitumen based	Protective coating on wood and steel, protects the surface from heat, water, weathering and insect attack suitable for hidden and underground steel and wood work
Tarothane	Two component coating made of high quality Tar with polyurethane. Doesn't require recoating year after year. No primer required for concrete surfaces.	Protective coating on wood and steel, protects the surface from heat, water, weathering and insect attack suitable for hidden and underground steel and wood work
Epoxy paint	Two component	On wood and metal (can be done on aluminium also)
Chemistic AC-101 Chemistic EP-TAR-202	Single component protective coating Anticorrosive tar epoxy coating	For metal and wood Used for protection of underground steel and wood work. It is also used for protection of underground concrete pipes and tanks.

12.4.3 Repair and Maintenance of Wall Paper

Wall papers are used for decorative purposes. The main defects in wall paper include peeling, blistering, appearance of oil stain and torn wall paper. Following steps should be taken for their repair:

- Remove the damage and peeled off section properly by cutting along the corners and joints. Apply a multipurpose adhesive on repair area. Paper should be then rolled firmly along the prepared surface with suitable adhesive.

- Blistering occurs due to differential expansion of the wall paper. Inject the gluing paste into the centre of each blister by means of a needle, allowing the paper to absorb it for five minutes and then flatten the blister. Alternatively cut horizontally and vertically through the centre of the blister with sharp knife, apply the adhesive paste inside the four corners and then push back the flaps firmly and flatten the area.

- For torn wall paper, patching is done in the damaged area. All the damaged and loose wall paper should be cut off. Replace and fit piece of required size of paper over the cut area to match the surrounding pattern of the paper and paste it.

- Greasy spots are cleaned by fuller's earth mixed with spirit. Washing of heavy papers can be done by mild detergents in water.

12.5 DEFECTS IN DOORS, WINDOWS, JOINERY AND FIXTURES

12.5.1 Door and Window Shutters

A well made door should last for the economic life of the building. Loose or worn hinges make the door to sag and not close properly. Worn hinges are required to be taken out and replaced with new hinges. Many times the screw holes are required to be shifted slightly upwards or downwards. If the depth of the menter permit, longer screw may be used. Aternatively the screw should be taken out and the screw hole be cleaned and refilled by a mixture of sawdust and glue. Once the glue sets, the screw can be tightened again.

If hinges, bolts and lock screws are loose, these pose a security risk to the house. The screws of hinges, bolts and fittings should be oiled and tightened from time to time.

External doors, which do not have the sill may allow rainwater through the gap between flooring and the door bottom. To reduce this problem, screw a triangular piece of wood on the door in order to throw off bulk of water. The fillet piece could be so fitted that it just grazes the floor. This piece can be taken out once the monsoon is over.

Warped Door

There is not much solution possible for a highly warped door and the shutter is to be replaced. Sometimes the shutters can be taken out and subjected to steam or shutters are kept under heavy loads placed on the plane surface.

Rotten Door Frame

Door frames which are embedded in floors rot quickly. After some time the wood may simply disappear. Instead of replacing the entire frame, a good piece could be fitted in place of rotten piece with a lap joint.

Warping Due to One Side Lamination or Paint

Where one side of wooden door shutter is painted or fixed with a laminate and the other side is not, the door may warp. As far as possible the door should have similar finish on both faces. Where lamination is put up, it should be on both sides although the thickness of both laminations may differ.

Joint Between Door and Masonry

Many defects arise due to poor joining details at the masonry and door frame. Penetration of water can be easily detected by damp or discoloured patches on the inside wall face around opening. A mastic seal between door frame and masonry face prevents transfer of moisture from outside to inside.

Moisture may also be absorbed near the lintel face and cause a problem especially when sun shade does not have a proper throating and drip course. Bathroom and WC doors get continuously water splashings and therefore should have metal lining of aluminium or GI sheet fixed with brass screws to protect shutter from water.

12.5.2 Door Closers

Hydraulic door closers are used where doors are expected to remain closed but not locked. Hydraulic door closers require periodic maintenance to ensure efficient operation and long life. An improperly installed door closer shortens its useful life and also places severe strain on the door frame. Every six months, all screws, bolts and nuts should be examined and tightened. This should include all bracket screws, arm bracket screws, open nuts and screws and adjustable arm screws.

A well-adjusted closer can swing the door fast till within 12 cm of closed position, and if a latch is provided it may close the latch.

12.5.3 Broken Glass / Fixtures

A cracked glass can be temporarily repaired by patching it with stick-tape. A broken glass should be replaced immediately. If glass is broken due to vandalism it might be better to replace it by reinforced glass or by fibre glass.

If the glass breaks because of its thinness, a thicker glass should be installed. Thin glass pane of large dimension may break even in high wind. When window or door glazing breaks due to forceful closing of shutter, it indicates weak shutter and it requires strengthening specially at corners. The excessive moisture absorption by wooden panel results in swelling of wood which may be due to poor paint. Such wooden frame needs to be painted after sealing of wood pores.

12.5.4 Joint Paints and Joinery

The doors, windows and external joinery may suffer from the following defects:

- Timber decay and surface deterioration of joints
- Paint deterioration
- Distorted joinery
- Rusting of tower bolts & latches, etc

Timber Decay and Surface Deterioration

Water enters via open joints, defective putty and wet walls. The decay in wood specially near wet masonry contact surface is due to wet rot. For a localised defect, cut away the affected wood and fit a new piece of wood with glue and suitable lap joint where general condition of joinery is poor, old putty should be removed, paintwork scrapped and the surface coated with primer. All holes and joints should be filled with water insoluble filler or mixture of glue and saw dust and repainted.

Paint Deterioration

This is due to poor workmanship, improper quality of paint or use of unseasoned timber. If failure of paint is in small areas, apply primer after scrapping old paint and repaint. Where paint failure is extensive, scrap all paint work and putty. Take out glass and apply primer, repaint and replace glass and fix putty.

Distorted Joinery and Lack of Fit

This can happen due to poor craftsmanship and poor painting maintenance. Clean the frame joints, apply epoxy resin glue and allow it to set. Strengthen the corners by fixing angles of steel. In extreme case of poor joinery, it may be cheaper to replace shutter completely.

Rusting

Rusting in steel window frames is caused by rain water contact and reaction with steel. The putty and glass panes should be removed. The steel frame should be derusted by scrubbing and chemical washing. Apply rust inhibitor on the steel surface. Apply suitable primer coat and glazing. Where frame is badly rusted, it may be economical to replace the frame. Removal of rust without taking off glass panes may result in temporary cure and the rust may reoccur soon. Rust must be fully scrapped by removing glass panes. Apply proper coating of antirust primers prior to painting.

12.6 REPAIR OF WALL PANELLING AND FALSE CEILING

Wall panelling, false ceiling and floors are provided for the purpose of decorative finishes.

Wall Panelling

Wall panelling on wooden or other frame work, may suffer from problems of shrinkage or loose wooden gutties. Dampness in the substrate wall develops fungus. Wall panelling becomes a breeding place for insects. Fumigation of space behind panelling is necessary. Pressure treatment with insecticide by suitable technique is required for timber frame work to avoid rotting.

False Ceiling

False ceiling connections of steel may corrode. Where false ceiling is fixed by wooden gutties in RCC, the gutties can become loose due to moisture. The false ceiling should be inspected regularly. Where incandescent bulbs are provided within false ceiling, the heat build up may become a fire hazard. Long stretches of false ceiling should have fire safety compartmentation. All electrical wiring within false ceiling must be run in steel conduits for proper insulation to minimise fire hazard. Broken or damaged pop tiles must be replaced by tiles of the same design. Dampness patches must be cleaned and painted with original paint. Damaged insulation behind plaster of paris tiles must also be replaced and repaired.

12.7 REPAIR OF ALUMINIUM ELEMENTS

Annodised aluminium is widely used in building. Even bare aluminium surfaces get protected in course of time by a thin tough film of an oxide which is formed on exposure to atmosphere.

Aluminium surfaces require cleaning with clean water every 2 to 6 weeks depending on exposure conditions. Use mild soap to remove dirt after loosening by solvent cleaner. A good clear methacrylate lacquer coating protects aluminium surface for one year. The coating can be done by spray or by brush. Waxing also facilitate protection. Aluminium should not be cleaned with alkalis or acids or strong detergent powders.

12.8 CLEANING OF GLASS

The dust and other pollutants settles on the glass. The dust particles get attached to the glass by electro static force. This force is generated when the glass is wiped off by a cloth. It is easier to clean glass face which is inside room. Glazing should be cleaned when in shade for drying slowly and to eliminate streaks. No acidic substances should be used for cleaning glass windows. Glass is cleaned by warm water, detergent, ammonia or lime.

The procedure for cleaning of glazing commences with dusting with soft brush. Starting from top, the glass is washed with sponge by taking straight overlapping strokes back and forth. Remove hardened spots of paints by scrapping. Dry the window by pulling and squeezing blade of rubber. The rubber squeeze makes a perfect contact with glass. Ammonia and lime are used to clean very dirty glass. Ammonia fumes can be strong and care should be taken to avoid suffocation. Very effective window cleaning fluid can be prepared by fixing equal parts of paraffin, methylated spirit and water shaken together in an airtight jar. Fine cloth soaked in solution is used first and latter cleaned with clean coarse cloth or crumpled newspaper. Cleaning of glass is an important step in enhancing its serviceable life and maintenance of buildings. Glass fixing putty should be repaired and replaced as and when necessary for stability of glass panes.

12.9 REPAIR OF FLOORS

Various type of defects in floors require proper remedial measures for effective occupation of buildings. The techniques depend on the type of defects as described subsequently.

12.9.1 Floor Surface Preparation for Repair

When the floor topping is to be replaced over the existing base concrete, the concrete surface must be properly prepared. Proper surface preparation of concrete involves three basic steps:

(a) Removal of contamination

(b) Removal of bulk

(c) Cleaning of concrete surface

(a) Removal of Contamination

Grease, wax, oil or sealers may impair proper bond of top floor coat with the base material. The surface preparation technique for repair may involve little removal of the substrate concrete while complete removal of contamination.

Presence of such contamination may be determined by dropping a small amount of muriatic acid onto the substrate and watching whether a reaction occurs. No reaction indicates that contaminants are present. If oil contaminating has penetrated into the concrete surface, it may be detected by raising the temperature of a small area to about 66°C with a heating lamp. Presence of the contaminants is indicated if oil appears or the area becomes" greasy' to touch. Typical contamination removal would include scrubbing with detergents such as **trisodium phosphate** or other commercial degreasers.

(b) Removal of Bulk

Various methods may be employed to prepare hardened concrete surface. Chipping is the most common technique. Use of a square tip chisel is recommended. Other methods used include hydrodemolition, **scrubbing, refomilling** and other mechanical means. Some of these techniques, such as hydro-demolition, may include the final three steps of surface preparation.

It is important to select a method that is effective enough to get the job done properly but should not be so aggressive as to damage the sound concrete. Various methods may be used for floor space preparation for application of products in thinner sections or where the substrate is not severely damaged. These methods are ranked according to performance as under:

● Shot blasting
● Sand blasting
● Water blasting
● Acid etching

Shotblasing is the preferred method for removal of thinner sections of concrete. Steel shots strike the concrete with high velocity to crush top layer. Follow mechanical cleaning with vacuum cleaner.

Aggressive Sandblasting may be used in place of shotblasting for concrete surface applications. Here steel shots are replaced with hard silicasand. Use the same procedure as shotblasting. It requires greater care during cleanup.

High-pressure **Waterblasting** using pressure over 8,000 kg/m^2 may be sufficient for some applications. Thorough rinsing of the substrate is necessary to remove wetted laitance. Water blasting with pressure below 8,000 kg/m^2 is insufficient for most applications. This helps in removal of loose laitance and other surface contaminants in substrate concrete.

Acid Etching should be used only when there is no other alternative applicable. Acid etching should be followed by scrubbing and flushing with adequate amounts of clean water to remove residual chemicals and, the fine dust produced by the etching, which may act as a bond breaker. Check for complete removal of acid by testing with moist pH paper. pH value should be more than 10.

(c) Cleaning of Concrete Surface

The final and the most important step of surface preparation is the **cleaning** of the concrete surface. All loose particles and dust must be removed prior to placement of the patching repair material. This is best done with a pressure washer using water of approximately 3,000kg/m^2. This water jet removes contamination and loose fine material to make the concrete surface ready to receive new floor with an uncontaminated, bondable toppings.

The prepared surface is dried before the application of floor topping. A bonding agent must be scrubbed on to the prepared surface with a clean, stiff bristle broom or brush prior to the application of the topping. Do not apply bond coat on more area than that can be covered with the topping before the bond coat dries out. The bonding coat should be **tacky** prior to the application of toppings layer.

12.9.2 Common Defects in Concrete floors

The problems such as random cracking, scaling, popouts and slab settlement can be controlled by good construction practices during original concrete floor construction. The roots of the problem can be eliminated by adopting general guidelines. These general guidelines for each of the root causes are stated in subsequent paras.

(a) Shrinkage Cracks

Concrete shrinks when it hardens. If a slab is not free to move when it tries to contract, tensile stresses are developed and the slab cracks. Concrete cracking is minimised by reducing shrinkage, by reducing the panel size and by providing suitable joints in the slab. Shrinkage is reduced by:

- Use of largest possible maximum size aggregate in concrete;
- Using low slump concrete with as much aggregate as possible and adding only enough water to provide the needed workability;
- Wet curing concrete for a longer period before allowing it to dry;

Shrinkage cracking can be further controlled by:

- Spacing joints 2.5 to 3 m apart for a 100 mm thick slab and 3 to 4 m apart for125 mm thick slab;
- Forming or sawcut joints to a depth of at least one fourth of the slab thickness;
- Placing joints at restraint corners;
- Isolating slabs from foundations, walls or other structural elements.

(b) Craze Cracking

Craze cracks are caused due to shrinkage of the rich paste in top surface of a slab. The cracks are unsightly but seldom affect durability or wear resistance of the surface. They do not move progressively. The cracking pattern is most noticeable after a rain when the surface first starts to dry. If the concrete contained calcium chloride, crazing is likely to be more pronounced. Crazing can be prevented by:

- Starting curing the concrete as soon as possible. Avoid fast drying;
- Not using sloppy concrete mix with the slump range from 75 mm and above;
- Not finishing concrete when bleed water is still on the surface;
- Not sprinkling dry cement and sand on the surface to blot excessive bleed water;
- Not using a pitbug unless the concrete slump is less than 75mm;
- Avoiding steel troweling which brings more paste to the surface.

(c) Settlement

Settlement sometimes causes cracking but can also create tripping hazards or drainage problems. Sidewalks that slope towards the foundation may form pond of water, increasing the chances of leaking basements. Inadequate soil compaction is the most usual cause for settlement. Settlement can be prevented by:

- Replacing top soil and organic material with a layer of coarse sand before placing concrete;
- Compacting cohesive sub-grade soil by ramming and granular sub-grade soil by vibrating;
- Using dense graded granular soil base and compacting it before placing concrete in poor soils;
- Draining rain run off away from the slab base to avoid washout of soil or base material;
- Thoroughly compacting back fill areas.

(d) Surface Scaling

Scaling is most likely to occur during the first or second winter after the floor concrete has been placed. Prevention requires high quality concrete and good construction practices. To prevent scaling:

- Use only air entrained concrete;
- Slope paving so that water drains readily and avoid creating low spots during finishing;
- Strike off the loose concrete, treat floor if necessary, and apply a broomed finish;
- Avoid finishing concrete surface having rain or bleed water and avoid adding dry cement or cement sand mixture to blot the surface water;
- Cure the concrete longer with wet hessian or curing compound after the concrete has dried for at least 28 days.

(e) Popouts

Popouts are pock (needle) marks in the concrete surface caused by internal pressure from particles near the surface. Common causes of popouts are absorptive or chemically reactive particles near the surface. To minimise the popout by absorptive and reactive aggregate :

- Use bug on the surface before floating to push larger particles further beneath the surface;
- Use concrete with smaller coarse aggregate size;
- Use concrete of **low water-cement ratio** to reduce permeability and increase the strength for more resistance to the forces causing popouts.
- Use concrete prepared with non-reactive aggregate;
- Wet cure the concrete by fog spray or ponding if only reactive aggregates are available in the area and rinse the concrete surface with water after curing.

12.9.3 Repairing Joints in Industrial Floors

Joints and cracks are the main problems in industrial floors. Random cracks can be attributed to inadequate slab curing and restraints to slab contraction. Large joint spacing, late sawing of joints and improper mix design also leads to slab cracking. Upward slab edge, corner curling and warping occur in some floors when concrete near the slab surface contracts in comparison to the slab bottom. Upward curling and warping occurs in some floors at joints and causes irreversible cracking. Upward slab deformations may be as large as 20mm near the joints and result in cracks with the slab losing contact with the base as far as 600-800mm from the joints. Curling is noticeable when a loaded truck crosses the joint. Slab deflections can be detected by simply standing on the joint. If slab curling is not corrected, loaded trucks cause tensile stresses in the top of the slab leading to cracks parallel to joints or edges.

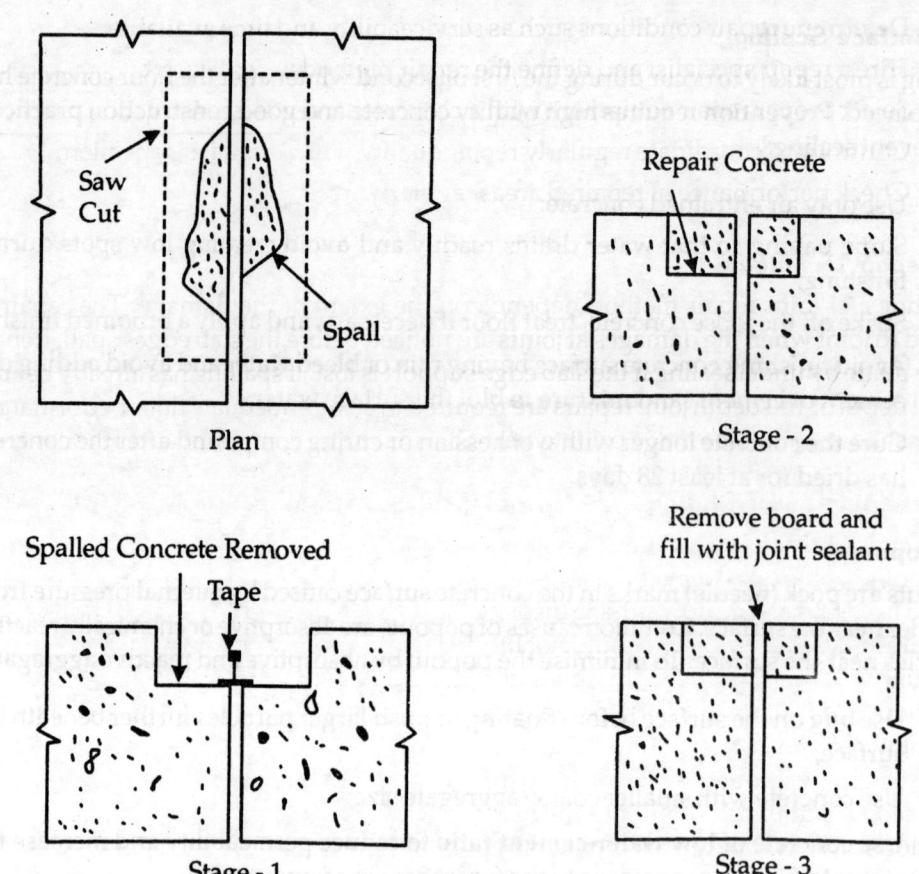

Fig. 12.2: Floor Joint Repair

The upper shoulder at keyed construction joint may also some times crack. This distress occurs due to wheel loads crossing the keyed joints. This load transfer effectiveness of the keyed joint gets diminished due to joint widening associated with slab contractions and the joint shoulder losing contact with the adjoining slab.

Concrete slab edges at joints and cracks can spall when subjected to truck traffic. Once spalling develops, slab edges continue to deteriorate under repeated loads specially when there is a loss of lateral support due to damaged or missing sealant.

(a) Steps for Floor Joint Repair

Joint cracks require a systematic repair approach by first treating the cause of cracking and then restoring the floor surface. The systematic repair approach involves the following steps

- Evaluate the causes and extent of distress;
- Determine traffic/occupancy condition of the repair area;

- Determine repair conditions such as serviceability and time available;
- Hire a repair specialist and define the repair methods;
- Select proper repair equipment with a competent crew;
- Control and coordinate regularly repair quality with adequate inspection;
- Check performance of repaired areas regularly.

(b) Repair Techniques

The choice of joint repair method depends on the extent of the damage. The repair is limited to joint when the damages at joints are noticed before the slab edges spall. Repair can be made by undersealing if the slab edge support is lost. If spalling has already begun, partial depth or full depth joint repairs are required to ensure adequate floor performance. Under sealing should be made prior to partial joint repairs or joint and crack sealing.

Undersealing and Grinding

When curling results slab deflections greater than 0.04mm, slab undersealing should be done prior to other joint repairs. Laser levels or dial gauges supported by Benkelman beam may be used to measure upward slab displacement caused by excessive undersealing grout pressure and volumes. Slabs should not lift higher than the existing adjacent floor levels.

To under seal a slab edge, drill holes in the slab about 12mm from the joint (or crack) and about 600-900mm apart. Prepare a grout with easy flowability by using a mixture of cement water and fly ash. Pump the grout under low pressure to prevent excessive slab lifting. Pump the grout until it comes up from an adjacent hole. Then put a wood plug in the hole and pump the grout into the next adjacent hole. Continue until pumping has been done at all the holes. This should provide continuous and uniform slab support. Fill in drilled holes by gravel concrete mix using expansive grout cement.

If the curled edges have created a hump at the joint, level the floor by diamond grinding the slab surface near the joint. Follow this with flat surface grinding to smoothen the rough surface. Repair of joints can be undertaken after undersealing all the slab edges.

Joint Cleaning, Routing and Resealing

Clean the joints and reform these joints using plastic or metal "T" strips. Metal "T" may cause minor spalling. Sealant can be moved with a joint plaugh or a router and for small projects by manual methods. Next use a diamond saw cutter to abrade the bottom and the vertical edges of the joint sealant reservoir. This is a quick method that causes the least slab edge spalling.

After sawing to at least one fourth of slab thickness, remove sawing dust using air blast in the joint. Remove all previous sealant and sandblast vertical edges. Place bond backer tape on the bottom of the sealant reservoir to prevent sealant loss into the control joint crack. Do not use the backer rod when the joint is exposed to hard rubber urethane casters or steel wheel traffic. The flexible backer rod decrease the lateral slab edge support, ꞏ ꞏ ꞏ Introduce the sealant when the temperatures are contraction before the sealant gains enough cohesive and adhesive strength to withstand joint movement without tearing or losing bond.

When metal key is left in place, joints are to be resealed and the procedure is different. Slab contraction widens the joint so much that the slab edge shoulders and keys are no longer in contact. Consequently the slab edge directly under the truck traffic does not transfer the loads across the joint. To re-establish contact and load transfer, allow the joint sealant to flow into and completely fill the space between the joint shoulder and key.

Repairing Spalled Joints and Cracks

Take concrete cores at spalled joints and examine the cores concrete and core hole walls. If the concrete has deteriorated to less than half the slab thickness, partial depth repairs are sufficient. Slab with damage extending into the bottom half will require full depth repairs.

Partial depth repairs consist of sawing round the boundary of the spalled area to a minimum depth of 40 mm. Minimum repair width is 50 mm inward from the joint or the crack edge. Remove the concrete between the saw cuts with a diamond grinder, scarifier, scrubber, or light chipping hammer. After removing the concrete debris and vacuuming, use air pressure to remove all dust from the repair area. Check the quality of the prepared surface by performing the bond pull off test. If lower **pull off** values are obtained, sandblast the surface before air-blasting.

Use either, high strength pea gravel concrete prepared from high monomer methyl methacrylate concrete mix or an epoxy aggregate mix as the repair material. The polymer materials are usually used for partial depth repairs only because of high cost. The polymers set quickly so workers have less time to place them. They gain strength quickly, bond well to the concrete and have good impact and abrasion resistance. Ensure good ventilation of area when polymers are used. This not only ensures the safety of the workers, but also surethe work area is well ventilated helps avoid leaving an unpleasant odour.

Before placing the repair material, form the sealant reservoir by placing a 20mm or thinner board or plastic strip in alignment with the joint to the full depth of the repair material. Apply a bond breaker to the form before placing the concrete. The form is removed ahead of sawing the sealant reservoir before resealing.

When using a cementitious repair material, wet the surface to a saturated surface dry condition. Next brush cement slurry admixed with bonding aid onto the prepared surface. Then immediately place the repair concrete without allowing the bonding slurry to dry. Most epoxy and methacrylate repair materials require bonding agents as described in the specific product data sheets for the respective materials. These should be applied to a **dry surface**.

Full depth repairs consist of sawing to full depth around the distressed joint or crack, compact the affected area and level the sub-grade. Load transfer dowels may be needed depending on slab thickness and magnitude of applied loads and observed objectionable slab edge deflections.

To establish load transfer between the old and new repair concrete, drill horizontal dowel holes 300 mm apart up to slab mid depth to a minimum horizontal depth of 150 mm. Dowel diameter should be 6mm for each 25 mm of slab thickness. The drilled holes should be of diameter large enough to accommodate deformed rebars that will be inserted into them.

Replace the spalled area with 35 MPa (35 N/mm²) concrete having 30 mm maximum size coarse aggregate and water cement ratio of 0.45. One side of the joint repair concrete should be placed and allowed to cure for at least 24 hours before placing the other side. This creates a plane of weakness at the joint ensuring that subsequent slab contractions do not cause random cracking at any other location. Use internal vibrators to consolidate the concrete when placing the second side. When full depth repairs are needed at joints requiring load transfer devices, use the repair scheme shown in Fig. 12.3. Install plain dowel bars instead of deformed rebars.

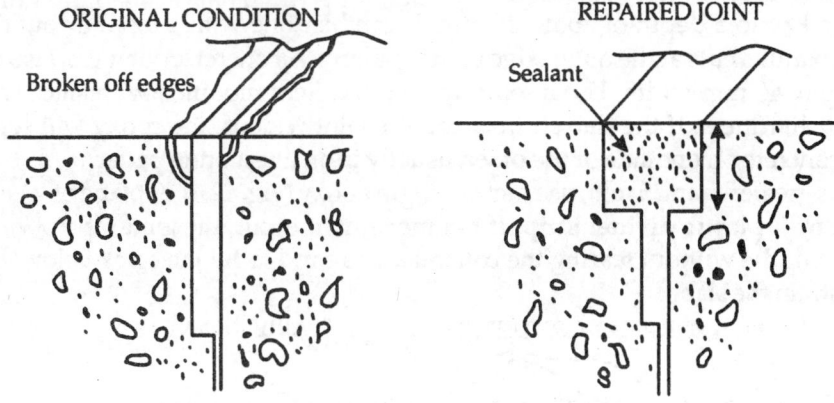

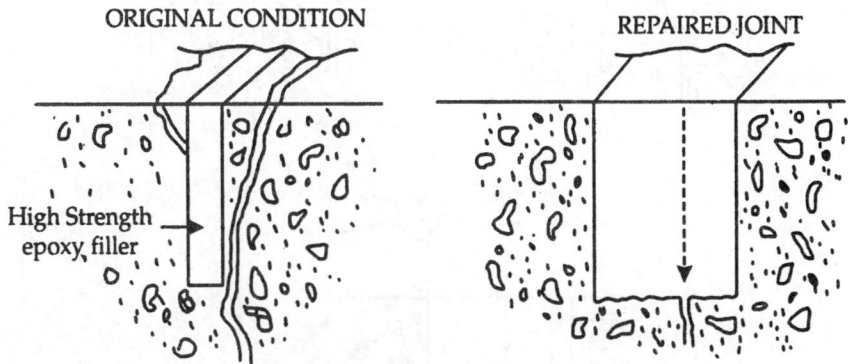

Fig. 12.3: Floor Joint Repair

To make a joint with smooth dowel load transfer, grease the dowels and place them in a dowel basket. Greasing the dowels allows horizontal movement of the slabs. After the concrete is poured, the form can be easily removed. The smooth dowels should be straight and perpendicular to the formed joint face in both the horizontal and vertical plane.

Crooked dowels cause restraints that can result in new cracking. Placing welded wire fabric is optional and can usually be omitted in all but the worst service conditions.

(c) Example of Floor Joint Repairs

Industrial floors have special requirement and the joints in floors must be kept in good condition. Some joint defects appear early in construction but can be repaired before occupancy. The various problems in industrial floor are described in subsequent paragraphs.

Problem (i) Metal Keys Problem

Joints create cantilevered concrete nose that may break off after the joint opens as any wheel load crosses the concrete batter.

Remove key to a depth of about 25 mm. It can be done with a torch or outlet saw cutting to parallel cuts at the outer edge of the spalling. A concrete router can also create the slot up to 40 mm width. Use a semi rigid epoxy, field modified with silica sand to increase its hardness. If the slab shrinks and the joint widens, the epoxy will separate from the concrete on one side or the other, usually in an alternating pattern.

For slots greater than 25 mm, use high strength epoxy (Fig. 12.4). A plane of weakness is created by a **plastic divider strip**. If the movement occur, the joint repair will split down the middle without tearing the concrete. Use sand to fill the crack below. Epoxy would restrain the slab.

ORIGINAL CONDITION   REPAIRED JOINT

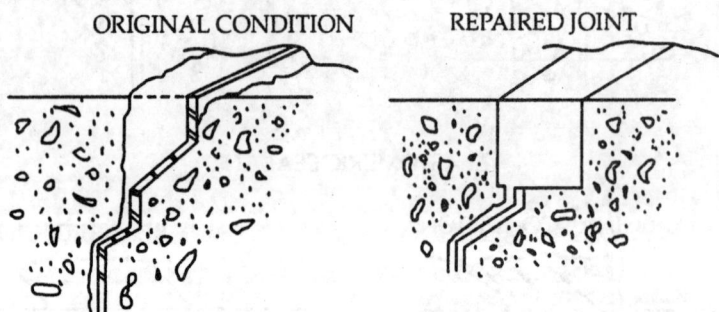

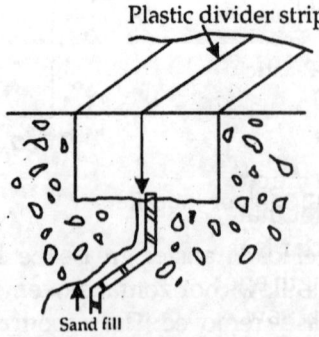

Fig. 12.4: Metal Key Left in Place

Problem (ii)

Plastic crack inducing strips may be out of plumb during finishing. This creates a cantilevered concrete nose that breaks off under traffic movement. Determine the position of the strip with the most severe damage which is usually on the cantilevered nose. Make a saw cut just beyond the spalling area and then another about 15 mm onto the nose. A cut minimum of 12 mm to 25 mm width is preferred. Remove the concrete between the cuts.

Fill the slot with a semi rigid epoxy mortar. Fill the joint flush to the surface or slightly higher and finish level.

Problem (iii)

Joint edges spall under hard wheel traffic because soft elastomeric sealants provided do not protect the joints from impact. Such sealants only keep dirt and moisture out. Create a vertical slot by saw cutting or routing a cut of at least 12 mm depth and 20 mm to 25 mm width.

Fill the joint with semi rigid epoxy mortar (as shown in Fig. 12.5) after cleaning, washing and drying. The epoxy may separate from the concrete if the slab moves at joints.

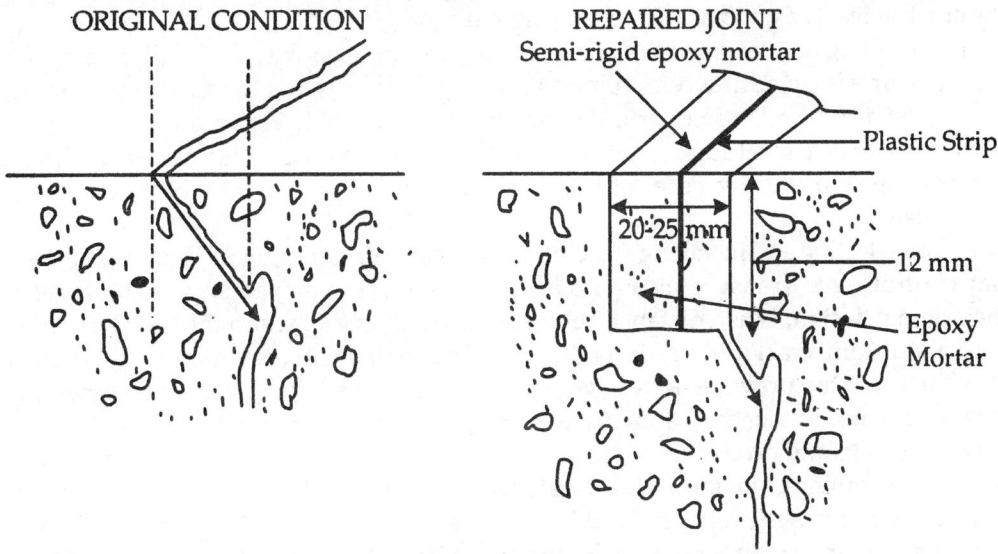

ORIGINAL CONDITION   REPAIRED JOINT

Fig. 12.5: Plastic Crack Inducing Strip

Problem (iv)

When the joint filler is stronger than the concrete, stress cracks due to shrinkage can occur on each side. Saw cut (or rout) a slot that is as deep or deeper than the original cut. In most cases, a semi rigid epoxy mortar is the best repair material, especially if the slot is less than 40 mm wide. A high strength epoxy mortar can be used for wider slot repairs, but a plastic divider strip (or a new saw cut control joint through the repair) must be provided to prevent shrinkage stress buildup.

12.9.4 Repair of Cracks in Industrial Floors

Cracks occur in floors on various counts:

(a) Shrinkage cracks

(b) Crazing and Map cracks

(c) Plastic shrinkage

(a) Shrinkage Cracks

Cracks in industrial floors are caused by restraints in slab to contraction. When slab is restrained, cracks drying shrinkage or contraction from cooling causes cracking. These cracks are usually widest near the floor surface and becomes narrower with depth. Shrinkage cracks in slabs reinforced with welded wire fabric often terminate at or near the level of reinforcement

If no significant crack edge spalling has occurred, structural epoxy pressure grouting can be done to restore structural slab integrity. For structural epoxy grouting, small holes are drilled along the crack at a spacing equal to the floor thickness. A well selected and properly installed structural epoxy system will provide flexural and tensile strength at the crack at least as good as that of the original concrete.

If no spalling has occurred at slab edges, the structural epoxy should fill the cracks to the slab surface. In addition to doing pressure grouting, workers can also use surface gravity feed to install epoxy at the widest part of the crack. Grind the surface of the hardened epoxy to meet adjacent surface levels and match as closely as possible to adjacent surface appearance. Take cores of the repaired crack to confirm that the crack is filled completely.

Before selecting and installing structural epoxy for slab crack repairs, examine adjacent slab control joints to ensure accommodation of contraction movements as envisaged in the original design. When distance between active control joint and the cracks are more than about 15m, cracks at location other than the joints accommodate slab contraction movements. This prevents excessive tensile stresses from building up at other slab locations. Thus, a structural epoxy-grouting repair should not be used when distances between cracks and proactive control joints exceed 15 m. In that case the crack should be repaired by routing grooving and sealing for accommodating slab contraction movements.

Routing and sealing repairs should be made at cracks not requiring structural epoxy pressure grouting repair. Route cracks about **12 mm to 38 mm deep** and **20 mm to 25 mm** wide depending on width of crack and minor surface spalls present. The routed sealant reservoir should be square or rectangular in cross section. Routed sealant reservoir width to depth ratio should be about 1 : 1.

Install a bond backing tape at the bottom of the crack sealant reservoir. Remove all dust and debris from sealant reservoir walls before installing the sealant.

(b) Crazing and Map Cracking

Crazing and map cracking typically occur soon after the concrete has set and which can usually be prevented by **adequate curing**. These cracks are extremely narrow and are often only noticeable when the floor has been coated with a translucent urethane or epoxy or when the floor dries after cleaning. Craze cracks terminate at depths less than 12 mm below the surface and do not affect floor performance.

(c) Plastic Shrinkage Cracks

Plastic shrinkage cracks are caused by rapid loss of moisture from the slab surface while the concrete is still plastic. The cracks usually penetrate only partially into the slab. These are often parallel to one another and range in length from a few mm to several metres. No slab movements occur at plastic shrinkage cracks because these cracks are narrow and seldom reach bottom. These areas subjected to the pneumatic tire traffic need no repairs as long as the crack edges have not spalled. Floors exposed to solid tire or caster traffic, fill the cracks with a low viscosity structural epoxy after removing all dust and debris from the crack.

12.9.5 Cleaning of Floors

The floors require every day cleaning and mopping to keep them clean. Many times stains and scratches on the floors spoil their appearance. Storm dust containing injurious particles should be removed as early as possible. It is easy to remove fresh stains. Following techniques may be used for removal of different types of stains:

Ink Stains

Fresh ink stains are removed by scrubbing and washing with Cotton/Rag dipped in ammonia till the stain disappears. Wash with clean water.

Old and deep stains are removed by applying a paste of sodium perborate mixed with whiting (marble powder) and leaving it for 15 minutes and then cleaning. If the stains are not removed they will be reduced in intensity, then repeat the treatment. Sodium perborate can be replaced by either of the following chemicals: **Ammonium Cxalate, Oxalic acid, Citric acid, Sodium citrate** and **Hydrogen peroxide**.

Dirt and Rust stains generally occur in bath rooms because of using iron buckets. Light and fresh stains can be removed by applying paste prepared with **Sodium Citrate, glycerine** and **water** in the ratio of **1 : 6 : 6** along with whiting. Apply this paste and leave for 15 to 20 minutes. Clean and wash with fresh water. Repeat the process if stains are not fully removed.

Deep and Old Rust stains are removed by applying paste of **Sodium Hydro Sulphate** with **Sodium Citrate** in equal parts. Apply this paste on the stained surface only for 15 minutes. This paste should not be left for more than an hour, otherwise it leaves a black mark which can be removed by applying fresh Hydrogen peroxide.

Fresh Oil Stains on the floor can be removed by mopping the stain and then rubbing with **hydrated lime** and then with **petrol or Benzene**.

Old and deep oil stains can be removed by placing a thick cloth saturated with **acetone and amylacetate** over the stain. Put whiting over it to allow the process of evaporation of chemicals. Clean up after 20-30 minutes. Repeat the process in case stains are not cleaned. Instead of **acetone and amylacetate, carbon tetrachloride** can also be used.

Paint Varnish Stains can be removed by ready made paint remover. Use of paint remover is the cheapest and best. Alternatively **use carbon tetra chloride** with **Amyl-acetate**. In case of a wooden surface, soften the paint with a blow lamp and then softly remove with an iron scraper.

Bitumen Stain can be removed by softening with mild heat by blow lamp. Sometimes bitumen melts and drips through an expansion joint. Scrap the bitumen stains after softening it with **heated kerosene oil**. The remaining light stains should be cleaned by **carbon tetrachloride**.

Floor Stains can be removed by scrubbing with a mix of **Tri Sodium Phosphate + hydrogen paroxide + ammonia** in equal ratio. Alternatively apply sodium hydrosulphite with whiting for 20-30 minutes and then clean and wash with water.

Tea-Coffee Stains can be removed by scrubbing with paste of **Glycorine** with water in 1:4 ratio mixed with **whiting** after keeping the paste over affected surface for 15-20 minutes. Then wash the floor with clean water.

12.10 SUMMARY

Finishes form an integral part of any building and constitute major component of the building cost. Apart from aesthetic considerations, finishes play considerable role in protection and enhancing the service life of buildings. Life of finishes is limited and requires repair and replacement at regular interval for maintaining the building in serviceable condition. Finishes are applied to walls, floors, roof terraces, ceilings and joinery for protection from deterioration and creating barrier to environmental forces.

Repair of wall finishes comprises of mortar joint pointing, plastering, rendering, distempering and painting etc. Damages are repaired by cleaning and sealing the pores by forming water proof films on the wall surface to stop environmental forces from entering the body of walls. Water proof cementitious grouts are used for treating wall surfaces. Surfaces are first prepared well before repair treatment. In case of deep damages, injection techniques are adopted for repair. Mask grouting can also be used in case of highly deteriorated joints.

Efflorescence on the wall surface (plastered or unplastered) can be removed by eliminating either the source of soluble salts causing efflorescence or sealing the moisture movement paths or using the solutions to dissolve the salts. Proprietary masonry cleaning solutions should be used to remove efflorescence after careful preparation of walls. The simplest technique to remove efflorescence may be to clean the surface with brushes and washing with clean water. In case of severe efflorescence, appropriate chemical cleaning agents (acidic or alkaline) may be used to wash the surface.

Protective coatings form water repellent films and moisture movement barrier. These coating protect the substrate structure from environmental attack. These coatings include **silicones, silanes** are selected based on the requirements of substrate structure, aesthetics, environment, service life and cost. Protective coatings are generally applied on dry surface when the temperature of coatings range between 10 to 30°C. Silicones generally form good protective coating. Some of the protective coatings form transparent film to maintain the originality of the substrate surface.

Some of the protective coatings are also decorative and colourful in appearance. Decorative coatings can also be prepared by mixing suitable colouring agents at the time of application or may be available ready mixed. These coatings are cement paint, oil free urethanes, oil/modified urethanes, acrylics, catalyzed epoxies, latex emulsions, alkyds, two compound urethanes, epoxyesters, polyesters, vinyl, phenolics and chlorinated rubber. Each of these decorative coatings have specific characteristics which must be studied to suit specific application in a specific situation.

Repair of wall finishes include repair of plaster work. Patch repair of plaster work should be done by cleaning and preparing the base, applying a bonding coat and placing the water proof final plaster to match the original surface. Efflorescence should first be removed by rubbing with brushes and then washing with dilute acid solution. Painted surface can be repaired by first removing and cleaning loose and cracked paint, paint curling and peelings by scrapping and brushing. Suitable primer coat should be applied for effective bonding of new paint. Whenever, wall paper gets damaged, it should be repaired by applying new matching wall paper in shape, size and design with good adhesive.

Warping of door and window shutters can be set right to some extent by steaming these shutters. Steamed shutters are then pressed on a plane surface under sufficient loads. Wet rot affected door/window frames can be replaced by cutting and lapjoint. Most of the joinery defects develop by entry of moisture through gaps between masonry and wooden frames. These gaps are treated by drilling holes, injecting the antitermite solutions and sealing all the gaps to stop ingress of moisture. Door-window painting should be done after cleaning, rubbing and preparing the old paint damaged surface. Glass panes should be cleaned using glass cleaning liquids. Broken glasses should be immediately replaced along with damaged putty.

New floor finish should be laid after thoroughly preparing the existing floor substrate. by removing contaminated and damaged portion. Floor surfaces can be prepared by shot blasting, sandblasting, water blasting or acid etching depending on the extent and type of damage in floor. Concrete floors suffer from shrinkage cracks, crazy cracks, popouts, surface scaling and settlements. These cracks must be filled with epoxies after removing spallings, cleaning and preparing the surface. Industrial floors are specially repaired by treating the damaged joints and removing contaminated and loose floor materials. Joints are repaired by cleaning, routing and resealing. Joints can be partially repaired or treated by full depth sawing cleaning and sealing properly. Floor cleaning regularly and removing various types of stains and contamination plays critical role in maintaining serviceability of floors in buildings.

QUESTIONS

12.1 Describe importance of finishings in building maintenance.
12.2 List the type of building finishing elements.
12.3 List wall finishes.
12.4 Describe repair techniques of wall mortar joints.
12.5 Explain grout injection method for masonry repair.

12.6 Explain mask grouting techniques for masonry joint repair.

12.7 Explain repair of joints in structural slab.

12.8 Explain method of removal of efflorescence in masonry.

12.9 List various chemical solutions for treating efflorescence.

12.10 Describe importance of protective and decorative coatings.

12.11 Write short notes on

- Efflorescence triangle
- Selection criteria for protective coatings
- Decorative cement paints
- Acrylics
- Urethanes
- Epoxy esters
- Polyesters
- Latex emulsions
- Alkyds
- Chlorinated rubber.

12.12 Describe special features of internal wall plaster repair.

12.13 Describe special features of paint repair.

12.14 Explain removal of floor contamination.

12.15 List common defects in floors.

12.16 Explain method of industrial floor joint repair any one technique.

12.17 Explain repair techniques for floor defects.

12.18 Describe importance of floor cleaning in building serviceability.

12.19 Explain methods of cleaning various floor stains.

12.20 Explain repair techniques of crazy cracks in floors.

Chapter 13

REPAIR OF
BUILDING JOINTS

13

REPAIR OF BUILDING JOINTS

LEARNING OBJECTIVES

After studying this chapter, the learner understands the repair of building joints and will be able to:

- **Describe** functions of joints in buildings and importance of their repair;
- **Explain** structural behaviour of building elements at the joint;
- **Explain** the process of sealing the joint;
- **Describe** the precautions in sealing of joints.

13.1 INTRODUCTION

Joint in a building structure may be defined as a **discontinuity** in the component located in a predetermined position between either similar or dissimilar materials. Joints are essential parts of any building structure to provide desired structural behaviour under various type of forces. These joints are provided at various stages of construction to serve specific purpose. Any restraint to expansion and contraction induce stresses in structural members caused by differential temperatures and moisture conditions. Depending on the material and the length of the element, the expansion/contraction joints are located at certain points to allow expansion or contraction movement freely without inducing severe stresses. The distance between these joints in any building is decided during design stage considering material properties viz. coefficient of expansion, modulus of elasticity, compressive and tensile strengths. This distance between consecutive expansion joints should not exceed 30 metre in masonry and concrete building elements. Any building longer than 30 m will require expansion joint to permit free movement due to temperature changes. The joints are capable of:

- Allowing for shrinkage, contraction and other movements without causing excessive tensile stresses in the components;
- Allowing for expansion and other movements without causing severe compressive stresses in the components;
- Accommodating shear movements;
- Accommodating the allowable deviations from design in dimensions of the component in the building;
- Allowing for temporary interruptions in the progress of construction.

Joints may either be flexible (movable) or rigid type. Apart from accommodating structural movements, the joints are also required to provide water barrier or seal between the two sides. Joint seals generally get deteriorated and require replacement.

These joints, apart from their structural behaviour, are also required to provide appropriate barrier to flow of environmental forces viz. water, moisture and air across the joint. Slide joints are provided at the support in 1/3 rd span points of continuous RCC slabs and beams so as to avoid cracking of supporting masonry structure. Slide joints accommodate RCC roof slab movement due to temperature changes.

In due course of use, the joints get damaged either by weathering effects or collection of debris making them ineffective. Sometimes mortar joints in masonry walls fail due to weathering effect viz. changes in temperature, moisture condition, attack of pollutants and running cracks along masonry joints. These cracks in mortar joints require to be repaired immediately to avoid further damages in walls.

Joints are an inevitable part of any structure. The size of the structure, type of building material and method of construction determine the number and type of joints in a structure. These can be categorized as expansion joints, joints between dissimilar materials, joints between precast elements and floor joints. Type of joints are formed during construction of a building. These joints may stop functioning as envisaged during construction due to damage and deterioration. The basic type of joints are:

- Construction Joint
- Expansion Joint
- Contraction Joint
- Isolation Joint

Construction Joints are provided between concrete lifts, or continuous slab construction in which one part is allowed to harden before the next is placed. This type of joint occurs due to interruption in concreting job.

Expansion Joints are provided to permit expansion and contraction freely without restraints and causing cracks.

Contraction Joint are used to permit shrinkage contraction in concrete slab without causing random cracks.

Isolation joints are provided to prevent random cracking between structural members of different thickness and shapes subjected to different loads. Isolation joint is used between concrete slabs of different shapes or thickness and around columns, footings and peripheries of buildings to isolate variable load bearing sections.

Poor detailing and improper sealing of joints often contribute significantly to the distress in structure. Sealing of joints may appear to be trivial in comparison to constructing and repairing joints. Joint construction requires proper attention of engineers for designing, detailing and specification drafting. Correct sealing applies to new joints as well as repairing of old non-functional joints. Any defect in repairing joints would lead not only to bad joints but also to structural distress.

13.2 SEALING OF JOINTS

Joints are the areas most susceptible to leaks in buildings and structures. To repair and establish the joint seal that lasts its expected life, select the right quality of sealant, prepare the joint properly and install the sealant correctly. Repair treatment for sealing of joints comprises of following necessary steps:

(a) Choosing a suitable sealant
(b) Preparing the Joint properly
(c) Installing sealant at proper temperature
(d) Providing bond breaker tape for shallow joints
(e) Providing backer rods for deep joints
(f) Backer rod placement
(g) Application of the sealant
(h) Field moulded hot applied sealant
(i) Field moulded cold applied sealants

(a) Choosing a suitable sealant

The flexible sealants are able to deform to accommodate joint movements. Rigid materials or those that harden with age or at low temperatures may fail when the joint opens and closes. If the sealant material shrinks after it is installed, tensile stresses develop causing it to spilt or lose bond with the joint face.

The type of sealant available for construction industry have already been discussed in detail in Chapter-6 on Materials for Repair. The physical properties listed for various materials can help users to choose a sealant that will perform well for a given situation of environment and loads.

(b) Preparing the Joint

If the existing sealant has performed well for the given environment then fill the same sealant in small gaps and cover with hard or soft sealants. However, if the failure is extensive, **remove and replace** the sealant completely.

Remove the sealant with hand tools from the groove or plough the joint with suitable mechanical tools on larger projects. If the joint needs to be widened to improve its shape factor, saw cut the sealant reservoir of required size.

Remove loose aggregate and embedded foreign material from the joint. Fill minor edge spalls with a repair mortar (this is especially important when using compression seals).

Sand blast the faces of the grooved joint to remove contaminants and to clean the pores of the concrete. Then remove any remaining dust by air blasting or vacuuming. These steps are important to develop a good bond with field moulded sealant and to provide uniform contact of compression seals. Solvents intended to remove oil contaminants may sometimes carry the contaminants further into the pores of the concrete. Careful **cleaning** with solvents, brushes, or oil free compressed air is important to obtain good **adhesion** to the sides of a joint. Solvent cleaning is required to remove many contaminants from nonporous substrates (such as metals and glass). Porous masonry substrates can be **cleaned by** applying **solvent oil**, free **compressed air**, saw cutting, mechanical scraping, grinding, or sand blasting. Grinding, however, can drive oily contaminants further in to the substrate, while **sand blasting** can **aggressively erode** the surface of joints in porous masonry materials.

In addition to **cleaning** of joint, application of **surface primer** on certain substrates is necessary to assure **good adhesion**. Sealant manufacturers usually recommend specific primers for each sealant and type of substrate. Some sealant material can not be substituted because of **chemical incompatibility**. As a general rule, joint faces must be clean and dry since the sealant has to bond to the concrete surface. Better results are achieved if the sealant is installed in a dry joint.

(c) Installing the sealant at proper temperature

Joints open and close with changing temperatures. As the temperature increases, the concrete expands and the joint closes. As the temperature decreases, the joint opens. If possible, install the sealant when the temperature is close to the average annual temperature. This minimizes the stress due to temperature changes on the sealant during its service life.

Sealant in a rigid control joint should be installed over a **bond breaker tape**, while sealant in an open expansion joint should be installed over a sponge backer rod.

(d) Providing bond breaker tape for shallow joints

The mortar in a rigid control joint must be raked out to the proper depth while it is still plastic. Bond breaker tape prevents adhesion to the back of three sided shallow joints, which can cause splitting failure in the sealant. Polyethylene tape is the most commonly used material, but it can be difficult to handle in cold weather. Butyl tape is a good substitute for easier handling at lower temperatures.

(e) Providing backer rods for deep joints

Deep joints require placement of a backer rod to **control sealant depth** and to provide a **firm surface** against which the sealant can be applied and tooled. Backer rods can be made up of either open cell **polyurethane foam**, or closed cell expanded **polyethylene foams**.

Backer rods should be sufficiently **stiff and immovable** to absorb the pressure applied during tooling of the sealant and large enough to maintain compression during the full cycle of joint expansion and compression. **Closed cell** backer rods should be sized one quarter to one third wider than the joint, so these develop at least 25% compression. Open cell backer rods should be sized atleast **one-half wider** than the joint, so these develop at least 50% compression.

(f) Backer rod placement

It is critical to install backer rod to the correct depth in the joint. If the rod is positioned too deep, high sealant stresses cause adhesive failure. Many technicians judge placement depth visually, but tools designed to control placement can assure proper width to depth ratios for better sealant performance. Tools should be smooth surfaced to avoid puncturing or tearing closed cell rods.

(g) Application of the sealant

Multi component sealants require job site mixing to ensure proper curing and development of physical properties. Both under mixing and over mixing can lead to sealant and joint failure, so manufacturer's instructions should be followed carefully.

The nozzle of the caulking gun should be plastic or a standard metal tip with an extrusion width equal to that of the joint. The angle of the nozzle tip should be kept parallel to the plane of the joint surface to avoid over or under filling the joint. Holding the nozzle tip tightly to the surface also prevents the sealant from squeezing out. The sealant must fill the joint completely.

All sealant joints must be tooled. Tooling should be used to seal the joint by pressing the sealant firmly against the joint sides eliminating voids and air pockets to create a clean surface. Wetting agents such as solvents, spirit solutions, and water are not recommended. Wet tooling can inhibit the curing process if solvent wetting agents are used. It can also cause excessive dirt pickup if solvents or soaps are used. Because wet tooling products frequently are misused and misapplied, most manufacturers suggest the sealant use with dry tooling.

Successful sealant application requires both good workmanship and the right specification of the backer rod or bond breaker tape for the specific job. Joint sealant can be applied either field moulded hot or field moulded cold.

(h) Field moulded hot applied sealant

Each manufacturer has recommended a proper **pour temperature** as well as a safe heating temperature, which should not be exceeded. The safe heating temperature usually is 20 degree above the recommended pour temperature. Subjecting the sealant to temperatures above the safe heating limit results in impairing setting of the compound which may ultimately result in poor field performance.

These materials are usually heated in double boiled type melting kettle equipped with a suitable agitation system in the sealant-melting chamber. The kettle also contain a positive pressure delivery and re-circulation system and a recording thermometer. The inner tank should be oil jacked and the temperature of the heat transfer oil should be thermostatically controlled. The application pipes and nozzles should be insulated.

Hot poured materials are normally suitable for installation in horizontal joints only. They can be placed in vertical joint, but adequate dams are necessary to prevent the sealant from flowing out from the bottom before it cools and sets. Fill horizontal joints slightly below the slab surface.

(i) Field moulded cold applied sealants

The equipment used to install field moulded cold applied sealants is the hand operated caulking gun. The sealant is supplied prepackaged in cartridges to suit the gun. Sometimes the chamber or cartridge is loaded on the job from bulk containers, or in the case of two component materials, they are filled with the compound after mixing.

Mix two component sealants thoroughly to ensure proper curing and uniform properties. Patches of sealant that do not harden due to improper mixing are required to be replaced with properly mixed material..

Small quantities of two component sealants can be mixed manually with a broad bladed putty knife while significant quantity of material requires mechanical mixing. For small batches, hand held electric drills fitted with paddle blades can be used for mixing. For larger projects, more sophisticated equipment is available in which the two components are brought by individual pipe lines to the nozzle where they are mixed in a small chamber before they are extruded.

A skilled operator is needed to apply sealant in a joint reservoir. Hold the gun nozzle at a 45-degree angle and move steadily along the joint to apply a uniform bead without dragging or tearing the sealant or leaving unfilled spaces. In large joints, several runs may be needed, building up the sealant in roughly triangular edges at each run.

For nonsag sealants, tool the material after the joint has been filled. Tooling of sealant ensures intimate contact with the joint faces, removal of any trapped air, consolidation of the sealant material, and provision of seal with uniform appearance.

Two component sealants have a limited working life, especially on hot days. Once the accelerator is mixed in, the curing reaction starts. Therefore, limit the batch size to what can be used within the **pot life** of the sealant.

To provide uniform contact with the joint faces, compression seals require a uniform joint width and straight, clean, smooth joint faces. To ease installation of the seal, apply a neoprene or other lubricant in a bead to the upper edge of each joint face. Apply the lubricant with hand pressure applicator immediately ahead of inserting the seal so that it does not dry out.

Position the seal vertically over the joint to press down and move the seal into the joint by using a hand roller. Do not twist the seal or fold it over on itself. A small amount of stretching (upto 5%) may occur as the seal is forced in. Do not willfully stretch and lengthen the seal to ease installation. Stretching the seal may develop cracks and crevices which may impair the effectiveness of the seal. Install the seal in as long a continuous length as possible. If field splices cannot be avoided, make them in the least critical location.

13.3 SAFETY PRECAUTIONS IN JOINT SEALANTS

There are certain hazards in using joint sealants and must be considered before installation.

- Hot applied materials can cause serious burns or fire hazard if the flammable materials are spilled.

- Excessive breathing of fumes or skin contact with coal tar compounds may cause irritation.

- Cold applied materials (other than emulsion) and primers may contain flammable solvents. Such containers should be kept closed and also away from flames. Working areas must be well ventilated.

- Many elastomeric sealants contain toxic chemicals. Avoid skin, eye or internal contact with such materials. Protective gloves and sometimes aprons are necessary for working with such materials.

- Solvents used in cleanup or during sealant curing may be restricted since these are atmospheric pollutants even though these may be non-hazardous.

- Hazards in joint sealing can be minimized by adopting correct practices.

13.4 SUMMARY

Joint in a building represents a discontinuity and serves specific purpose of free movement without causing excessive stresses. Joints are located at suitable places. Joints are normally spaced at less than 30m to avoid excessive stresses due to contraction or expansion.

Apart from structural behaviour, these joints are also required to provide barrier to flow of moisture, water and gases. Generally seal of the joints get deteriorated with time and needs to be repaired or replaced based on its condition.

Sealing of joints is done after removing the damaged seal and preparing the joint surface properly. A suitable sealant material is selected for the purpose and applied step by step. Joint surface is prepared by sand blasting the groove, cleaning and using bond breaker for unbounded shallow joints and Backer rods for deep joints. Sealant is applied at suitable temperature specified by the manufacturer. Sealants are field moulded hot or cold applied depending on the situations and the material used. Hot poured materials are normally suited in horizontal joints. Field moulded cold applied sealants use hand operated caulking gun. The sealant is often supplied prepackaged in cartridges. Two component materials are mixed just before application in a special mixing equipment. Cartridges should not be kept filled beyond the pot life of the mixed material. Seal should be applied uniformly and without twist or stretch.

Precautions should be observed while using hot applied materials. Proper gloves, masks, aprons and goggles should be used. Flammable materials should not be spilled and containers should be kept with closed lids. Avoid contact with toxic chemicals. Cleaning solvents are atmospheric pollutants and should be used with restriction. Hazards can be avoided by simple precautions and use of correct practices in joint sealing.

QUESTIONS

13.1   Describe the main functions of joints in buildings.

13.2   List different types of joints.

13.3   List steps in repair treatment for sealing the joints.

13.4   Explain the preparation of joints for repair sealing.

13.5   Write short notes on:

(a) Temperature for sealant installation

(b) Backer rod for deep joint sealing.

13.6   Explain field moulded hot applied sealant technique.

13.7   Explain field moulded cold applied sealant technique.

13.8   Describe briefly precautions necessary in joint sealing.

Chapter 14

REPAIR OF WATER SUPPLY & SANITARY SYSTEMS IN BUILDINGS

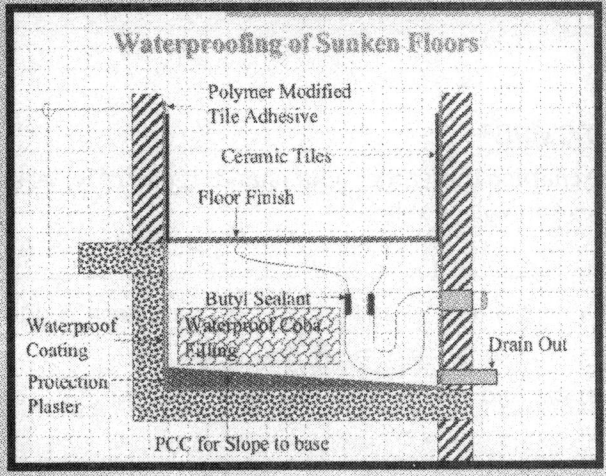

Waterproofing of Sunken Floors

Polymer Modified Tile Adhesive

Ceramic Tiles

Floor Finish

Butyl Sealant

Waterproof Coba Filling

Waterproof Coating

Protection Plaster

Drain Out

PCC for Slope to base

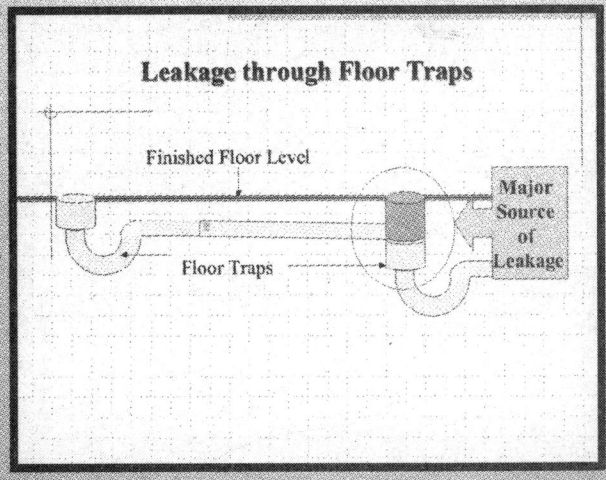

Leakage through Floor Traps

Finished Floor Level

Major Source of Leakage

Floor Traps

14

REPAIR OF WATER SUPPLY AND SANITARY SYSTEMS IN BUILDINGS

LEARNING OBJECTIVES

After studying this chapter, the learner understands repair of water supply and sanitary systems in buildings and will be able to:

- **Explain** repair consideration in water supply fixtures in buildings;
- **Explain** repair of sanitary fixtures in buildings;
- **Describe** repair and maintenance of underground and overhead tanks in Building.

14.1 INTRODUCTION

The maintenance of Public Health Services in buildings requires thorough knowledge about the design, specification and the material used for providing effective services. Deteriorations in buildings mostly occur due to ill functioning of the Public Health Services. The faults in these services occur due to defective system and misuse. It is necessary to carry out routine maintenance of various services to avoid failures of the system. The common problems in water supply and sanitary system are as follows:

- Low Pressure
- Leakage in G.I Pipes
- Overflowing Cisterns
- Blocked Waste Pipes
- Blocked Drains and leakage
- Damaged or cracked China Ware.

14.1.1 Low Pressure

Low pressure is the result of blockage in the ferrule, meter, main stopcock or wrong fitting of stopcock and meter, using undersized main pipe. The ferrule and the meter should be cleaned if blocked. The stopcock can also be checked and repaired. Wrong fitting of the stopcock and the meter can also be set right.

Internal leakage can be checked by closing all the taps and observing the movement of the meter needle. If it is moving then it will clearly indicate internal leakage in the line, which can then be located and set right by replacing defective section. If it is in walls or roofs, it will dampen the wall or the roof. First check the pipe line under ground for leakage and then check rest of the pipe under the floors and locate correctly. Locating this defect and setting it right will also save the structure from damage, as the leakage of pipes under the floor will result in seepage of water directly to the foundations. If the pressure is less because of undersized main then it has to be replaced completely to increase the pressure.

14.1.2 Leakage in G.I Pipes

If the dampness occurs all the time then it indicates the leakage in some portion of G.I Pipe. The point where the dampness is maximum indicates the nearness to the location of the damaged pipe. If the dampness is in a wall, it may descend towards the point of most prominent leakage. Dampness in the roof is observed either in the form of falling drops or a continuous stream depending upon the extent of damage and the water pressure at the point of leakage. Such defects can be avoided by proper leakage testing of pipelines before embedding them in the wall or the slab. The determination of the fault can be simplified by fixing the pipeline on the surface of the wall. The dampness near C.I fitting normally occurs on roof due to leakage through C.I fitting. It is also normally below the water closet or the floor trap. These defects occur due to seepage of water through the gap between the floor and the 'P' trap or the floor trap. This leakage can be repaired and plugged by applying a suitable sealant in the gap. This type of leakage is most common in majority of buildings due to in proper sealing of gap between floor and the trap at the time of original construction.

14.1.3 Overflowing Cisterns

Overflow from the cistern or the overhead tank occurs due to a damaged ball cock or due to blockage of the ball valve itself. The ball may develop a crack or leak, which results in filling of the ball with water and thus resulting in ill functioning. The ball may also get disconnected from the copper rod due to the damage of its holding slot. This requires replacement of the ball or the whole ball assembly. In case the ball is okay and the fault is in the ball cock then the ball cock is set right.

Fault can be due to the presence of some grit in the valve, damaged washer or erosion of the seat. The grit can be removed, the washer can be replaced and if necessary the defective seating can also be replaced. Malfunctioning of these float walls result in loss of water and cause severe dampness in walls and roof slab and thus weakening the building structure.

14.1.4 Blocked Waste Pipes

Blocked waste pipe results in the delayed/slow emptying of the fitting to which the waste pipe is connected. Normally cleaning eyes/plugs are provided in the pipe for cleaning and the same should be opened to remove foreign matter. To decrease the chances of the occurrence of this defect, a strainer should be fixed at the entry point of waste pipe. Flow should be checked in the waste pipe at regular intervals.

14.1.5 Blocked Drains and Leakage

Blocked drains are evident by the leakage of water from the inspection chamber. These blockages are removed with the help of cleaning rods by sanitary workers.

14.1.6 Damaged or Cracked China Ware

Damaged chinaware fittings give a very shabby look in addition to the leakage problem. Chinaware fittings can only be replaced if these are cracked. Washbasins are mostly fixed on iron clamps embedded in wall upto very small depth. A slight displacement can break them. The support angles should invariably be embedded adequately (200-250mm) in the wall so as to make them more stable. Whenever supporting fixtures are loose, these should be immediately repaired and reset.

14.2 MAINTENANCE OF PIPES

For repair jobs, a plumber should carry a repair kit. Repair kit should contain rubber paddings, clamps, spanner pipe wrench, sockets, nipples, sealing paste, sockets, stoppers, unions, etc. The pipe line requiring replacement of any section must have union for easy and quick replacement. Union has reverse threads in both sides. After cutting the defective or damaged section, the pipe can be unscrewed easily. Union joint facilitates proper tightening of replaced pipe section. While designing water supply lines, provision of unions facilitate repair and replacement.

In case of CI sanitary pipeline, the joints are filled with lead to close the gap in collar and it is difficult to insert a new pipe in some section. Collar can be utilized in forming lead joint on both sides of collar. However, collar does not give a proper joint and as such another method to insert a new pipe is to melt the existing lead joints both from above and below the pipe piece or cut the defective pipe. Where a new fitment is needed, then a long socket joint pipe is arranged which has double the ordinary depth, then a fitment is placed into the position and all joints are caulked with lead.

14.3 REPAIR OF TAPS

Most of the problems which relate to taps are on account of heavy use and the abuse to which taps are subjected. When a tap does not completely close to stop water, there are three possible causes:

- A defective handle not capable of pressing washer properly against the seat.
- The deteriorated or broken washer not capable to block the opening properly.
- Seat is worn out or broken.

A steady drip which comes out drop by drop can waste as much as 9000 litres of water every three months. A stream of less than 0.9 mm thick can waste as much as 1,40,000 literes every three months while 3mm thick stream can waste upto 15 lakh litres every 3 months. These losses of water from leaking taps indicate the gravity of the problem and loss of essential resource.

A good quality tap has a life of about 10 years and in most of cases it is the washer which deteriorates and its replacement costs very little. When the tap stem becomes defective, the handle should be removed and threaded properly. The stem should be checked again for proper fitting. Stem is replaced if worn out beyond repair. When washer is to be replaced, it is enough to take out the upper part of the tap alongwith the washer spindle assembly. Fix new washer to the spindle assembly with screw.

When seat of tap is worn out, it is necessary to replace the tap, as resetting is very difficult and costly too.

A good tap should normally run for one-lakh operations of turning on and turning off. Many times the tap is noisy when water is running. There are four possible causes (a) the washer is loose (b) the packing has deteriorated (c) the stem handle is worn out and (d) seat base is worn out. Check and set right the defective item.

Many times top handle is defective and the handle rotates freely without changing the rate of water flow. Replace the defective handle.

14.4 REPAIRING WC CISTERNS

Generally cistern parts get deteriorated and requires frequent repair and setting. Parts that usually require repair are flush valve, the intake valve and float. The problem is either of water flowing continuously or water is not flowing at all. Continuously flowing water occurs on account of **improperly adjusted ball valve** or punctured float ball.

A ball valve is expected to shut off water intake valve when water level reach to the normal level mark just below overflow out let in the tank. Ball cock should be adjusted by bending the rod connected to the valve. Where the ball of the float valve has developed a leak, the float ball should be replaced or the puncture of the ball be sealed after removing leak water.

Some times WC do not flush which indicates that bell type or syphon type arrangement is out of alignment. The bell which induces the sypon, jumps out of seat which requires resetting and placing back in position. When syphonic type flushing cistern is either broken or worn out severely, the cistern should be replaced.

14.5 CLEANING CLOGGED DRAINS

Drains may get clogged by large size objects dropped in to them or by accumulation of grease, dirt or other matter. If the obstruction is in a trap, the trap cover can be removed and cleared. If the obstruction is elsewhere in the pipe, cleaning can be done by using extendable rods attached with flushing rubber cups in front. This rod should be moved to and fro in the clogged drains and flushed with water. If the clogging is more hard and difficult, use augers with steel cable.

Clean out augers with long, flexible, steel cables commonly called, snakes. These 'snakes' may be run into drain partly to break the obstruction and to hook on to pull out the obstructing objects. Augers are available in various lengths and diameters and should be selected according to the size of the drain.

Small obstructions can sometimes be forced down or drawn up by use of an ordinary rubber cup force. Flushing of grease and soap with hot water may be carried out for 10 minutes. Sand and dirt sometimes clog the floor trap and drains. Remove the strainer and take out much of the sediment around the floor trap manually and flush the drain with clear water. If more pressure is needed, use a garden hose to clean the clogging. Push and fix warp clothes around the hose into the drain to prevent flush back flow of water.

Garden hose, augers, rubber force cups and other tools used in direct contact with sewage are subjected to contamination and should be handled carefully with protective gloves and boots.

14.6 MAINTENANCE OF TRAPS

The maintenance problems of traps arise due to defective design, inappropriate seal, and defective installations.

Design defect of traps may be caused due to construction partition or due to loose cover or improper type of trap. The proper trap has a 'U' shaped pipe. It consumes more material during casting process and decreases depth of seal. However, it creates new pockets or voids in the trap, which do not get cleaned easily with the flow of water. Thus, this type of trap is neither self cleaning nor satisfying the basic requirements of absence of cavities and internal projections in traps.

The top cover of trap is to be screwed on or made an integral part of the trap. Loose covers on the trap will promote a tendency to put lot of undesirable material inside the trap. Trap has wider mouth inlet of more than 100 mm, which should be brought down to 75 mm at the outlet. Additional connecting pipe piece between the trap and the main pipe need to be cleaned frequently. It is therefore necessary to screw the cover instead of fixing.

Water Seal of a trap is the minimum vertical distance between the weir crest of the trap and the minimum normal inside water level. In order to reduce the cost of the trap, water seal is reduced but this may result in self syphonage. Water seal should be 38 mm in two-pipe system and 40 mm in single stack system. Inadequate water seal create leakage problems.

To check the seal, trap is laid with its month horizontally. Then water is poured till it comes to the level of first bend (A). Depth from the bottom of the trap to this water (level A) is measured. Water is then continued to be poured till it just flows out from the outlet (B level). Once water just flows out measurement is again taken and the difference between A level and B level is the water seal.

Defective Installation is the main cause of leakage. The defective floor installations can be summarised as:

- Installation where trap is seated too deep from the floor level;
- Installation where one or more pipe enters into floor trap;
- Installation where trap is overflooded.

The level of trap is determined by the level of connection of the horizontal piping. The present practice is to locate the trap below the floor and seat it into a mass of concrete. Seating of trap on concrete alone may not stop leakage. Therefore, a watertight lead joint should be provided upto the floor level in CI trap.

When one or more pipe empties into the floor trap and sometimes one of the horizontal pipe from other fitments may not reach the trap at all due to short length. This should be avoided to avoid leakage. Some times due to too much inflow to the trap and small outlet, it may be overflowing. This should be avoided by designing proper sizes of outlet and inlet.

14.7 DEFECTIVE DOWN TAKE DRAIN PIPES

Down take drain pipes should be nonperforated and nonleaking. Metal pipes are favoured as these are nonperforated and nonleaking. Mass of spongy material in the vicinity of the leaks may be caused due to bimetallic corrosion. Remedial measures will consist of painting of the inside of gutter and downpipes with bitumen paint. Otherwise these pipes may have to be replaced.

In case of plastic pipes, bowing during hot season is a major problem. Generally it is reversible. However, if joints starts leaking, these need to be repaired by using appropriate leak proof sealing adhesive. Bowing of plastic pipes can be straightened at average temperatures and joints checked for leak proofing.

14.8 MAINTENANCE SURVEY

Even the best building ever constructed will get ruined in due course of time if not properly maintained. A delay for sometime is tolerable in case of buildings, but in case of services, it can lead to bigger hazards, including to danger to building life. Therefore regular and periodic surveys are necessary for maintaining building serviceability. Proper inspection is time consuming job. Proper way to start inspection is to obtain drawing according to which services have been laid. All service lines should be shown by different colours for systematic checking and inspection. Site for inspection must be prepared for testing of various services. Prepare a check list for specific inspection and necessary test schedule. The checklist should be comprehensive, systematic, and simple to follow. All occupants must be informed of the inspection and report making for the purpose of maintenance of services. The site must be available and accessible for inspection and testing. A comprehensive report should be prepared after inspection, testing and interviews with the users of services.

14.9 REPAIR AND MAINTENANCE OF OVERHEAD AND UNDERGROUND WATER TANKS IN BUILDINGS

14.9.1 Introduction

Overhead tanks are provided in buildings to give uninterrupted water supply to the occupants. The size of tank depends upon the number of occupants, per capita water requirement and number of water closets it has to serve. The tanks may be constructed from different materials and in different shapes. The most common shapes are rectangular and circular. The outlets from the tanks should always be provided with antisyphonage pipe to avoid air locking. The most common materials for the construction of simple over head or under ground tanks in buildings are:

- Masonry,
- R.C.C,
- Mild steel, and
- Moulded plastic.

Both OH and UG tanks may suffer from following problems :

- Seepage in bottom/side walls
- Cracks in brick masonry or R.C.C work
- Rusting in concrete reinforcement
- Appearance of dampness on tank walls
- Corrosion of MS tanks
- Uncontrolled overflow
- Cracks or puncture in plastic tanks

14.9.2 Remedial Measures in RCC/Masonry Tanks

R.C.C and masonry tanks even with the hairline cracks require immediate attention. Whenever cracks appear on face not in contact with water, it should be cleaned with wire brush. Place 25 mm thick layer of ferrocement lining all round with rich cement mortar after sealing the cracks with epoxy or cementitious grouts.

Paint two coats of solvent free epoxy resin after preparing the inside surface of the tank. Care has to be taken to ensure that the material is not reactive to potable water.

For repairing **wide cracks**, chip off the loose concrete. The surface of the crack is cleaned with dilute hydraulic acid and rusting is cleaned with sandblasting. Then provide two layers of guniting with steel welded mesh (1.20mm gauge) having a thickness of not more than 30mm in case of face not in contact with water. In case of wide cracks in brickwork, the surface should be carefully prepared. Then the stitching with reinforcing bars is done by putting the bar across the crack as discussed earlier under repair of cracks. The filling is done by non-shrinkable cement mortar.

In case the tank require strengthening it should be provided with reinforcement or welded wire mesh along the inside/outside of walls. All the treated surface should be gunited with water proof cement mortar.

14.9.3 Galvanised Cold Water Tanks

Galvanised tanks may have leakage either from the bottom or the sides with signs of rust. The point of leakage is usually quite obvious from the external examination of the tank. The visual examination will confirm this by the presence of rust. Not all the rusted areas will leak but the number of rust stains should be noted to control the remedial measures and treatment.

Sometimes water may leak out in the form of a very fine jet, difficult to see, though its existence may be found by moving the hand in space near the damp place. The inside of the leak hole in tank should be inspected for signs of iron filings after emptying. These metal filings may be around the leak or have fallen from the tank wall. Such defect occurs because of acids or soft waters which may eat away zinc coating. The attack may be general or may be at isolated places. Pieces of foreign metal, such as iron filings, metal tools may cause localised corrosion. Bimetallic corrosion may be caused if dissimilar metals are used in the fittings connected to the tank, especially those which are in contact with water. Ready made sealing materials are also available for repair of minor leaks.

Remedial Work much depends upon the extent of the rusting and particularly the number of rusted spots which are not actually leaking. A single leak may be patched and the whole interior of the tank then painted with a suitable nontoxic bitumen paint. Systems have been developed for lining a tank with a plastic sheeting in the form of a bag and these can often provide a remedy for badly corroded tanks needing replacement. These plastic lining adheres to steel tank walls without leaving any air pocket between the two. These linings should be attached using nontoxic epoxies and thermo setting plastics.

14.9.4 Plastic Tank

Cracks, holes or damages can be treated using nontoxic thermo setting plastics to seal these openings. If these defects are due to structural failures then inadequacy of structural design should first be set right by applying steel rings or vertical stiffeners to strengthen. Generally, small cracks and punctures are repaired by sealing with nontoxic epoxy injection and plastic paints. Plastic tanks are generally circular in shape and are strengthened using plastic mouldings of box ring or vertical sections. Simple rectangular sections may be used for additional vertical or ring stiffeners.

14.10 SUMMARY

The problems in Public Health Services are due to low pressure, leakage in pipes, tanks ill functioning of waste pipes, blocked drains and damaged chinaware. G.I. and Sanitary pipes should be repaired or replaced whenever there are heavy leakages. In case of taps, the problem is due to defective handle or broken washer or wornout seat. W.C. can be repaired by removing the problem in flush valve, intake valve or float ball. Clean out augers, rubber cups, cleaning acids and other tools are used to remove blockages. The maintenance problem in traps arise out of defective design, inadequate water seal and defective installation. Locating leaks and systematic surveys should be carried out for maintenance of Public Health Services. Problem in underground and overhead tanks are due to cracks, overflow and currosion. Repair should be carried out using nontoxic epoxy resins, ferrocement and other waterproofing materials. Badly corroded MS tanks need to be replaced or repaired thoroughly with plastic lining and other coatings.

QUESTIONS

14.1 List four most common problems in Public Health Services from maintenance consideration.

14.2 Describe repair of leaking G.I. pipe.

14.3 Describe the possible causes of defects in water taps alongwith their remedial measures.

14.4 Describe common problems in W.C. systems and explain how these can be removed.

14.5 Explain the use of various tools for removing blockage in drains.

14.7 Explain the repair of leaking plastic overhead tanks.

14.6 Describe the common problems of overhead and underground tanks.

14.8 Explain the remedial measures for hairline cracks on tankwall faces.

14.9 Describe the main problems in G.I. tank and their remedies.

14.10 Explain, in not more than 100 words, the importance of maintenance of services in buildings.

Chapter 15

COMMON STRENGTHENING TECHNIQUES

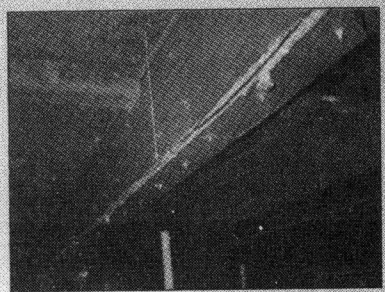

15

COMMON STRENGTHENING TECHNIQUES

LEARNING OBJECTIVES

After studying this chapter, the learner understands common strengthening techniques for buildings and will be able to:

- **Describe** importance of strengthening of building elements
- **List** various strengthening techniques
- **Describe** strengthening by providing additional interior reinforcement
- **Explain** the supplymentary strengthening techniques for R.C.C. elements in buildings
- **Explain** underpinning of foundation of distressed building structure.

15.1 INTRODUCTION

Structural members are strengthened to increase their load carrying capacity, stability and stiffness. All structural restorations are inherently affected by field conditions. The techniques for restoring strength must be workable with existing member sizes, reinforcement spacing and other typical site details. However, the decision on feasibility and suitability of structure to be strengthened should be considered in each particular case, depending on the service requirements and the actual field condition of the structure. The cost of strengthening should be compared with the remaining life after strengthening and cost of an entirely new structure assessed for total replacement.

Thus, it is imperative to perform structural analysis prior to repair of structural members to determine if the members are overloaded for the service loads. The analysis should consider both serviceability and strength. The analysis should include consideration of the causes of the structural degradation. The cause for existing distress should always be eliminated before repairing and strengthening.

Rehabilitation systems can not be classified into a cook book type of design procedures. The engineer should have the imagination to select and adopt the best of the several available strengthening options to remove the existing defects and weaknesses.

There are five basic approaches of structural strengthening that can be considered without completely demolishing and rebuilding the member or structure as a whole. These strengthening technique are:

- Internal strength restoration by adding interior reinforcement.
- Strengthening by adding exterior reinforcement(encased or exposed.)
- Strengthening by external post-tensioning
- Strengthening by use of jackets, brackets and Collars
- Strengthening by providing supplementary members

Each method is well suited for a particular set of field conditions. The subsequent sections describe construction techniques used for strengthening such as placement of reinforcement within existing concrete or placement of new additional reinforcement exterior to the existing member. In all cases, the objective is to enhance the strength or stiffness of the member to resist tension or compression caused by flexure, shear, torsion, and axial forces so that the strengthened element/structure meets the minimum requirements. The strength and serviceability is checked according to various type of building codes to suit the type of occupation.

15.2 ADDITIONAL INTERIOR REINFORCEMENT

Description

A common method of providing additional reinforcement across cracked surfaces is to install new dowels into holes drilled perpendicular to the crack surfaces (as shown in Fig 15.1). The entire length of the dowel is fixed to the concrete by using a bonding matrix. The structure must be shored and jacked if it is desired to relieve the member from the dead load stresses so that the new reinforcement will share the original dead load after

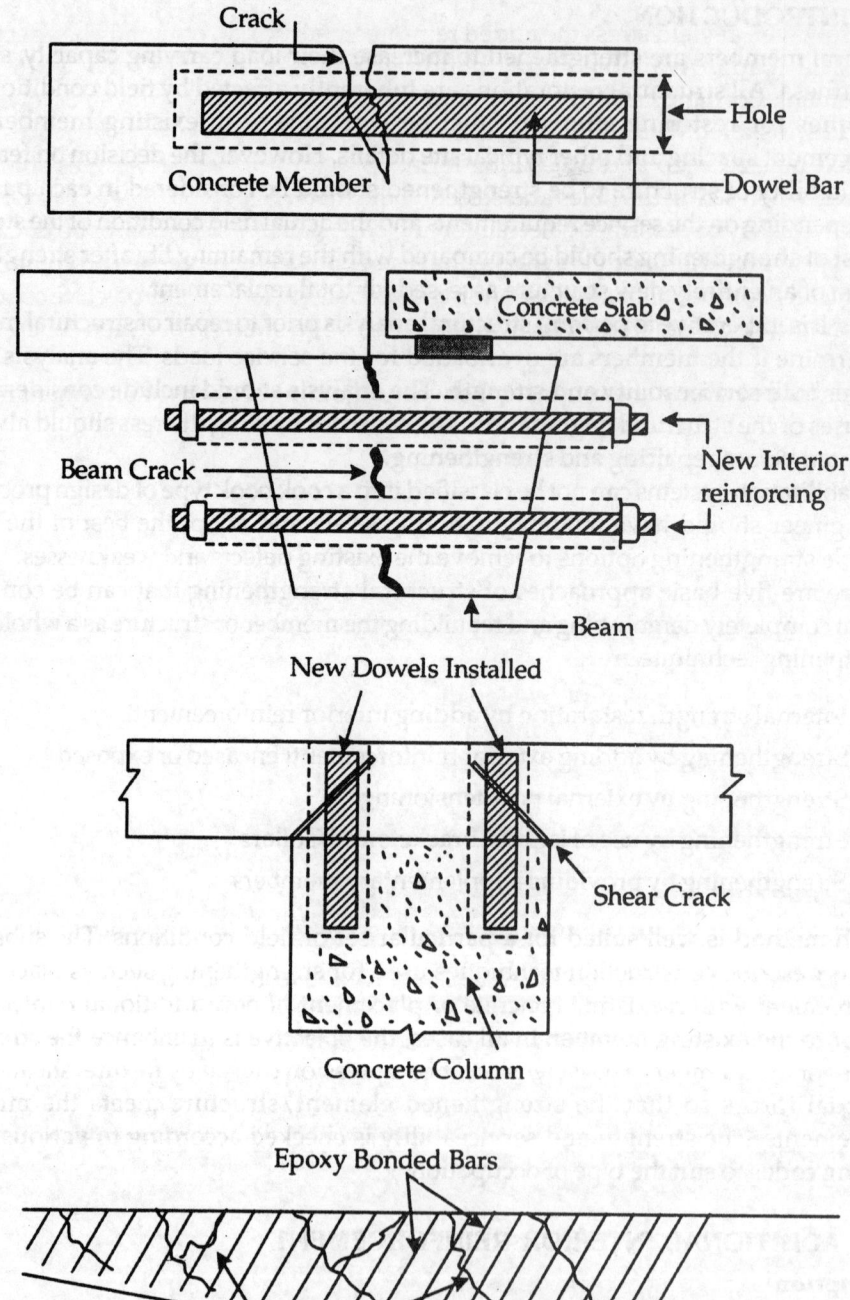

Fig. 15.1: Interior Reinforcing in Different Elements

repair. However, at yield stress the added reinforcement would optimally be effective in sharing full loads. Available epoxy bonding materials may be used with added reinforcement. Portland cement grouts, epoxy, and other chemical adhesives have been successfully installed within the annular space between the dowel bars and sides of the pre-drilled holes. Creep and other long-term behaviour of such adhesives should be considered when sizing the holes and selecting the materials.

The dowels may be deformed steel reinforcing bars, stainless steel rods, or bolts. Coating steel dowels with either **zinc galvanizing** or **fused epoxy** is acceptable if all components are chemically compatible with the bonding material. The effect of protective coating of dowels should be considered when evaluating the bond strength between concrete and dowel bars.

Dowels for providing shear transfer between adjoining sections of pavement may be placed in slots cut from the top to mid-depth of the adjoining sections. In one section dowel is **bonded** while in the other, the dowel is **un-bonded** using a sleeve or debonding agent.

Installation Details

The holes should be drilled approximately perpendicular to the crack. A core bit or a solid drill bit may be used, but core bits create a smooth surface inside the hole. The smooth surface is less effective for bonding to the concrete. The embodiment length on both sides of the crack must be sufficient to develop the required stress in the bar by bond strength. Epoxy bond length of 10 to 15 times the bar diameter is usually sufficient, but development lengths should be based on calculated design loads and bond stresses or on actual test results of mortars. The minimum development lengths given by the code are necessary. The final hole diameter should be 3 to 6 mm larger than the dowel diameter if epoxy bonding agents are used. The hole diameter should be at least 50 mm larger than the bar diameter for cement mortar grout. This provides an annular space that is adequate to allow for compaction of the mortars. However, epoxy viscosity is critical in selecting hole diameter for bar installation. Follow manufacturer's instructions and conduct trial installations in order to determine the best hole diameter and depth. A vacuum and wire brush will clean the inside of drill holes sufficiently. Carbide tipped drills with internal vacuum ports are available to save time blowing out the hole after drilling. Install sufficient bonding agent into the hole to fill the annular space, or approximately half full if epoxy is used. Insert the dowel immediately into the hole and displace the bonding agent. The viscosity of the bonding agent should be fluid enough to permit the agent to flow between the dowel and the hole.

Special Features

Internal reinforcement can strengthen concrete element cracked by flexural and shear stresses due to restrained volumetric changes. The procedure is simple and uses commonly available equipment. Epoxy injection is commonly used to fill all cracks after installation of the bolts and their adhesive, but before tensioning bolts.

Avoid cutting or puncturing internal reinforcing bars or conduits during the drilling operation. Non-destructive testing and design drawings can be used to determine the locations of embedded items. Heavily reinforced structural members may not permit drilling and such members should be strengthened by **external techniques**.

Sometimes constraints from the outside of the member may not permit drilling holes transverse to the crack. Do not install internal dowels in deteriorated concrete if the bond strength cannot be developed. The concrete strength must be evaluated for each installation. Clean the drill holes of concrete dust prior to the installation of reinforcement and bonding agent. If the hole is not thoroughly cleaned, the bonding matrix will adhere to dust and only a limited bond strength of the concrete will result.

15.3 EXTERIOR REINFORCING (ENCASED OR EXPOSED)

Description

Steel elements may be placed on the exterior of an existing concrete member (as shown in Fig. 15.2). The new reinforcement may be encased with Shotcrete, mortar or other material, or it may be left exposed and protected from corrosion with a protective coating. The reinforcement may be deformed bars, welded wire fabric, steel plate, steel rolled sections, steel strapping and specially fabricated brackets. For members damaged by overload, erosion, abrasion, or chemical attack, the deteriorated or cracked concrete must first be removed, and new reinforcement installed around and adjacent to the remaining concrete. Rusting of existing reinforcement must also be removed before treatment and placing additional rebars.

The new reinforcement is encased in conventionally placed concrete or in Shotcrete. Where the existing concrete is in good condition, the new reinforcement may be bonded directly to the existing concrete surface after preparing the surface and enclosing with stirrups especially in case of beams and columns.

Epoxy and other chemical adhesives as well as portland cement concrete may be used to bond the new reinforcement or it may be mechanically fastened to the existing reinforcement by welding.

Special Features

Placement of exterior reinforcement may be the most convenient method for repair and strengthening where obstructions limit access of equipment needed for placement of interior reinforcement. If surface repair and placement of epoxy mortar, plaster or shotcrete is necessary for concrete rehabilitation, placement of external reinforcement for strengthening may be accomplished in the same construction process.

External flexural, shear and torsion reinforcement for beams and girders may be provided by bonding deformed bars or plates to the surface of concrete girders with shotcrete, cast-in place concrete, or epoxy and polymer concrete. Anchors may be required in the repair method to ensure composite action.

Steel plates may be attached to existing girders using bolts. Structural adhesive requires adequate surface preparation for both the steel and the concrete. Select the appropriate adhesive to bond steel to concrete properly. Sandblasting of both the steel and concrete surface is the best method of surface preparation, while surface cleaning with solvents or high pressure water blasting is adequate.

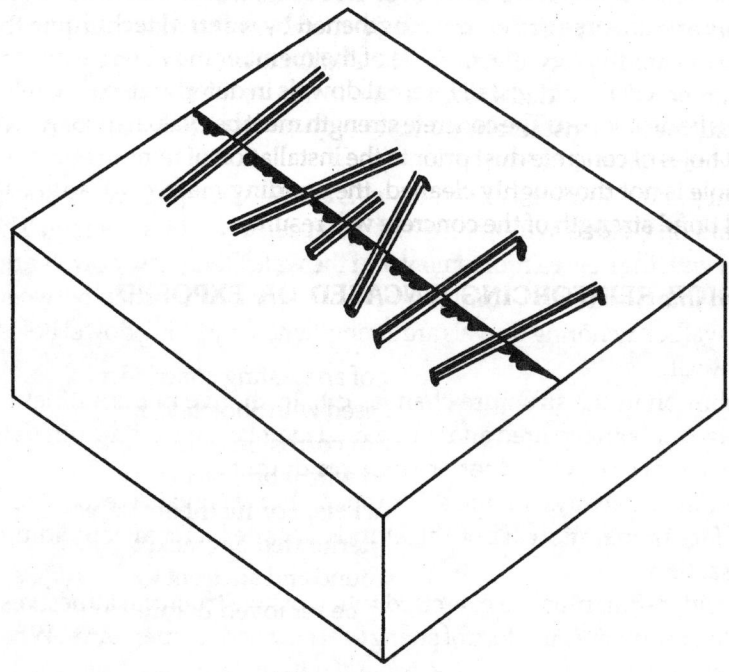

External Steel Plate

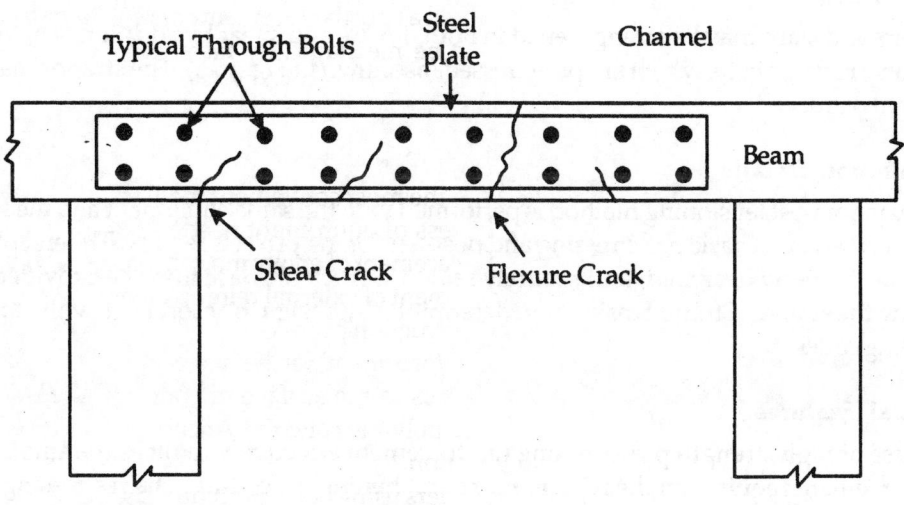

Steel Strapping

Fig. 15.2: Exterior Reinforcing

Beams, girders, columns, and walls can be strengthened by placement of longitudinal reinforcing bars and stirrups or ties around the members and then encasing the members with shotcrete or cast-in place concrete. The shotcrete bonds the new reinforcement to the existing members. The added shotcrete also increases the size of the member and adds strength and stiffness. Consideration must be given to additional dead weight of the member resulting from such additions.

Both masonry and reinforced concrete walls can be strengthened by adding external reinforcing bars or welded wire fabric and by application of shotcrete. The shotcrete develops bond with the new reinforcement and the wall. Often new dowels are embedded in holes drilled into perimeter columns and beams for a connection between the frame members and walls. Anchoring dowels are strengthened, by filling dowel holes along with shotcreting of wall.

Load distribution in the structure changes due to change in the stiffness of repaired members. Design for both repaired and unrepaired members including foundations should be checked for actual or modified service load conditions.

External reinforcing always occupies extra space that was available for other uses before the repair. Surface preparation of both steel and concrete is critical if bonding is required for composite action.

Careful consideration must be exercised when using structural adhesives, especially epoxy, due to their softening and loss of strength at elevated temperatures. Where required, appropriate fire protection must also be provided to such repaired elements.

15.4 EXTERIOR POST TENSIONING

Description

Girders and slabs may be strengthened in both flexure and shear by addition of external tendons, rods, or bolts which are pre-stressed (as shown in Fig. 15.3). The strands may be straight or curved.

Installation Details

The exterior post tensioning method is performed with the same equipment and the same design criteria that basic pre-stressing and post-tensioning projects require. The end plates must be properly designed, and securely seated. Sufficient space must be provided for pulling the strands. Strand tensions are determined by design in accordance with various building codes.

Special Features

The use of high strength pre-stressing reinforcement effectively reduces the amount of reinforcement required in the repair and strengthening procedure. The restressing also provides beam camber which helps to reduce deflection. The addition of a single post tensioning arrangement increases both shear and flexural strengths. The shear resistance is increased due to concrete compression according to the profile of the strands. The flexural strength is increased due to the addition of longitudinal reinforcement with pretension.

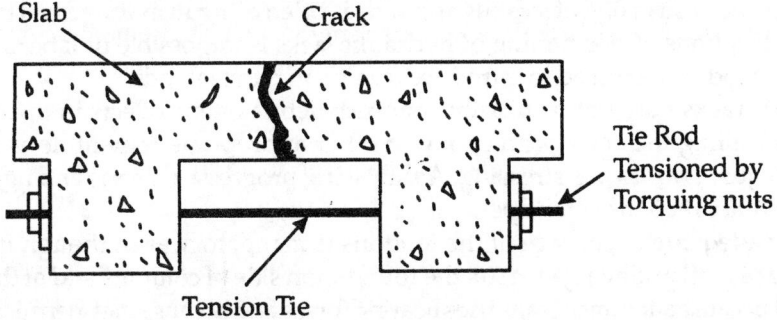

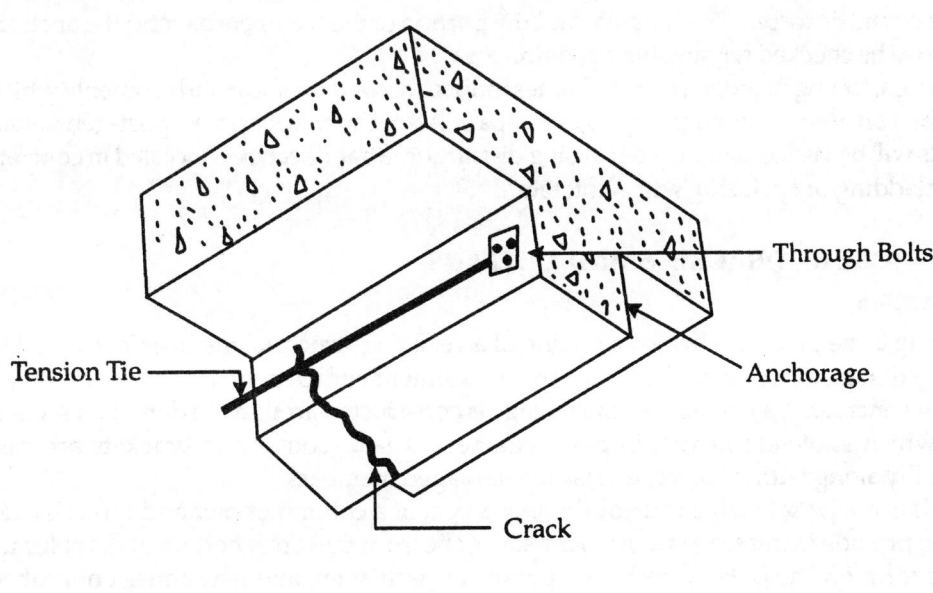

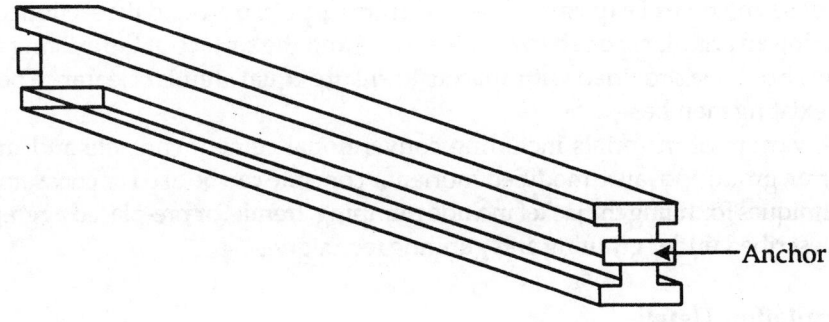

Fig. 15.3: External Prestressing

Placement and restressing of strands and rods is often difficult in congested buildings and interior locations. If positioning of hydraulic jacks is impossible or laborious, turn buckles or nut rods can be used to introduce tension in the steel rods.

Closing of cracks may not be possible because debris often collects in open cracks. Cleaning and filling the crack with epoxy or other appropriate cementitious material must be done just prior to pre-stressing. As stressing progresses, debris and aggregates already in contact keep the crack open.

Providing adequate anchorage of the tendons is mandatory even though it may be difficult. Use of steel grillage system on the foundation side of columns and at the end of beams is best because such anchorage uses bearing forces and because they permit stressing of the entire length of a member including end connections. Bolted or bonded side anchors may be used as an alternate anchorage. The steel and concrete in anchors must be designed for the eccentric forces. The concrete and the portion of the members beyond the anchors must also be checked for suitable strength.

Post tensioning in indeterminate frames induces secondary shear and moments which must be considered when planning the repair. Also, the new exposed post- tensioned strands will be visible, and may be causing distraction. Strands can be concealed in concrete or by cladding or enclosing with shotcrete.

15.5 JACKETS, BRACKETS AND COLLARS

Description

Jacketing is the process whereby a section of an existing structural member is restored to original dimensions or increased in size by encasement with cement or polymer modified cement concrete. A steel reinforcement cage is constructed around the damaged section onto which shotcrete or cast-in-place concrete is laid. Sometimes brackets are cast externally along with jackets to encase the damaged members.

Collars are jackets which surround only a part of a column or pier and typically are used to provide increased support to the slab or beam at the top or bottom of the column.

The form for the jacket may be temporary or permanent and may consist of timber, corrugated metal, precast concrete, rubber, fiberglass, or special fabric, depending on the purpose and type of exposure. The jacket form is placed around the section to be repaired, creating an annular space between the jacket and the surface of the existing member. The form should be provided with spacers to ensure equal annular clearance between it and the existing member.

A variety of materials including conventional cement concrete and mortar, epoxy mortar, grout, and latex-modified mortar or concrete can be used as encasement material. Techniques for filling the jacket include pumping, tremie, or pre-placed aggregate concrete as described under grouting and guniting techniques.

Installation Details

The form-work techniques and the decision to use permanent or temporary forms are important details in jacketing. Timber or cardboard forms may be used as temporary or permanent forms. Corrugated steel forms are easy to assemble and are adequate as temporary or permanent forms. Permanent fiberglass, rubber and fabric forms have gained wide acceptance because they also provide resistance to chemical attack during repair and even after the repair is complete.

Special Features

Jacketing is particularly applicable in repair of deteriorated columns, piers, and piling where all or a portion of the section to be repaired is under water. The method is applicable for protecting concrete, steel, and timber sections against further deterioration as well as for strengthening. Permanent forms are advantageous in marine environments where added protection from weathering, abrasion, and chemical pollution is desired. Collars are effective in providing new capitals on existing columns for supporting slab floors. The collar provides increased shear capacity for the slab, and it decreases the effective length of the column. Collars may help to satisfy architectural constraints better than jacketing the column for its full height.

Jackets, brackets and collars require that all deteriorated concrete be removed first. Clean the existing reinforcement, prepare the surface and repair the cracks. This preparation is required so that the newly placed materials bond well with the existing structure to act as monolithic structure. Because jackets are often under water, such preparation is expensive and difficult. Nevertheless the applicability of jackets and collar is widespread and is generally cost-effective, especially if the alternative is replacement of the deteriorated member.

Jackets, bracket and collars may occupy space that was available for other uses before repair. This approach was used by the author in strengthening of overhead service reservoir supporting shaft and foundation. Collar sleeve rings and vertical stiffeners of concrete were used. Epoxy adhesives and welding of reinforcement was used for monolithic action of the shaft.

15.6 SUPPLEMENTARY MEMBERS

Description

Supplementary members (as shown in Fig. 15.4) are new columns, beams, or braces which are installed to support and assist damaged structural members.

Installation Details

The new supplementary members may be timber, steel, concrete, or masonry. The new member must be correctly constructed, wedged, or anchored in position so that the loads are partly or fully transferred to the supplementary (new) members without changing deformation stress pattern.

Special Features

This repair method can be used if none of the other strengthening techniques is adequate for repair or if the structural configuration precludes use of other techniques. Supplementary members are quickly installed and therefore are suitable as temporary emergency repair solutions. Typically new members are installed to support seriously

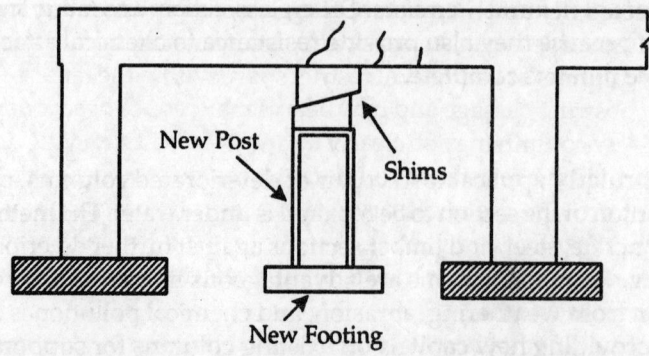

New Post

Shims

New Footing

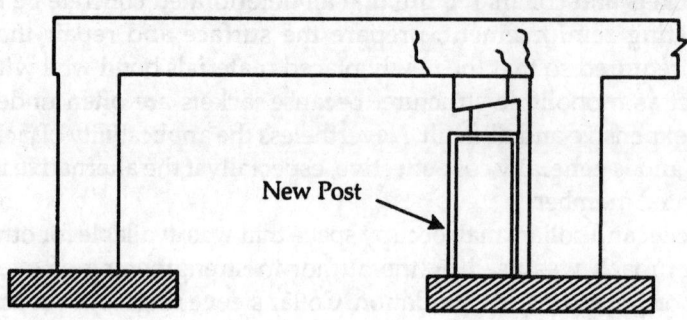

New Post

Anchor for Stability

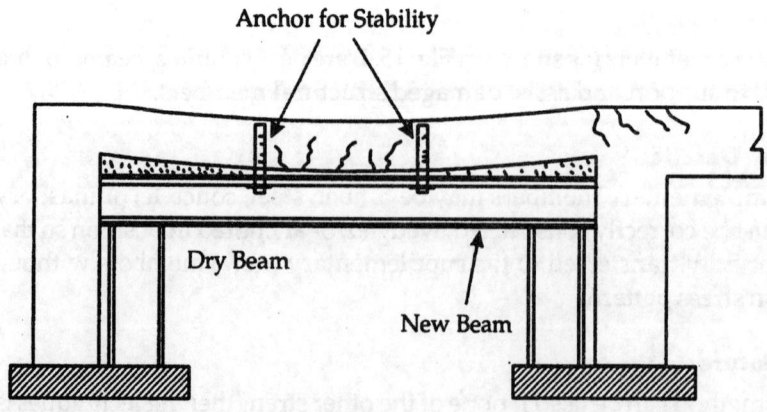

Dry Beam

New Beam

Fig. 15.4: Supplementary Members

cracked and deflected flexural members. Often, use of supplementary members may be most economic alternative because of easy and urgent installation.

Installing new columns or beams will restrict space within the repaired column bay. A new column may obstruct passage and new beams may reduce headroom. **Aesthetically** the new beam or new column may be clearly noticeable and distracting. Cross bracing or other means of providing resistance to lateral forces may be necessary if the original structure does not provide the necessary resistance. Such bracing may further restrict interior space utilisation. Loads and stresses in the existing structure may not be relieved unless special procedures are used to transfer these loads from existing damaged structures to the new additional elements.

The supplementary members may cause **redistribution** of loads and forces that may overstress an existing nearby member and hence require a detailed analysis.

15.7 UNDERPINNING

Underpinning may be defined as placing of new permanent structural element by supporting and removing existing structural element. In this method the existing loads of the damaged structural element are transferred to a temporary supporting structure installed at the place. After relieving the damaged element of any load and ensuring proper safety of the structure, the damaged portion is removed or strengthened by repair.

A new structural element is installed in place of removed damaged elements. The interface of the existing structure with the replaced element should be prepared as per surface preparation techniques. Suitable bonding materials are used for developing proper bond and interaction.

Placing of the reinforcement may be done as per design and techniques described earlier in repair of various structural elements. Techniques of concreting and materials of concrete may be adopted as described in earlier sections.

Underpinning should be done only under supervision of a competent structural engineer. Safety precautions should be taken for the structure, workers and occupants wherever underpinning is undertaken.

15.8 SUMMARY

Sometimes the structural elements deteriorate and **loose strength**. This makes the structural element **unsafe and inadequate** to carry the desired loads. Based on the extent of damage, such elements are either replaced completely or repaired and strengthened. Strengthening of structural elements can be carried out by any of the basic techniques of restoring internal strength and reinforcement, encased exterior reinforcement, external post tensioning, providing external collars, brackets and jackets, and providing supplementary members.

Before undertaking strengthening of structural element, remove the damaged concrete and steel corrosion, if any. Prepare the surface for proper bond with the repair material. Use one or combination of the strengthening technique and add suitable reinforcement as per design requirements. Anchor this additional reinforcement to the existing member by coating with suitable bonding material or by welding. The section can be enlarged by adding cementitious mortars or concretes admixed with suitable bonding agents and or epoxy resins.

The section of repaired and strengthened members generally become large and may cause redistribution of loads in strengthened members, may consume certain space and needs rescanning of space utilization. Adequacy of repaired members must be ensured by proper analysis. Service conditions of structural elements must be checked before undertaking repair and strengthening. Epoxy resins and other chemicals used for bonding must be checked for its adverse effect, if any.

Suitable and well designed strengthened elements may serve for long time. Many repaired and strengthened building structures provide highly economical service.

QUESTIONS

15.1 Explain the importance of strengthening in buildings.

15.2 List basic techniques of strengthening building elements.

15.3 Explain the technique of adding internal reinforcement for strengthening.

15.4 Explain the technique of adding external reinforcement for strengthening.

15.5 Explain the technique of post tensioning for strengthening of existing building elements.

15.6 Explain the technique of jacketing for strengthening of building columns with example.

15.7 Explain the technique of providing supplementary members to strengthen the existing columns in buildings.

GLOSSARY OF TERMS

A + B Contract
Cost plus-time bidding process where each contractor includes a time cost bid along with their construction bid and the contractor selected has the lowest combined bid total.

Absolute Specific Gravity
The ratio of the weight to a vacuum of a given volume of material at a stated temperature to the weight referred to a vacuum of an equal volume of gas-free distilled water at the same temperature.

Absolute Volume
The volume of the solid particles of loose granular material excluding voids.

Absolute Volume (of ingredients of concrete of mortar)
The displacement volume of an ingredient of concrete or mortar, in the case of solids, the volume of the particles themselves, including their permeable or impermeable voids but excluding space between particles, in the case of fluids, the volume which they occupy.

Absorbed Moisture
The moisture held in a material and having physical properties not substantially different from those of ordinary water at the same temperature and pressure.

Absorbed Water
Water held on surfaces of a material by physical and chemical forces, and having physical properties substantially different from those of absorbed water or chemically combined water at the same temperature and pressure.

Absorption
The amount of water absorbed under specific conditions, usually expressed as a percentage of the dry weight of the material, the process by which the water is absorbed.

Acceleration
Increase in rate of hardening or strength development of concrete.

Accelerator·
An admixture which, when added to concrete, mortar, or grout, increases the rate of hydration of hydraulic cement, shortens the time of set, or increases the rate of hardening or strength development.

Acrylic
Polymer and copolymers of the esters of acrylic and methacrylic acids. One of a group formed by polymerizing the esters or amides of acrylic acid

Adhesion Loss
The loss of bond between a joint sealant material and the concrete joint face noted by physical separation of the sealant from either or both joint faces.

Adhesives
The group of materials used to join or bond similar or dissimilar materials, for example, in concrete work, the epoxy resins.

Admixture	A material other than water, aggregates, and portland cement (including air entraining portland cement, and portland blast furnace slag cement) that is used as an ingredient of concrete and is added to the bath before and during the mixing operation.
Adsorption	Development at the surface of a solid of a higher concentration of a substance than exists in the bulk of the medium, especially in concrete and cement technology, formation of a layer of water at the surface of a solid, such as cement, or aggregate, or of air-entraining agents at the air-water boundaries, the process by which a substance is adsorbed.
Aggregate	Granular material, such as sand, gravel, crushed stone, crushed hydraulic-cement concrete, or iron blast furnace slag, used with a hydraulic cementing medium to produce either concrete or mortar.
Aggregate Blending	The process of intermixing two or more aggregates to produce a different set of properties, generally, but not exclusively, to improve grading.
Aggregate Gradation	The distribution of particles of granular material among various sizes, usually expressed in terms of cumulative percentages larger or smaller than each of a series of sizes (sieve opening) or the percentages between certain ranges of sizes (sieve openings). See also "Grading".
Aggregate Interlock	The projection of aggregate particles or portion of aggregate particles from one side of a joint or crack in concrete into recesses in the other side of the joint or crack so as to effect load transfer in compression and shear and maintain mutual alignment.
Aggregate Lightweight	Aggregate of low density, such as: (a) expanded or sintered clay, shale, slate, diatomaceous shale, perlite, vermiculite, of slag, (b) natural pumice, scoria, volcanic cinders, tuff and diatomite, (c) sintered fly ash or industrial cinders, used to produce lightweight concrete.
Aggregate, Angular	Aggregate particles that possess well defined edged formed at the intersection of roughly planar faces.
Aggregate, Cement Ratio	See Cement-Aggregate Ratio
Aggregate, Coarse	See Coarse Aggregate
Aggregate, Dense-graded	Aggregates graded to produce low void content and maximum weight when compacted.
Aggregate, Fine	See Fine Aggregate
Aggregate, Gap-graded	Aggregate so graded that certain intermediate sizes are substantially absent.
Aggregate, Heavyweight	Aggregate of high density, such as barite, magnetite, hematite, limonite, ilmenite, iron, or steel, used to produce heavyweight concrete.
Aggregate, Maximum Size	See Maximum Size of Aggregate
Aggregate, Nominal Maximum Size	In specifications for and descriptions of aggregate, the smallest sieve opening through which the entire amount of the aggregate is permitted to pass, sometimes referred to as "maximum size (of aggregate)"
Aggregate, open-graded	Concrete aggregate in which the voids are relatively large when the aggregate is compacted.
Agitating Speed	The rate of rotation of the drum of blades of a truck mixer when used for agitation of mixed concrete.

Agitating Truck	A vehicle in which freshly mixed concrete can be conveyed from the point of mixing to that of placing, while being agitated, the truck body can either be stationary and contain an agitator of it can be a drum rotated continuously so as to agitate the contents.
Agitation	The process of providing gentle motion in mixed concrete just sufficient to prevent segregation or loss of plasticity.
Agitator	A device for maintaining plasticity and preventing segregation of mixed concrete by agitation.
Air Content	The amount of air in mortar or concrete, exclusive of pore space in the aggregate particles, usually expressed as a percentage of total volume of mortar or concrete.
Air Void	A space in cement paste, mortar or concrete filled with air, an entrapped air void is characteristically 1 mm or more in size and irregular in shape, an entrained air void is typically between 10 m and 1 mm in diameter and spherical (or nearly so).
Air-Entraining	The capabilities of a material or process to develop a system of minute bubbles of air in cement paste, mortar, or concrete during mixing.
Air-Entraining Agent	An addition for hydraulic cement or an admixture for concrete or mortar which causes air, usually in small quantity, to be incorporated in the form of minute bubbles in the concrete or mortar during mixing, usually to increase its workability and frost resistance.
Air-Entraining Cement	A cement that has an air-entraining agenda added during the grinding phase of manufacturing.
Air-Entrainment	The inclusion of air in the form of minute bubbles during the mixing of concrete or mortar.
Air-Meter	A device for measuring the air content of concrete and mortar.
Air-Water Jet	A high velocity jet of air and water mixed at the nozzle, used in clean up of surfaces of rock or concrete, such as horizontal construction joints.
Algae	Black brown or green seaweed growth on moist roofs and walls.
Alkali-Aggregate Reaction	Chemical reaction in mortar or concrete between alkalis (sodium and potassium) released from portland cement or from other sources, and certain compounds present in the aggregates, under certain conditions, harmful expansion of the concrete or mortar may be produced.
Alkali-Carbonate Reaction	The reaction between the alkalies (sodium and potassium) in portland cement binder and certain carbonate rocks, particularly calcite dolomite and dolomitic limestones, present in some aggregates, the products of the reaction may cause abnormal expansion and cracking of concrete in service.
Alloy Steel	Special steel, containing one or more added elements with ferrous which impart some special property to the steel.
Alternate Lane Construction	A method of constructing concrete roads, runways, or other paved areas, in which alternate lanes are placed and allowed to harden before the remaining immediate lanes are placed
Angle of Repose	The angle between the horizontal and the natural slope of loose material below which the material will not slide.
Anodize	Coating with a thin protective film by electrolytic action.

Apron Lining
A board used to form a protective finish at the edge of the floor, round a stair well or other similar opening.

Area of Steel
The cross-sectional area of the reinforcing bars in or for a given concrete cross section.

Artificial Turf Drag
Surface texture achieved by inverting a section of artificial turf that is attached to a device that allows control of the time and rate of texturing.

Asphalt
A brown to black bituminous substance that is chiefly obtained as a residue of petroleum refining and that consists mostly of hydrocarbons.

Automatic Batcher
A batcher equipped with gates or valves which, when actuated by a single starter switch, will open automatically at the start of the weighing operation of each material and close automatically when a designated weight of each material has been reached, interlocked in such a manner that (1) the charging mechanism cannot be opened until the scale has returned to zero (2) the charging mechanism cannot be opened if the discahrge mechanism is opened, (3) the discharge mechanism cannot be opened if the charging mechanism is opened (4) the discharge mechanism cannot be opened until the designated weight has been reached within the allowable tolerance, and (5) if different kinds of aggregates or different kinds of cements are weighed cumulatively in a single batcher, interlocked sequential controls are provided.

Axle Load
The portion of the gross weight of a vehicle transmitted to a structure or a pavement through wheels suporting a given axle.

Backer Rod
Foam cord that inserts into a joint sealant reservior and is used to shape a liquid joint sealant and prevent sealant from adhering to or flowing out of the bottom of the reservoir.

Backup Wall
A wall to support a facing material.

Bag (of cement)
A quantity of cement; 42.6 kg in the United States, 39.7 kg in Canada; portland or air-entraining portland cement, or as indicated on the bag for other kinds of cement.

Ball Test
A test to determine the consistency of fresh concrete by measuring the depth of penetration of a steel ball. The apparatus is usually called a kelly ball

Bar
A member used to reinforce concrete, usually made of steel.

Bar Chair
An individual supporting device used to support or hold reinforcing bars in proper position to prevent displacement before or during concreting.

Bar Spacing
The distance between parallel reinforcing bars, measured center to center of the bars perpendicular to their longitudinal axis.

Bar Support
A rigid device used to support or hold reinforcing bars in proper position to prevent displacement before or during concrete placing.

Barrel (of cement)
A unit of weight for cement: (170.6 kg) net, equivalent to 4 US bags for portland or air-entraining portland cements, or as indicated by the manufacturer for other kinds of cement. (In Canada, 158.8 kg. net per barrel).

Base
A subfloor slab or "working mat," either previously placed and hardened or freshly placed, on which floor topping is placed in a later operation; also, the underlying stratum on which a concrete slab, such as a pavement, is placed.

Base Course	A layer of specified select material of planned thickness constructed on the subgrade or subbase below a pavement to serve one or more functions such as distributing loads, providing drainage, minimizing frost action, or facilitating pavement construction.
Batch	Quantity of concrete or mortar mixed at one time.
Batch Plant	Equipment used for batching concrete materials.
Batch Weights	The weights of the various materials (cement, water, the several sizes of aggregate, and admixtures) that compose a batch of concrete.
Batched Water	The mixing water added to a concrete or mortar mixture before or during the initial stages of mixing.
Batching	Weighing or volumetrically measuring and introducing into the mixer the ingredients for a batch of concrete or mortar.
Beam Test	A method of measuring the flexural strength (modulus of rupture) of concrete by testing a standard unreinforced beam.
Benkelman Beam	Static deflection measuring tool equipped with dial gauges able to detect slab deflection to 0.025 millimeters.
Binder	See Cement Paste
Bitu-mastic Compound	A water sealer containing bitumen, which remains pliable enough to accommodate movement in joints.
Bitumen	Any of various mixtures of hydrocarbons (as tar) often together with their non-metallic derivatives that occur naturally or are obtained as residues after heat-refining petroleum
Bituminous	Resembling, containing or impregnated with bitumen.
Blanking Band	A plastic scale, or computer-generated scale, 1.7 inches wide and 21.12 inches long representing a length of 0.1 miles on a profilograph trace. The opaque blanking strip, running the length of the scale and located at its midpoint, covers the profile trace. Typically, a bandwidth of 0.0 to 0.2 in. is used.
Blast Furnace Slag	The non-metallic by-product, consisting essentially of silicates and aluminosilicates of lime and other bases, which is produced in a molten condition simultaneously with iron in a blast furnace.
Bleeding	The self-generated flow of mixing water within, or its emergence from, freshly placed concrete or mortar.
Bleeding Rate	The rate at which water is released from a paste or mortar by bleeding, usually expressed as cubic centimeters of water released each second from each square centimeter of surface.
Blemish	Any superficial defect that causes visible variation from a consistently smooth and uniformly colored surface of hardened concrete. (See also Bug Holes, Efflorescence, Honeycomb, Laitance, Popout, Rock Pocket, Sand streak.)
Blended Cement	See Cement, Blended
Blended Hydraulic Cement	See Cement, Blended
Blistering	The irregular rising of a thin layer of placed mortar or concrete at the surface during or soon after completion of the finished operation.

Bond	The adhesion of concrete or mortar to reinforcement or other surfaces against which it is placed; the adhesion of cement paste to aggregate.
Bond Area	The interface area between two elements across which adhesion develops or may develop, as between concrete and reinforcing steel.
Bond Beam	A beam that ties an upper and lower section of wall together giving them added stiffness. A continuous course of brick or block that is reinforced with longitudinal bars and grouted so it acts as a structural beam.
Bond Breaker	A material used to prevent adhesion of newly placed concrete from other material, such as a substrate.
Bond Hardness	The support (bond strength) that the metal matrix in a diamond saw blade segment provides to each diamond that is embedded within the matrix.
Bond Strength	Resistance to separation of mortar and concrete from reinforcing steel and other materials with which it is in contact; a collective expression for all forces such as adhesion, friction due to shrinkage, and longitudinal shear in the concrete engaged by the bar deformations that resist separation.
Bond Stress	The force of adhesion per unit area of contanct between two surfaces sufh as concrete and reinforcing steel or any other material such as foundation rock.
Bonded Concrete Overlay	Thin layer of new concrete (2-4 inches) placed onto slightly deteriorated existing concrete pavement with steps taken to prepare old surface to promote adherence of new concrete.
Bonding Agent	A substance applied to an existing surface to create a bond between it and a succeeding layer, as between a bonded overlay and existing concrete pavement.
Bottom Chord	Lower horizontal member of a steel joist or truss.
Box Frame	A structural building system that uses shear walls to resist lateral loads. The floor and roof systems transfer vertical live and dead loads to the load bearing masonry shear walls, lateral loads act on diaphragms that carry lateral loads to masonry shear walls for transfer to the foundation. For the system to work, floor wall connections must be adequate.
Box Out	To form an opening or pocket in concrete by a box-like form; Used for manholes, drainage inlets and other in-pavement objects.
Breeze	Usually understood to mean clinker but has sometimes been used to refer to cock breeze.
Broom	The surface texture obtained by stroking a broom over freshly placed concrete. A sandy texture obtained by brushing the surface of freshly placed or slightly hardened concrete with a stiff broom.
Bug Holes	Small regular or irregular cavities, usually not exceeding 15 mm in diameter, resulting from entrapment of air bubbles in the surface of formed concrete during placement and compaction.
Bulk Cement	Cement that is transported and delivered in bulk (usually in specially constructed vehicles) instead of in bags.
Bulk Density	The mass of a material (including solid particles and any contained water) per unit volume, including voids.
Bulk Specific Gravity	The ratio of the weight in air of a given volume of a permeable material (including both permeable and impermeable voids normal to the material)

at a stated temperature to the weight in air of an equal volume of distilled water at the same temperature.

Bulking Factor   Ratio of the volume of moist sand to the volume of the sand when dry.

Bull Float   A tool comprising a large, flat, rectangular piece of wood, aluminium, or magnesium usually 20 cm wide and 100 to 150 cm long, and a handle 1 to 5 m in length used to smooth unformed surfaces of freshly placed concrete.

Burlap   A coarse fabric of jute, hemp, or less commonly flax, for use as a water-retaining cover for curing concrete surfaces; also called Hessian.

Burlap Drag   Surface texture achieved by trailing moistened coarse burlap from a device that allows control of the time and rate of texturing.

Butt Joint   A plain square joint between two concrete slabs.

Calcareous   Containing calcium carbonate, or less generally, containing the element calcium.

Calcine   To become quick powdery lime by the action of heat.

Calcium Chloride   A crystalline solid, $CaCl_2$; in various technical grades, used as a drying agent, as an accelerator of concrete, a deicing chemical, and for other purposes.

Calcium Lignosulfonate   An admixture, refined from papermaking wastes, employed in concrete to retard the set of cement, reduce water requirement and increase strength.

Caliche   Gravel, sand, or desert debris cement by porous calcium carbonate of other salts.

California Bearing Ratio   The ratio of the force per unit area required to penetrate a soil mass with a 19.4 sq cm circular piston at the rate of 1.27 mm per min to the force required for corresponding penetration of a standard crushed-rock base material; the ratio is usually determined at 2.5 mm penetration.

California Profilograph   Rolling straight edge tool used for evaluating pavement profile (smoothness) consisting of a 25-ft frame with a sensing wheel located at the center of the frame that senses and records bumps and dips on graph paper or in a computer.

Capillary   In cement paste, any space not occupied by unhydrated cement or cement gel (air bubbles, whether entrained or entrapped, are not considered to be part of the cement paste).

Capillary Absorption   The action of surface tension forces which draws water into capillaries (i.e., in concrete) without appreciable external pressures.

Capillary Flow   Flow of moisture through a capillary pore system, such as concrete.

Capillary Space   In cement paste, any space not occupied by anhydrous cement or cement gel. (Air bubbles, whether entrained or entrapped, are not considered to be part of the cement paste.)

Capillary Transmission   Passage of water or other fluid through capillaries, either by capillarity or under hydraulic pressure; capillary flow.

Carbide-Milling   Surface removal or sawing done with carbide milling machine; Machine used blade or arbor equipped with carbide-tipped teeth that impact and chip concrete or asphalt.

Carbonation   Reaction between carbon dioxide and the products of portland cement hydration to produce calcium carbonate.

Carcass	Framing in position before addition of covering.
Casement Door	A hinged door or pair of doors almost wholly glazed.
Casement Window	A window frame with glass in which one or more parts are hinged to open
Cast Stone	Simulated stone, made from concrete cast in unit with artificially coloured surface.
Cast-in-Place	Concrete placed and finished in its final location.
Catalyst	A substance whose presence increase the rate of a chemical reaction. (In some cases the catalyst is consumed and regenerated, in other cases the catalyst seems not to enter into the reaction, but functions by virtue of some other characteristic).
Cavity Wall	A double leaf wall with a continuous air space (generally 25 mm or more thick) separating the facing leaf from the backup leaf. Either or both leaves may be designed as load bearing. The air space may be insulated to increase the walls thermal efficiency.
Cellular Concrete	Lightweight concrete made by introducing large numbers of air cells into the mix.
Cement	See Portland Cement
Cement Content	Quantity of cement contained in a unit volume of concrete or mortar, ordinarily expressed as pounds, barrels, or bags per cubic yard.
Cement Factor	See Cement Content
Cement Lime Mortar	Mortar made with a proportion of slaked lime added to the cement also known as (gauge mortar)
Cement Paste	Constituent of concrete consisting of cement and water.
Cement, Blended	A hydraulic cement consisting essentially of an intimate and uniform blend of granulated blast-furnace slag and hydrated lime; or an intimate and uniform blend of portland cement and granulated blast-furnace slag cement and pozzolan, produced by intergrinding Portland cement clinker with the other materials or by blending Portland cement with the other materials, or a combination of intergrinding and blending.
Cement, Expansive	A special cement which, when mixed with water, forms a paste that tends to increase in volume at an early age; used to compensate for volume decrease due to drying shrinkage.
Cement, High-Early-Strength	Cement characterized by producing earlier strength in mortar or concrete than regular cement, referred to in the United States as "Type III."
Cement, Hydraulic	Cement that is capable of setting and hardening under water, such as normal portland cement.
Cement, Normal	General purpose portland cement, referred to in the United States as "Type I."
Cement, Portland-Pozzolan	A hydraulic cement consisting essentially of an intimate and uniform blend of portland cement or portland blast-furnace slag cement and fine pozzolan produced by intergrinding portland-cement clinker and pozzolan, by blending portland cement or portland blast-furnace slag cement and finely divided pozzolan, or a combination of intergrinding and blending, in which the pozzolan constituent is within specified limits.

Cement-Aggregate Ratio The ratio, by weight or volume, of cement to aggregate.

Cementitious Having cementing properties.

Cementitious Materials Substances that alone have hydraulic cementing properties (set and harden in the presence of water). Includes: ground granulated blast furnace slag, natural cement, hydraulic hydrated lime, and combinations of these and other materials.

Central Mixer A stationary concrete mixer from which the fresh concrete is transported to the work.

Central-Mixed Concrete Concrete that is completely mixed in a stationary mixer from which it is transported to the delivery point.

Chair See Bar Support

Chalking A phenomenon of coatings, such as cement paint, manifested by the formation of a loose powder by deterioration of the paint at or just beneath the surface.

Charging Introducing, feeding, or loading materials into a concrete or mortar mixer, furnace, or other container or receptacle.

Checking Development of shallow cracks at closely spaced but irregular intervals on the surface of mortar or concrete.

Chipping Treatment of a hardened concrete surface by chiseling away a portion of material.

Chute A sloping trough or tube for conducting concrete, cement, aggregate, or other free-flowing materials from a higher to a lower point.

Cladding The external non-load bearing covering to the frame wall of a building.

Cleaving Lines Natural lines of the weakness along which stone can be broken most easily.

Clinker A stage in the manufacture of cement in which the ingredients are fused into small pieces by heat.

Coarse Aggregate See Aggregate, Coarse

Coefficient of Thermal Expansion Change in linear dimension per unit length or change in volume per unit volume per degree of temperature change.

Cohesion Loss The loss of internal bond within a joint sealant material; noted by a noticeable tear along the surface and through the depth of the sealant.

Cohesiveness The property of a concrete mix which enables the aggregate particles and cement paste matrix therein to remain in contact with each other during mixing, handling, and placing operations; the "stick-togetherness" of the concrete at a given slump.

Cold Joint A discontinuity produced when the concrete surface hardens before the next batch is placed against it.

Colloidal Mixer Grout mixing device that uses a high velocity blade to shear or separate cementitious particles in order to break surface tension and enable complete contact between the particles and mixing water.

Column An upright (vertical or near vertical) load bearing member whose length on plan is not more than four times it's width and mainly carries compressive load.

Combined Aggregate Grading Particle size distribution of a mixture of fine and coarse aggregate.

Compacting Factor The ratio obtained by dividing the observed weight of concrete which fills a container of standard size and shape when allowed to fall into it under standard conditions of test, by the weight of fully compacted concrete which fills the same container.

Compaction The process wherby the volume of freshly placed mortar or concrete is reduced to the minimum practical space, usually by vibration, centrifugation, tamping, or some combination of these; to mold it within forms or molds and around embedded parts and reinforcement, and to eliminate voids other than entrained air. See also Consolidation.

Compressible Insert Board used to separate a partial-depth patch from an adjacent slab, usually consisting of a 12 mm thick Styrofoam or compressed fiber material that is impregnated with asphalt.

Compression Seal See Preformed Compression Seal

Compression Test Test made on a specimen of mortar or concrete to determine the compressive strength; in the United States, unless otherwise specified, compression tests of mortars are made on 50 mm cubes, and compression tests of concrete are made on cylinders 152 mm in diameter and 305 mm high.

Compressive Mortar Capacity of a material to withstand compressive load. The maximum compressive load (in Newtons) that a masonry unit (in square mm) can carry. The resulting strength is expressed in (Newton) per square mm. Compressive strength also can be calculated on net cross sectional area.

Compressive Strength The measured resistance of a concrete or mortar specimen to axial loading; expressed as pounds per square inch (psi) of cross-sectional area.

Concentric Load Load placed along the centerline of the resisting member. Also see eccentric load.

Concrete A composite material that consists essentially of a binding medium in which is embedded particles or fragments of relatively inert material filler. In portland cement concrete, the binder is a mixture of portland cement and water; the filler may be any of a wide variety of natural or artificial aggregates.

Concrete Bond The adhesion of two concrete surfaces together.

Concrete Spreader A machine designed to spread concrete from heaps already dumped in front of it, or to receive and spread concrete in a uniform layer.

Concrete, Normal-weight Concrete having a unit weight of approximately 2400 kg/m³ made with aggregates of normal weight.

Concrete, Reinforced Concrete construction that contains mesh or steel bars embedded in it.

Consistency The relative mobility or ability of fresh concrete or mortar to flow. The usual measures of consistency are slump or ball penetration for concrete and flow for mortar.

Consolidate Compaction usually accomplished by vibration of newly placed concrete to minimum practical volume, to mold it within form shapes or around embedded parts and reinforcement, and to reduce void content to a practical minimum.

Consolidation The process of inducing a closer arrangement of the solid particles in freshly mixed concrete or mortar during placement by the reduction of voids, usually by vibration, centrifugation, tamping, or some combination of these

	actions; also applicable to similar manipulation of other cementitious mixtures, soils, aggregates, or the like. See also Compaction.
Construction Joint	The junction of two successive placements of concrete, typically with a keyway or reinforcement across the joint.
Continuously Reinforced Pavement	A pavement with continuous longitudinal steel reinforcement and no intermediate transverse expansion or contraction joints.
Contract	Decrease in length or volume. (See also Expand, Shrinkage, Swelling, and Volume Change.)
Contraction Joint	A plane, usually vertical, separating concrete in a structure of pavement, at a designated location such as to prevent formation of objectionable shrinkage cracks elsewhere in the concrete. Reinforcing steel is discontinuous.
Control Joint	See Contraction Joint
Co-polymerisation	Polymerisation of two or more dissimilar monomers.
Core	A cylindrical specimen of standard diameter drilled from a structure or rock foundation to be bested in compression or examined petrographically.
Corner Break	A portion of the slab separated by a crack that intersects the adjacent transverse or longitudinal joints at about a 45° angle with the direction of traffic. The length of the sides is usually form 0.3 meters to one-half of the slab width on each side of the crack.
Course	In concrete construction, a horizontal layer of concrete, usually one of several making up a lift; in masonry construction, a horizontal layer of block or brick. See also Lift.
Cover	In reinforced concrete, the least distance between the surface of the reinforcement and the outer surface of the concrete.
Crack Saw	Small three-wheeled specially saw useful for tracing the wandering nature of a transverse or longitudinal crack; usually contains a pivot wheel and requires a small diameter crack sawing blade.
Cracking	The process of contraction or the reflection of stress in the pavement.
Crazing	Minute surface pattern cracks in mortar or concrete due to unequal shrinkage or contraction on drying or cooling.
Cross Section	The section of a body perpendicular to a given axis of the body; a drawing showing such a section.
Cross Sectional Area Gross	The total area of cross-section perpendicular to the direction of the load, including areas within cells.
Cross Sectional Area Net	The net area is average contact area of cross section and is equal to gross area minus the cellular spaces.
Crushed Gravel	The product resulting from the artificial crushing of gravel with a specified minimum percentage of fragments having one or more faces resulting from fracture. See also Coarse Aggregate.
Crushed Stone	The product resulting from the artificial crushing of rocks, boulders, or large cobblestones, substantially all faces of which possess well-defined edges and have resulted from the crushing operation.
Crusher-run Aggregate	Aggregate that has been broken in a mechanical crusher and has not been subjected to any subsequent screening process.
Cryolite	A mineral comprising of sodium aluminium fluoride

Cubic Yard	Normal commercial units of measure of concrete volume, equal to 27 cubic feet.
Cure	Maintenance of temperature and humidity for freshly placed concrete during some definite period following placing and finishing to ensure proper hydration of the cement and proper hardening of the concrete.
Curing	The maintenance of a satisfactory moisture content and temperature in concrete during its early stages so that desired properties may develop.
Curing Blanket	A built-up covering of sacks, matting, Hessian, straw, waterproof paper, or other suitable material placed over freshly finished concrete. See also Burlap.
Curing Compound	A liquid that can be applied as a coating to the surface of newly placed concrete to retard the loss of water or, in the case of pigmented compounds, also to reflect heat so as to provide an opportunity for the concrete to develop its properties in a favourable temperature and moisture environment. See also Curing.
Damp	Either moderate absorption or moderate covering of moisture; implies less wetness than that connoted by "wet," and slightly wetter than that connoted by "moist." See also Moist and Wet.
Damp Proof Course	A layer or sheet of material placed within a wall, column, or similar construction at the junction of super structure and foundation to prevent the passage of moisture (A Damp proof course is damp proof membrane within a floor or a wall)
DBI	A dowel bar inserter that places the load transfer bars into plastic concrete as part of the paving operation.
Dead Load	The fixed static load represented solely by the weight of walls, partitions, roofs, floors, and other permanent constructions including finishing.
Deformed Bar	A reinforcing bar with a manufactured pattern of surface ridges that provide a locking anchorage with surrounding concrete.
Deformed Reinforcement	Metal bars, wire, or fabric with a manufactured pattern of surface ridges that provide a locking anchorage with surrounding concrete.
Density	Mass per unit volume; by common usage in relation to concrete, weight per unit volume, also referred to as unit weight.
Density (dry)	The mass per unit volume of a dry substance at a stated temperature. See also Specific Gravity.
Density Control	Control of density of concrete in field construction to ensure that specified values as determined by standard tests are obtained.
Design Strength	Load capacity of a member computed on the basis of allowable stresses assumed in design.
Deterioration	(1) Physical manifestation of failure (e.g., cracking delamination, flaking, pitting, scaling, spalling, staining) caused by environmental or internal autogenous influences on rock and hardened concrete as well as other materials; (2) decomposition of material during either testing or exposure to service. See also Disintegration and Weathering.
Diamond Grinding	The process used to remove the upper surface of a concrete pavement to remove bumps and restore pavement rideability; also, equipment using many diamond-impregnated saw blades on a shaft or arbor to shave the surface of concrete slabs.

Disincentive	Deduction in payment resulting from a measured quality lower than specified for full payment.
Dispersing Agent	Admixtures capable of increasing the fluidity of pastes, mortar or concretes by reduction of inter-particle attraction.
Distress	Physical manifestation of deterioration and distortion in a concrete structure as the result of stress, chemical action, and/or physical action.
Dolomite	A mineral having a specific crystal structure and consisting of calcium carbonate and magnesium carbonate in equivalent chemical amounts (54.27 and 54.73 per cent by weight, respectively); a rock containing dolomite as the principal constituent.
Dowel	(1) A steel pin, commonly a plain round steel bar, which extends into two adjoining portions of a concrete construction, as at a joint in a pavement slab, so as to transfer shear loads; (2) a deformed reinforcing bar intended to transmit tension, compression, or shear through a construction joint.
Dowel Basket	See Load-Transfer Assembly.
Down Pressure	The force that keeps the grinding head on a diamond grinding machine cutting through bumps in the concrete surface and prevents the grinding head from riding up and merely tracing the bump profile.
Drainage	The interception and removal of water from, on, or under an area or roadway; the process of removing surplus ground or surface water artificially; a general term for gravity flow of liquids in conduits.
Drive Packer	For slab stabilization or slab jacking, tapering metal nozzle that seats into an injection hole by tapping or standing on footplate. Usually most appropriate for small-diameter holes.
Drop Hammer	Impact-type pavement breaking equipment.
Dry Mix	Concrete, mortar, or plaster mixture, commonly sold in bags, containing all components except water; also a concrete of near zero slump.
Dry Mixing	Blending of the solid materials for mortar or concrete prior to adding the mixing water.
Dry Process	In the manufacture of cement, the process in which the raw materials are ground, conveyed, blended, and stored in a dry condition. See also Wet Process.
Dry Rot	A type of decay of timber caused by dry rot fungus, Merulius lacrymans.
Drying Shrinkage	Contraction caused by drying.
Dry-Rodded Volume	The volume that would be occupied by an aggregate if it were compacted dry under the standardized conditions used in measuring unit weight of aggregate.
Durability	The ability of concrete to remain unchanged while in service; resistance to weathering action, chemical attack, and abrasion.
Dynamic Load	A variable load; *i.e.*, not static, such as a moving live load, earthquake, or wind.
Dynamic Loading	Loading from units (particularly machinery) which, by virtue of their movement or vibration, impose stresses in excess of those imposed by their dead load.
E. J. Sealant	A compressible material used to exclude water and soiled foreign materials from joints.

Early Strength Strength of concrete developed soon after placement, usually during the first 72 hours.

Early-Entry Dry Saw Lightweight saw equipped with a blade that does not require water for cooling and that allows sawing concrete sooner than with conventional wet-diamond sawing equipment.

Eccentric Load Load offset from the center line of the supporting member.

Econocrete Portland cement concrete designed for a specific application and environment and, in general, making use of local commercially produced aggregates. These aggregates do not necessarily meet conventional quality standards for aggregates used in pavements.

Edge Beam A stiffening beam at the edge of a slab.

Edge Form Formwork used to limit the horizontal spread of fresh concrete on flat surfaces, such as pavements or floors.

Edger A finishing tool used on the edges of fresh concrete to provide a rounded edge.

Effective Height The height of a member that is assumed when calculating the slenderness ratio. If a wall is laterally supported top and bottom, its effective height is the distance between these supports. If the wall is free at the top, the effective height is assumed to be twice the height above the bottom support.

Effective Thickness The thickness of a member that is assumed when calculating the slenderness ratio. For unreinforced solid walls it is the actual thickness. For cavity walls loaded on only one leaf, it is the actual thickness of the loaded leaf.

Effective Width The width of a wall that is assumed, in flexural computations, to work with reinforcing bars.

Efflorescence Deposit of calcium carbonate (or other salts), usually white in colour, appearing upon the surface or found within the near-surface pores of concrete. The salts deposit on concrete upon evaporation of water that carries the dissolved salts through the concrete toward exposed surfaces.

Element A part of a building or structure having its own functional identity. Examples are foundation, floor, roof, and wall

Emulsion A two-phase liquid system in which small droplets of one liquid (the internal phase) are immiscible in, and dispersed uniformly throughout a second continuous liquid phase (the external phase).

End-Result Specification Specification that requires the contractor to take the entire responsibility for supplying each item of construction. The highway agency's responsibility is to either except or reject the final product or apply a price adjustment that compensates for the degree of compliance with the specification.

Entrained Air Round, uniformly distributed, microscopic, non-coalescing air bubbles entrained by the use of air-entraining agents; usually less than 1 mm in size.

Entrapped Air Air in concrete that is not purposely entrained. Entrapped air is generally considered to be large voids (larger than 1 mm).

Epoxies Synthetic polymer which are condensates of epicloro hydrin and a suitable polyhydroxyl material, most commonly used polyhydroxyl material is bisphenol-A.

Epoxy Resins	A class of organic chemical bonding system used in the preparation of special coatings or adhesive for concrete or as binder in epoxy resin mortars and concrete.
Evaporable Water	Water set in cement paste present in capillaries or held by surface forces; measured as that removable by drying under specified conditions. See also Nonevaporable Water.
Expanding Rubber Packer	For slab stabilization or slab jacking, nozzle containing an expandable rubber sleeve that expands from injection pressure to fill the injection hole during injection of stabilizing material.
Expansion	Increase in length or volume. See also Autogenous Volume Change, Contraction, Moisture Movement, Shrinkage, and Volume Change.
Expansion Joint	An opening or gap between adjacent parts of a building structures or concrete work which allows for safe and inconsequential relative movement of those parts, as caused by thermal variations or other conditions. See Isolation Joint
Expansion Joint Filler	A compressible material used to fill a joint to prevent the infiltration of debris and to provide support for sealants.
Expansion Joints Cover	A protective cover placed over and spanning a joint. It may be prefabricated or field fabricated. Designed to flex with the movements of the joints without loss of protection to the joint.
Expansion Sleeve	A tubular metal covering for a dowel bar to allow its free longitudinal movement at a joint.
Exposed Aggregate	Surface texture where cement paste is washed away from concrete slab surface to expose durable chip-size aggregates for the riding surface.
Exposed Concrete	Concrete surfaces formed so as to yield an acceptable texture and finish for permanent exposure to view. See also Architectural Concrete.
External Vibrator	See Vibration.
Extruded	Shaped or moulded by forcing through a die by pressure.
Extruded Mortar Joint	Expansion T backer rod or back up any material or substance placed into a joint to be reduce its depth and/or to inhibit sagging of the sealant.
False Set	The rapid development of rigidity in a freshly mixed Portland cement paste, mortar, or concrete without the evolution of much heat, which rigidity can be dispelled and plasticity regained by further mixing without addition of water; premature stiffening, hesitation set, early stiffening, and rubber set are terms referring to the same phenomenon, but false set is the preferred designation.
Fascia	A long and relatively narrow upright face at the eaves or cornice or over a shop front.
Fast-Track	Series of techniques to accelerate concrete pavement construction.
Faulting	Differential vertical displacement of a slab or other member adjacent to a joint or crack.
Fibrous Concrete	Concrete containing dispersed, randomly oriented fibers.
Field-Cured Cylinders	Test cylinders cured as nearly as practicable in the same manner as the concrete in the structure to indicate when supporting forms may be removed, additional construction loads may be imposed, or the structure may be placed in service.

Final Set A degree of stiffening of a mixture of cement and water greater than initial set, generally stated as an empirical value indicating the time in hours and minutes required for a cement paste to stiffen sufficiently to resist to an established degree, the penetration of a weighted test needle; also applicable to concrete and mortar mixtures with use of suitable test; procedures. See also Initial Set.

Final Setting Time The time required for a freshly mixed cement paste, mortar, or concrete to achieve final set. See also Initial Setting Time.

Fine Aggregate Aggregate passing the 3/8-inch (9.5 mm) sieve and almost entirely passing the No. 4 (4.75 mm) sieve and predominantly retained on the No. 200 (75 mm) sieve.

Fineness Modulus It is a measure of relative fineness or coarseness.

Finish The texture of a surface after compacting and finishing operations has been performed.

Finishing Leveling, smoothing, compacting, and otherwise treating surfaces of fresh or recently placed concrete or mortar to produce desired appearance and service.

Finishing Machine A power-operated machine used to give the desired surface texture to a concrete slab.

Fixed Form Paving A type of concrete paving process that involves the use of fixed forms to uniformly control the edge and alignment of the pavement.

Flash Set The rapid development of rigidity in a freshly mixed Portland cement paste, mortar, or concrete, usually with the evolution of considerable heat, which rigidity cannot be dispelled nor can the plasticity be regained by further mixing without addition of water; also referred to as quick set or grab set.

Flashing A strip of impermeable sheet material fixed so as to protect a joint or surface from the entry of rainwater.

Flexible Pavement A pavement structure that maintains intimate contact with and distributes loads to the subgrade and depends on aggregate interlock, particle friction, and cohesion for stability; cementing agents, where used, are generally bituminous materials as contrasted to portland cement in the case of rigid pavement. See also Rigid Pavement.

Flexural Strength A property of a material or structural member that indicates its ability to resist failure in bending. See also Modulus of Rupture.

Float A tool (not a darby) usually of wood, aluminium, or magnesium, used in finishing operations to impart a relatively even but still open texture to an unformed fresh concrete surface.

Float Finish A rather rough concrete surface texture obtained by finishing with a float.

Floating Process of using a tool, usually wood, aluminium, or magnesium, in finishing operations to impart a relatively even but still open texture to an unformed fresh concrete surface.

Floor Slab A slab forming the continuous load bearing structure of floor and spanning between supports or laid on the ground.

Flow (1) Time dependent irrecoverable deformation. See Rheology. (2) A measure of the consistency of freshly mixed concrete, mortar, or cement paste in

	terms of the increase in diameter of a molded truncated cone specimen after jigging a specified number of times.
Flow Cone Test	Test that measures the time necessary for a known quantity of grout to completely flow out of and empty a standard sized cone; usually used in slab stabilization to determine the water quantity necessary for stabilization grout.
Flux	A substance used to promote fusion.
Fly Ash	The finely divided residue resulting from the combustion of ground or powdered coal and which is transported form the fire box through the boiler by flu gasses; Used as mineral admixture in concrete mixtures.
Form	A temporary structure or mold for the support of concrete while it is setting and gaining sufficient strength to be self-supporting.
Form Oil	Oil applied to interior surface of formwork to promote easy release from the concrete when forms are removed.
Foundation	Construction to support and spread loads applied to the supporting soil.
Framing	Timber used in the structural work of a building.
Free Moisture	Moisture having essentially the properties of pure water in bulk; moisture not absorbed by aggregate. See also Surface Moisture.
Free Water	See Free Moisture and Surface Moisture.
Frog	A depression in the wide surface ot a brick provided for bonding.
Full-Depth Patching	Removing and replacing at least a portion of a concrete slab to the bottom of the concrete, in order to restore areas of deterioration.
Full-Depth Repair	See "Full-Depth Patching."
Furring	The building or leveling of a surface.
Gap-graded Concrete	Concrete containing a gap-graded aggregate.
Grading	The distribution of particles of granular material among various sizes, usually expressed in terms of cumulative percentages larger or smaller than each of a series of sizes (sieve openings) or the percentages between certain ranges of sizes (sieve openings).
Granolithic Finish	A surface layer of granolithic concrete which may be laid on a base of either fresh, green, or hardened concrete.
Gravel	Granular material predominantly retained on the 4.75 mm (No. 4) sieve and resulting from natural disintegration and abrasion of rock or processing of weakly bound conglomerate.
Green Concrete	Concrete that has set but not appreciably hardened.
Green Sawing	The process of controlling random cracking by sawing uniform joint spacing in early age concrete without tearing or dislocating the aggregate in the mix.
Grid	A rectangular network of lines used in planning or setting out of structure.
Grinding Head	Arbor or shaft containing many diamond blades on diamond grinding equipment.
Grooving	The process used to cut slots into a concrete pavement surface to provide channels for water to escape beneath tires and to promote skid resistance.
Gross Vehicle Load	The weight of a vehicle plus the weight of any load thereon.

Gross Volume (of concrete mixers)	In the case of a revolving-drum mixer, the total interior volume of the revolving portion of the mixer drum; in the case of an open-top mixer, the total volume of the trough of pan calculated on the basis that no vertical dimension of the container exceeds twice the radius of the circular section below the axis of the central shaft.
Ground Beam	A horizontal beam of RCC usually built at ground or plinth level supported by piles or pads.
Ground Bond	The adhesion of grout to masonry units and to reinforcement. To transfer stress from the masonry to the reinforcement good grout bond must be achieved.
Grout	A mixture of cementitious material and water, with or without aggregate, proportioned to produce a pourable consistency without segregation of the constituents; also, a mixture of other composition but of similar consistency. See also Neat Cement Grout and Sand Grout.
Grout Lift	The height to which grout is poured into cores, collar joints, or cavities without stopping.
Grout Pour	The height of grout poured in a day. A grout pour may consist of one or more grout lifts.
Grouting Low Lift	Filling voids in masonry with grout as the wall is being built. For partially built walls grouting in pours not to exceed 1.2 m vertically.
Grout-Retention Disk	Small plastic disk that provides a barrier that prevents grout or epoxy from escaping from a dowel hole.
Hairline Cracking	Barely visible cracks in random pattern in an exposed concrete surface which do not extend to the full depth or thickness of the concrete, and which are due primarily to drying shrinkage.
Hardener	A chemical applied to concrete floors to reduce wearing and dusting.
Hardening	When portland cement is mixed with enough water to form a paste, the compounds of the cement react with water to form cementitious products that adhere to each other and to the intermixed sand and stone particles and become very hard. As long as moisture is present, the reaction may continue for years, adding continually to the strength of the mixture.
Harsh Mixture	A concrete mixture that lacks desired workability and consistency due to a deficiency of mortar.
Harshness	Deficient workability and cohesiveness caused by insufficient sand or cement, or by improperly graded aggregate.
Header	A transverse construction joint installed at the end of a paving operation or other placement interruptions. To a contractor, a header is the location at which paving will resume of the next day.
Heat of Hydration	Heat evolved by chemical reactions of a substance with water, such as that evolved during the setting and hardening of Portland cement.
Heavy-Weight Concrete	Concrete in which heavy aggregate is used to increase the density of the concrete; unit weights in the range of 165 to 330 pounds per cubic foot are attained.
High Density Polyurethane	Also, HDP; Polyurethane used in slab stabilization or slab jacking process.
High Range Water-Reducing Admixture	See Water-Reducing Admixture (high range).

High-Early-Strength Cement	See Cement, High-Early-Strength.
High-Early-Strength Concrete	Concrete that, through the use of high-early-strength cement or admixtures, is capable of attaining specified strength at an earlier age than normal concrete.
Holiday	An area in a diamond ground surface that is not ground because the head on the diamond grinding equipment does not cut deep enough to touch a low spot in the surface.
Honeycomb	Concrete that, due to lack of the proper amount of fines or vibration, contains abundant interconnected large voids or cavities; concrete that contains honeycombs was improperly consolidated.
Hooked Bar	A reinforcing bar with the end bent into a hook to provide anchorage.
Horizontal-Axis Mixer	Concrete mixers of the revolving drum type in which the drum rotates about a horizontal axis.
Hot-pour Sealant	Joint sealing materials that require heating for installation, usually consisting of a base of asphalt or coal tar.
Hydrated Lime	A dry powder obtained by treating quicklime with sufficient water to convert it to calcium hydroxide.
Hydration	The chemical reaction between cement and water which causes concrete to harden.
Hydraulic Cement	A cement that is capable of setting and hardening under water due to the chemical interaction of the water and the constituents of the cement.
Hydraulic Ram	Impact-type pavement breaking equipment.
Hydroplaning	To go out of steering control by skimming the surface of a wet road.
Impact Load	An imposed load whose effect is increased due to its sudden application.
In situ	Literally 'inplace,' referring to material or components that are cast or assembled in their permanent positions in a building or structure as distinct from being cast or assembled before installation.
Incentive	Additional payment to the contractor resulting from a measured quality higher than specified for full payment.
Inclined-Axis Mixer	A truck with a revolving drum that rotates about an axis inclined to the bed of the truck chassis.
Incompressibles	Small concrete fragments, stones, sand or other hard materials that enter a joint sealant, joint reservoir, or other concrete pavement discontinuity.
Initial Set	A degree of stiffening of a mixture of cement and water less than final set, generally stated as an empirical value indicating the time in hours and minutes required for cement paste to stiffen sufficiently to resist to an established degree the penetration of a weighted test needle; also applicable to concrete or mortar with use of suitable test procedures. See also Final Set.
Initial Setting Time	The time required for a freshly mixed cement paste to acquire an arbitrary degree of stiffness as determined by specific test.
Injection Hole	Hole drilled vertically through a concrete slab that is used to inject stabilizing grout underneath the slab or subbase layers.

Inlay	A form of reconstruction where new concrete is placed into an area of removed pavement; The removal may be an individual lane, all lanes between the shoulders or only partly through a slab.
Isolation Joint	A pavement joint that allows relative movement in three directions and avoids formation of cracks elsewhere in the concrete and through which all or part of the bonded reinforcement is interrupted. large closure movement to prevent development of lateral compression between adjacent concrete slabs; usually used to isolate a bridge.
Jamb	A vertical side member of a doorframe or door lining or window frame.
Jitterbug	A grate tamper for pushing coarse aggregate slightly below the surface of a slab to facilitate finishing. See also Tamper.
Joint	A plane of weakness to control contraction cracking in concrete pavements. A joint can be initiated in plastic concrete or green concrete and shaped with later process.
Joint Depth	The measurement of a saw cut from the top of the slab to the bottom of the cut.
Joint Deterioration	See Spalling, Compression.
Joint Filler	Compressible material used to fill a joint to prevent the infiltration of debris and to provide support for sealant.
Joint Plane	This position and direction of the bottom of joint in an individual layer, roughly perpendicular to the bedding plane
Joint Shape Factor	Ratio of the vertical to horizontal dimension of the joint sealant reservoir.
Joint, Construction	See Construction Joint.
Joint, Contraction	See Contraction Joint.
Joint, Expansion	See Expansion Joint.
Jointed Plane Pavement	Pavement containing enough joints to control all natural cracks expected in the concrete; steel tie bars are generally used at longitudinal joints to prevent joint opening, and dowel bars may be used to enhance load transfer at transverse contraction joints depending upon the expected traffic.
Jointed Reinforced Pavement	Pavement containing some joints and embedded steel mesh reinforcement (sometimes called distributed steel) to control expected cracks; steel mesh is discontinued at transverse joint locations.
Joist	One of a number of members spanning horizontally between supports to carry flooring or ceiling or both.
Kerb	A low barrier rising above the surrounding surface.
Key	Irregularity or indentation of a surface which creates a mechanical bond for plaster.
Keyway	A recess or groove in one lift or placement of concrete which is filled with concrete of the next lift, giving shear strength to the joint. See also Tongue and Groove.
Knock Down	To dismantle into separate components for ease of packing and transport.
Laitance	A layer of weak material containing cement and fines from aggregates, brought to the top of overwet concrete, the amount of which is generally increased by overworking and over-manipulating concrete at the surface by improper finishing.

Lap Splice	The distance one rebar overlaps another rebar.
Lateral Support	Structural members of the building that brace masonry walls against lateral loads. Floor beams, spandrel beams, roof beams or floors or roofs themselves can provide lateral support at vertical can spacings. Piers, columns, pilasters or sheer walls can provide lateral support at horizontal spacings.
Latex	Organic polymer particles dispersed in water.
Layer	See Course.
Lean Concrete	Concrete of low cement content.
Lean Mix	A concrete mix with a low proportion of cement to aggregates.
Life-Cycle Cost Analysis	The process used to compare projects based on their initial cost, future cost and salvage value, which accounts for the time value of money.
Lift	The concrete placed between two consecutive horizontal construction joints, usually consisting of several layers or courses.
Lintel	A beam over an opening in a wall.
Liquid Sealant	Sealant materials that install in liquid form and cool or cure to their final properties; rely on long-term adhesion to the joint reservoir faces.
Load	A force acting on a structure or a member causing deformation.
Load Bearing Concrete Masonry	Concrete block masonry made for load bearing applications.
Load Bearing Wall	A wall designed to support a load in addition to its own weight and wind pressure on its surface.
Load Transfer Efficiency	The ability of a joint or crack to transfer a portion of a load applied on side of the joint or crack to the other side of the joint or crack.
Load-Transfer Assembly	Most commonly, the basket or carriage designed to support or link dowel bars during concreting operations so as to hold them in place, in the desired alignment.
Longitudinal Broom	Surface texture achieved in similar manner as transverse broom, except that broom is pulled in a line parallel to the pavement centerline.
Longitudinal Joint	A joint parallel to the long dimension of a structure or pavement.
Longitudinal Reinforcement	Reinforcement essentially parallel to the long axis of a concrete member or pavement.
Longitudinal Tine	Surface texture achieved by a hand held or mechanical device equipped with a rake-like tining head that moves in a line parallel to the pavement centerline.
Lot	A defined quantity.
Map Cracking	(1) Intersecting cracks that extend below the surface of hardened concrete; caused by shrinkage of the drying surface concrete which is restrained by concrete at greater depths where either little or no shrinkage occurs; vary in width from fine and barely visible to open and well-defined. (2) The chief symptom of chemical reaction between alkalis in cement and mineral constituents in aggregate within hardened concrete; due to differential rate of volume change in different portions of the concrete; cracking is usually random and on a fairly large scale, and in severe instances the cracks may reach a width of 0.50 in. See also Checking, Crazing, and Pattern Cracking.

Masonry Construction of walls with stones, bricks, or blocks.

Maximum Size Aggregate The largest size aggregate particles present in sufficient quantity to affect properties of a concrete mixture.

Membrane Curing A process that involves either liquid sealing compound (e.g., bituminous and paraffinic emulsions, coal tar cut-backs, pigmented and non-pigmented resin suspensions, or suspensions of wax and drying oil) or non-liquid protective coating (e.g., sheet plastics or "waterproof" paper), both of which types function as films to restrict evaporation of mixing water from the fresh concrete surface.

Mesh The number of openings (including fractions thereof) per unit of length in either a screen or sieve in which the openings are 6 mm or less.

Mesh Reinforcement See Welded-Wire Fabric Reinforcement.

Method and Material Specification Specification that directs the contractor to use specified materials in definite proportions and specific types of equipment and methods to place the material.

Mix The act or process of mixing; also mixture of materials, such as mortar or concrete.

Mix Design See Proportioning.

Mixer A machine used for blending the constituents of concrete, grout, mortar, cement paste, or other mixture.

Mixer, Batch See Batch Mixer.

Mixer, Horizontal Shaft A mixer having a stationary cylindrical mixing compartment, with the axis of the cylinder horizontal, and one or more rotating shafts to which mixing blades or paddles are attached; also called Pugmill.

Mixer, Non-tilting A horizontally rotating drum mixer that charges, mixes, and discharges without tilting.

Mixer, Open-top A truck-mounted mixer consisting of a trough or a segment of a cylindrical mixing compartment within which paddles or blades rotate about the horizontal axis of the trough. See also Mixer, Horizontal Shaft.

Mixer, Tilting A rotating drum mixer that discharges by tilting the drum about a fixed or movable horizontal axis at right angles to the drum axis. The drum axis may be horizontal or inclined while charging and mixing.

Mixer, Transit See Truck Mixer.

Mixing Cycle The time taken for a complete cycle in a batch mixer; i.e., the time elapsing between successive repetitions of the same operation (e.g., successive discharges of the mixer).

Mixing Plant See Batch Plant.

Mixing Speed Rotation rate of a mixer drum or of the paddles in an open-top, pan, or trough mixer, when mixing a batch; expressed in revolutions per minute (rpm), or in peripheral feet per minute of a point on the circumference at maximum diameter.

Mixing Time The period during which the mixer is combining the ingredients for a batch of concrete., For stationary mixers, the time is measured from the completion of batching cement and aggregate until the beginning of discharge. For truck mixers, mixing is given in term of the number of revolutions of the drum at mixing speed.

Mixing Water	The water in freshly mixed sand-cement grout, mortar, or concrete, exclusive of any previously absorbed by the aggregate (e.g., water considered in the computation of the net water-cement ratio). See also Batched Water and Surface Moisture.
Mixture	The assembled, blended, commingled ingredients of mortar, concrete, or the like, or the proportions for their assembly.
Modular Size	A size which allows a material to a given module or standard measurement.
Modulus of Rupture	A measure of the ultimate load-carrying capacity of a beam, sometimes referred to as "rupture modulus" or "rupture strength." It is calculated for apparent tensile stress in the extreme fiber of a transverse test specimen under the load that produces rupture. See also Flexural Strength.
Moist	Slightly damp but not quite dry to the touch; the term "wet" implies visible free water, "damp" implies less wetness than "wet," and "moist" implies not quite dry. See also Damp and Wet.
Moisture Barrier	A vapour barrier.
Moisture Content of Aggregate	The ratio, expressed as a percentage, of the weight of water in a given granular mass to the dry weight of the mass.
Moisture-Free	The condition of a material that has been dried in air until there is no further significant change in its mass. See also Mass and Overdry.
Monomer	An organic liquid, of relatively low molecular weight, that creates a solid polymer by reacting with itself or other compounds of low molecular weight.
Mortar	Concrete with essentially no aggregate larger than about 3/16 inch.
Mortice and Tenon Joint	A joint in which a tenon on the end of one member is fitted into a mortice cut in the member.
Mud Balls	Balls of clay or silt ("mud").
Mullion	An intermediate vertical member in a window frame, doorframe or similar structure.
Must-Grind Bump	In a rideability specification, any bump exceeding a certain height in 25 feet (requirement may very between 0.3 and 0.5 in. bump height).
Natural Sand	Sand resulting from natural disintegration and abrasion of rock. See also Sand and Aggregate, Fine.
Neat Cement Grout	Grout consisting of Portland cement and water.
Neoprene	Synthetic rubber like plastic capable of resisting heat, light and heavy loads.
Nominal Maximum Size (of aggregate)	In specifications for and descriptions of aggregate, the smallest sieve opening through which the entire amount of the aggregate is permitted to pass; sometimes referred to as "maximum size (of aggregate)."
Non-agitating Unit	A truck-mounted container for transporting central-mixed concrete that is not equipped to provide agitation (slow mixing) during delivery. (Dump truck)
Non-air-entrained Concrete	Concrete in which neither an air-entraining admixture nor an air-entraining cement has been used.
No-Slump Concrete	Concrete with a slump of 6 mm or less. See also Zero-slump Concrete.
Open-Graded Subbase	Unstabilized layer consisting of crushed aggregates with a reduced amount of fines to promote drainage.

Ovendry	The condition resulting from having been dried to essentially constant weight, in an oven, at a temperature that has been fixed, usually between 221 and 239° F (105 and 115° C).
Overlay	The addition of a new material layer onto an existing pavement surface. See also Resurfacing
Over-Sanded	Containing more sand than would be required for adequate workability and satisfactory finishing characteristics.
Over-Vibrated	Concrete vibrated more than is necessary for good consolidation and elimination of entrapped air.
Over-Wet	The consistency of concrete when it contains more mixing water and hence is of greater slump than is necessary for ready consolidation.
Partial-Depth Patching	Patches for restoring localized areas of surface deterioration; Usually for compression spalling problems, severe scaling, or other surface problems that are within the upper one-third of the slab depth.
Partial-Depth Repair	See "Partial-Depth Patching"
Partially Reinforced Masonry	Masonry that is reinforced only where design analysis indicates likely development of tensile stress. Partially reinforced walls have no minimum steel area requirements, whereas fully reinforced masonry must have a minimum area and maximum spacing of steel.
Particle-Size Distribution	The division of particles of a graded material among various sizes; for concrete materials, usually expressed in terms of cumulative percentages larger or smaller than each of a series of diameters or the percentages with certain ranges of diameter, as determined by sieving.
Party Wall	A wall common to two buildings or two pieces of land.
Paste	Constituent of concrete consisting of cement and water.
Patina	A film formed on copper by exposure or by treatment with acids.
Pattern Cracking	Fine openings on concrete surfaces in the form of a pattern; resulting from a decrease in volume of the material near the surface, an increase in volume of the material below the surface, or both.
Pavement (Concrete)	A layer of concrete over such areas as roads, sidewalks, canals, airfields, and those used for storage or parking. See also Rigid Pavement.
Paving Train	An assemblage of equipment designed to place and finish a concrete pavement.
Pea Gravel	Screened gravel the particle sizes of which range between 3/16 and 3/8 inch in diameter.
Pencil Rod	A thin rod used as an anchor for stone or terra-cotta facing.
Percent Fines	Amount, expressed as a percentage, of material in aggregate finer than a given sieve, usually the No. 200 (75 mm) sieve; also, the amount of fine aggregate in a concrete mixture expressed as a percent by absolute volume of the total amount of aggregate.
Performance-Based Specification	Specification that describes the desired levels of fundamental engineering properties (for example, resilient modulus and/or fatigue properties) that are predictors of performance and appear in primary prediction relationships (i.e., models that can be used to predict pavement stress, distress, or performance from combinations of predictors that represent traffic, environmental, roadbed, and structural conditions.)

Performance-Related Specification	Specification that describes the desired levels of key materials and construction quality characteristics that have been found to correlate with fundamental engineering properties that predict performance. These characteristics (for example, strength of concrete cores) are amenable to acceptance testing at the time of construction.
Permeability	A measure of the rate of the passage of water or moisture concrete or other materials.
Permeable Subbase	Layer consisting of crushed aggregates with a reduced amount of fines to promote drainage and stabilized with Portland cement or bituminous cement.
Petroleum Coke	Residue left when petroleum is distilled to dryness.
Phasing	The sequences used by a contractor to build elements of a project.
Pie Tape	Tape used to measure the circumference of the grinding head blades on diamond grinding equipment.
Pig Iron	Basic iron from a blast furnace.
Pilaster	A built in column in a masonry wall, intended to stiffen the wall.
Pile	A support driven into or cast in situ in the ground for bearing transferring load to substrate of foundation.
Pitch	The angle of a roof slope to the horizontal in case of inclined roof.
Pitch Fibre Pipe	Pipe of wood fibre pulp impregnated under vacuum with coal-tar pitch to about 75% of its total weight.
Pitting	A localized disintegration taking the form of cavities at the surface of concrete.
Placement	The process of placing and consolidating concrete; a quantity of concrete placed and finished during a continuous operation; also inappropriately referred to all Pouring.
Placing	The deposition, distribution, and consolidation of freshly mixed concrete in the place where it is to harden; also inappropriately referred to as Pouring.
Plain Bar	A reinforcing bar without surface deformations, or one having deformations that do not conform to the applicable requirements.
Plain Concrete	Concrete without reinforcement.
Plain Pavement	Concrete pavement with relatively short joint spacing and without dowels or reinforcement
Plane of Weakness	The plane along which a body under stress will tend to fracture; may exist by design, by accident, or because of the nature of the structure and its loading.
Plastic	A condition of freshly mixed concrete such that it is readily remoldable and workable, cohesive, and has an ample content of cement and fines, but is not over-wet.
Plastic Consistency	Condition of freshly mixed cement paste, mortar, or concrete such that deformation will be sustained continuously in any direction without rupture; in common usage, concrete with slump of 3 to 4 inches (80 to 100 mm).
Plastic Cracking	Cracking that occurs in the surface of fresh concrete soon after it is placed and while it is still plastic.

Plastic Deformation	Deformation that does not disappear when the force causing the deformation is removed.
Plastic Shrinkage Cracking	Cracks, usually parallel and only a few inches deep and several feet long, in the surface(s) of concrete pavement that are the result of rapid moisture loss through evaporation.
Plasticity	That property of fresh concrete or mortar which determines its resistance to deformation or its ease of molding. or Ability of freshly mixed concrete to flow and get moulded into different shapes.
Plasticizer	A material that increases the plasticity of a fresh cement paste, mortar, or concrete.
Ply Wood	A product made up of plies and adhesives in which the plies are crossed to improve the strength properties.
Pneumatic	Moved or worked by air pressure.
Point Bearing	Occurs when a partial-depth patch is made without the compressible insert; also, slab expansion in hot weather forces an adjacent slab to bear directly against a small partial-depth patch and causes the patch to fail by delaminating and popping out of place.
Polyester	One of a large group of synthetic resins. Mainly produced by reaction of unsaturated dibasic acids with dihydroxy alcohol; commonly prepared for application by mixing with vinyl-group monomer and free radial catalysts at ambient temperature and used as binders for resin mortars and concrete, fibre laminates (mainly glass) adhesives, and the like.
Polymer	Polymer are long molecules of simple units called monomers. Monomers are generally organic compounds. Conversation of monomers into polymers is called polymerisation, which is effect either by heat radiation (Gamma or ultra violet) catalysts, etc.
Polymerisation	The reaction in which two or more molecules of the same substance (monomer) combine to form a compound containing the same elements, but of high molecular weight.
Polyol	A polhydric alcohol, i.e., one containing two or more hydroxyl groups.
Polysulfide	Synthetic polymers obtained by the reaction of sodium polysulfide with organic dichlorides.
Polyurethane	Reaction product of an isocyanate with any one of a wide variety of other compounds containing an active hydrogen group; used to formulate tough, abrasion resistant coatings.
Polyvinyl Acetate	Colourless, permanently thermoplastic resin; usually supplied as an emulsion or water dispersible powder characterized by flexibility, stability towards light, transparency to ultraviolet rays, high dielectric strength, toughness, and hardness; the higher the degree of polymerization, the higher the softening temperature; may be used in paints for concrete.
Poput	Pit or crater in the surface of concrete resulting from cracking of the mortar due to expansive forces associated with a particle of unsound aggregate or a contaminating material, such as wood or glass.
Porosity	The ratio, usually expressed as a percentage, of the volume of voids in a material to the total volume of the material, including voids.
Portland Cement	A commercial product which when mixed with water alone or in combination with sand, stone, or similar materials, has the property of

combining with water, slowly, to form a hard solid mass. Physically, portland cement is a finely pulverized clinker produced by burning mixtures containing lime, iron, alumina, and silica at high temperature and in definite proportions, and then intergrinding gypsum to give the properties desired.

Portland Cement Concrete
A composite material that consists essentially of a binding medium (Portland cement and water) within which are embedded particles or fragments of aggregate, usually a combination of fine aggregate course aggregate.

Portland-Pozzolan Cement
See Cement, portland-pozzolan.

Pozzolan
A siliceous or siliceous and aluminous material, which in itself possesses little or no cementitious value but will, in finely divided form and in the presence of moisture, chemically react with calcium hydroxide at ordinary temperatures to form compounds possessing cementitious properties.

Pozzolan-Cement Grout
Common slab stabilization grout consisting of water, portland cement and pozzolan; usually fly ash.

Preformed Compression Seal
Joint sealant that is manufactured ready for installation and is held in a joint by lateral pressure exerted against the reservoir by the seal after being compressed during installation.

Preservation
The process of maintaining a structure in its present condition and arresting further deterioration. See also Rehabilitation, Repair, and Restoration.

Pressure-Relief
Cut made in a concrete pavement to relieve compressive forces of thermal expansion during hot weather.

Process Control
Those Quality assurance actions and considerations necessary to assess production and construction processes so as to control the level of quality being produced in the end product. This includes sampling and testing to monitor the process but usually does not include acceptance sampling and testing.

Profile Index
Smoothness qualifying factor determined from profilograph trace. Calculated by dividing the sum of the total counts above the blanking band for each segment by the sum of the segment length.

Profile Line
On a profile trace, line drawn by hand on the field trace to average out spikes and minor deviations caused by rocks, texturing, dirt or transverse grooving.

Project Scoping
An early planning step in the development of a project where all project requirements are defined and a plan is developed to address them.

Proportioning `
Selection of proportions of ingredients for mortar or concrete to make the most economical use of available materials to produce mortar or concrete of the required properties.

Pugmill
A stationary mechanical mixer for blending cement and aggregate.

Pumping
The forceful displacement of a mixture of soil and water that occurs under slab joints, cracks and pavement edges which are depressed and released quickly by high-speed heavy vehicle loads; occurs when concrete pavements are placed directly on fine-grained, plastic soils or erodible subbase materials.

Punchout
In continuously reinforced concrete pavement, the area enclosed by two closely spaced transverse cracks, a short longitudinal crack, and the edge

	of the pavement or longitudinal joint, when exhibiting spalling, shattering, or faulting. Also, area between Y cracks exhibiting this same deterioration.
Quality Assurance	Planned and systematic actions by an owner or his representative to provide confidence that a product or facility meet applicable standards of good practice. This involves continued evaluation of design, plan and specification development, contract advertisement and award, construction, and maintenance, and the interactions of these activities.
Quality Assurance/ Quality Control Specification	Statistically based specification that is a combination of end result and material and method specifications. The contractor is responsible for quality control (process control), and the highway agency is responsible for acceptance of the product.
Quality Control	Actions taken by a producer or contractor to provide control over what is being done and what is being provided so that the applicable standards of good practice for the work are followed.
Raft Foundation	A supporting foundation slab continuous over the whole area covered by a building or structure.
Rafter	One of a number of inclined members of a sloping roof structure/truss to which a roof covering is fixed.
Random Crack	See Uncontrolled Crack.
Raveling	Displacement of aggregate or paste near the slab surface from sawing; normally indicates that concrete strength is too low for sawing.
Reactive-Aggregate	Aggregate containing certain silica or carbonate compounds that are capable of reacting with alkalis in portland cement, in some cases producing damaging expansion of concrete.
Ready-Mixed Concrete	Concrete manufactured for delivery to a purchaser in a plastic and unhardened state.
Rebar	Abbreviation for "reinforcing bar." See Reinforcement.
Rebound Hammer	An apparatus that provides a rapid indication of the mechanical properties of concrete based on the distance of rebound of a spring-driven missile.
Reconstruction	The process of removing an existing pavement from its grade and replacing it with a completely new pavement.
Recycled Concrete	Concrete that has been processed for use, usually as aggregate.
Recycling	The act of processing existing pavement material into usable material for a layer within a new pavement structure.
Reinforced Masonry	Masonry that is strengthened by the addition of reinforcing steel. Rebar grouted into the cores of masonry units or into the cavity between masonry leaves increase the walls ability to resist flexural tensile stresses.
Reinforced Concrete	Concrete containing adequate reinforcement (prestressed or not prestressed) and designed on the assumption that the two materials act together in resisting forces.
Reinforcement	Bars, wires, strands, and other slender members embedded in concrete in such a manner that the reinforcement and the concrete act together in resisting forces,.
Reinforcement, Transverse	Reinforcement at right angles to the longitudinal reinforcement; may be main or secondary reinforcement.

Relative Humidity	The ratio of the quantity of water vapour actually present to the amount present in a saturated atmosphere at a given temperature; expressed as a percentage.
Release Agent	Material used to prevent bonding of concrete to a surface. See also Bond Breaker.
Remoldability	The readiness with which freshly mixed concrete responds to a remolding effort, such as jigging or vibration, causing it to reshape its mass around reinforcement and to conform to the shape of the form. See also Flow.
Rendering	The application of a coat of mortar on wall surfaces by means of a trowel of float.
Repair	To replace or correct deteriorated, damaged, or faulty materials, components, or elements of a structure. See also Preservation, Rehabilitation, and Restoration.
Reservoir	The part of a concrete joint that normally holds a sealant material. Usually a widening saw cut above the initial saw cut.
Restoration	The process of reestablishing the materials, form, and appearance of a structure to those of a particular era of the structure. See also Preservation, Rehabilitation, and Repair.
Resurfacing	The addition of a new material layer onto an existing pavement surface for the purposes of correcting a functional factor, such as smoothness or texture.
Retaining Wall	A construction providing lateral support to the ground or mass of other material.
Retardation	Reduction in the rate of hardening or strength development of fresh concrete, mortar, or grout; i.e., an increase in the time required to reach initial and final set.
Retarder	An admixture that delays the setting of cement and hence of mixtures such as mortar or concrete containing cement.
Retempering	Addition of water and remixing of concrete or mortar which has lost enough workability to become unplaceable or unusable. See also Tempering.
Retrofit Dowel Bars	Dowels that install into slots cut into the surface of an existing concrete pavement.
Reveal	The vertical face revealed in the thickness of an opening or the depth of a recess.
Reverberatory Furnace	A furnace in which the flame is reflected from the roof to the material being smelted
Revibration	A second vibration applied to fresh concrete, preferably as long after the first vibration as the concrete will still respond properly.
Rheology	The science of dealing with flow of materials, including studies of deformation of hardened concrete, the handling and placing of freshly mixed concrete, and the behaviour of slurries, pastes, and the like.
Ribbon Loading	Method of batching concrete in which the solid ingredients, and sometimes the water, enter the mixer simultaneously.
Rich Mixture	A concrete mixture containing a large amount of cement.
Ridge Board	The longitudinal member at the apex or rigid of a roof truss.

Rigid Pavement	Pavement that will provide high bending resistance and distribute loads to the foundation over a comparatively large area.
Rock Pocket	A portion of hardened concrete consisting of a concentration of coarse aggregate that is deficient in mortar; caused by separation during placement or insufficient consolidation, or both; see honeycomb.
Rod	A specified length of metal, circular in cross section with one end rounded; used to compact concrete or mortar test specimens.
Rod, Tamping	A straight steel rod of circular cross section having one or both ends rounded to a hemispherical tip.
Rodability	The susceptibility of fresh concrete or mortar to compaction by means of a tamping rod.
Rodding	Compaction of concrete by means of a tamping rod. See also Rod, Tamping, and Rodability.
Rubblizing	A destructive procedure to break existing concrete pavement in place to fragments that range in size from 4 to 8 in.
Sack	See Bag.
Salt Glaze	Transparent glazing.
Sample	A group of units, or portion of material, taken from a larger collection of units or quantity of material, which serves to provide information that can be used as a basis for action on the larger quantity or on the production process; the term is also used in the sense of a sample of observations.
Sampling, Continuous	Sampling without interruptions throughout an operation or for a predetermined time.
Sampling, Intermittent	Sampling successively for limited periods of time throughout an operation or for a predetermined period of time. The duration of sample periods and of the intervals between are not necessarily regular and are not specified.
Sand	The fine granular material (usually less than 3/16 inch in diameter) resulting from the natural disintegration of rock, or from the crushing of friable sandstone.
Sand Grout	Grout mixture containing water, Portland cement, and sand.
Sand Streak	A streak of exposed fine aggregate in the surface of formed concrete caused by bleeding.
Saturated Surface-Dry	Condition of an aggregate particle or other porous solid when the permeable voids are filled with water but there is no water on the exposed surface.
Saturated Surface-dry (SSD) Particle Density	The mass of the saturated-surface-dry aggregate divided by its displaced volume in water or in concrete. (Also called Bulk Specific Gravity).
Saturation	(1) In general, the condition of the coexistence in stable equilibrium of either a vapour and a liquid or a vapour and solid phase of the same substance at the same temperature. (2) As applied to aggregate or concrete, the condition such that no more liquid can be held or placed within it.
Saw Blade, Abrasive	Concrete sawing medium that uses non-diamond abrasion elements. These blades do not need water to cool, but water is sometimes used.
Saw Blade, Diamond	Concrete sawing medium that uses industrial diamonds as the primary abrasion element., Blades are cooled with water to protect the host metal from melting and prematurely dislodging the diamonds.

Saw Cut	A cut in hardened concrete utilizing diamond or silicone-carbide blades or discs.
Sawed Joint	A joint cut in hardened concrete, generally not to the full depth of the member, by means of special equipment.
Sawing	Cutting of joints in hardened concrete by means of special equipment utilizing diamond or silicon carbide blades or discs; cut goes only part way through the slab.
Scaling	Flaking or peeling away of the near-surface portion of hydraulic cement concrete or mortar.
Scallop	Areas enclosed by profile line and blanking band.
Schmidt Hammer (trade name), Swiss Hammer, or Rebound Hammer	A device used to estimate the compressive strength of hardened concrete by measuring surface hardness.
Screed	(1) To strike off concrete lying above the desired plane or shape. (2) A tool for striking off the concrete surface, sometimes referred to as a Strikeoff.
Screed Guide	Firmly established grade strips or side forms for unformed concrete that will guide the strikeoff in producing the desired plane or shape.
Sealant	See Joint Sealant and Membrane Curing
Sealant Reservoir	The saw kerf or formed slot in which a joint sealant is placed. Many times this refers to a cut made to widen the original saw cut made for a contraction joint.
Sealing	The process of filling the sawed joint with material to minimize intrusion into the joint of water and incompressible materials.
Sealing Compound	See Joint Sealant and Membrane Curing
Secondary Sawing	The sawing that takes place to establish shape in the joint. Many times this shape is the reservoir of the joint.
Segregation	The tendency, as concrete is caused to flow laterally, for coarse aggregate and drier material to remain behind and for mortar and wetter material to flow ahead. This also occurs in a vertical direction when wet concrete is over-vibrated, the mortar and wetter material rising to the top. In the vertical direction, segregation may also be called stratification.
Semiautomatic Batcher	A batcher equipped with gates or valves that are separately opened manually to allow the material to be weighed but which are closed automatically when the designated weight of each material has been reached.
Separation`	The tendency, as concrete is caused to pass from the unconfined ends of chutes or conveyor belts, for coarse aggregate to separate from the concrete and accumulate at one side; the tendency, as processed aggregate leaves the ends of conveyor belts, chutes, or similar devices with confining sides, for the larger aggregate to separate from the mass and accumulate at one side; the tendency for solids to separate from the water by gravitational settlement. See also Bleeding and Segregation.
Set	The condition reached by a cement paste, mortar, or concrete when it has lost plasticity to an arbitrary degree, usually measured in terms of resistance to penetration or deformation. Initial set refers to first stiffening. Final set refers to attainment of significant rigidity.

Setting of Cement	Development of rigidity of cement paste, mortar, or concrete as a result of hydration of the cement. The paste formed when cement is mixed with water remains plastic for a short time. During this stage it is still possible to disturb the material and remix without injury, but as the reaction between the cement and water continues, the mass loses its plasticity. This early period in the hardening is called the "setting period," although there is not a well-defined break in the hardening process.
Setting Time	The time required for a specimen of concrete, mortar or cement paste, prepared and tested under standardized conditions, to attain a specified degree of rigidity.
Settlement	Sinking of solid particles in grout, mortar, or fresh concrete, after placement and before initial set. See also Bleeding.
Settlement Shrinkage	A reduction in volume of concrete prior to the final set of cementitious mixtures; caused by settling of the solids and decreases in volume due to the chemical combination of water with cement. See Plastic Shrinkage.
Shear Walls	A wall that resists horizontal forces (such as wind, blast, or earthquakes) applied in the plane of the wall. Shear walls may brace other walls perpendicular to them. The added loading offers greater resistance to overturning stresses. Also see lateral support.
Shrinkage	Decrease in length or volume.
Shrinkage Crack	Crack from restraint of volume reduction due to shrinkage or temperature contraction; usually occurring within the first few days after placement.
Shrinkage Cracking	Cracking of a slab due to failure in tension caused by external or internal restraints as reduction in moisture content develops.
Shrink-mixed Concrete	Ready-mixed concrete mixed partially in a stationary mixer and then mixed in a truck mixer.
Sieve	A metallic plate or sheet, a woven-wire cloth, or other similar device, with regularly spaced apertures of uniform size, mounted in a suitable frame or holder for use in separating granular material according to size.
Sieve Analysis	The classification of particles, particularly of aggregates, according to sizes as determined with a series of sieves of different openings.
Silicone	A resin, characterized by water-repellent properties, in which the main polymer chain consists of alternating silicon and oxygen atoms, with carbon-containing side groups; silicones may be used in joint sealing compounds, caulking or coating compounds, or admixtures for concrete.
Silicone Sealant	Liquid joint sealant consisting of silicone-based material.
Skid Resistance	A measure of the frictional characteristics of a surface.
Skirting	A finishing member along perimeter of a wall or other vertical surface where it meets the floor.
Slab Jacking	Process of injecting grout materials beneath concrete slabs in order to lift or elevate the slabs.
Slab Stabilization	Process of injecting grout materials beneath concrete slabs in order to fill voids without raising the concrete slabs.
Slag	Waste material resulting from the refining of an ore.
Sleeper Wall (dwarf)	A low wall supporting the joists of the lowest floor of a building.

Slenderness Ratio	The ratio of effective length or height to thickness.
Slip Form Paving	A type of concrete paving process that involves extruding the concrete through a machine to provide a uniform dimension of concrete paving.
Slipform	A form that is pulled or raised as concrete is placed; may move in a generally horizontal direction to lay concrete evenly for highway paving or on slopes and inverts of canals, tunnels, and siphons; or vertically to form walls, bins, or silos.
Slump	A measure of consistency of freshly mixed concrete, equal to the subsidence measured to the nearest 1/4 inch (6 mm) of the molded specimen immediately after removal of the slump cone.
Slump Cone	A mold in the form of the lateral surface of the frustum of a cone with a base diameter of 8 in (203 mm), top diameter 4 in (102 mm), and height 12 in (305 mm), used to fabricate a specimen of freshly mixed concrete for the slump test.
Slump Loss	The amount by which the slump of freshly mixed concrete changes during a period of time after an initial slump test was made on a sample or samples thereof.
Slump Test	The procedure for measuring slump.
Slurry	Mixture of water and concrete particles resulting from concrete sawing or grinding.
Soffit	An exposed under surface of beam including that of a ceiling
Solar Screen	A pieced wall, made from masonry or tile units, intended to act as a sunshade.
Solid Floor	A floor laid direct on the ground or over a continuous filling.
Solid Volume	See Absolute Volume.
Sounding	Process of tapping concrete slab surface with metal object, listening for tone from the impact, to determine areas of delamination.
Soundness	In the case of a cement, freedom from large expansion after setting. In the case of aggregate, the ability to withstand aggressive conditions to which concrete containing it might be exposed, particularly those due to weather.
Spalling, Compression	Cracking, breaking, chipping, or fraying of slab edges within 0.6 meter of a transverse joint.
Spalling, Sliver	Chipping of concrete edge along a joint sealant; usually within 12 millimeters of the joint edge.
Spalling, Surface	Cracking, breaking, chipping, or fraying of slab surface; usually within a confined area less than 0.5 square meters.
Specific Gravity	The ratio of the weight in air of a given volume of material at a stated temperature to the weight in air of an equal volume of distilled water at the same temperature.
Specific Gravity Factor	The ratio of the weight of aggregates (including all moisture), as introduced into the mixer, to the effective volume displaced by the aggregates.
Split Batch Charging	Method of charging a mixer in which the solid ingredients do not all enter the mixer together; cement, and sometimes different sizes of aggregate, may be added separately.
Spud Vibrator	A vibrator used for consolidating concrete, having a vibrating casing or head that is used by insertion into freshly placed concrete.

Standard Deviation   The root mean square deviation of individual values from their average.

Static Load   The weight of a single stationary body or the combined weights of all stationary bodies in a structure (such as the load of a stationary vehicle on a roadway); during construction, the combined weight of forms, stringers, joists, reinforcing bars, and the actual concrete to be placed. See also Dead Load.

Stationary Hopper   A container used to receive and temporarily store freshly mixed concrete.

Stile   A framed outer vertical member of a door or window frame.

Storage Hopper   See Stationary Hopper.

Straight-Edging   Process of using a rigid, straight piece of either wood or metal to strike off or screed a concrete surface to proper grade or to check the planeness of a finished surface.

Stratification   The separation of over-wet or over-vibrated concrete into horizontal layers with increasingly lighter material toward the top; water, laitance, mortar, and coarse aggregate will tend to occupy successively lower positions (in that order).

Strength   A generic term for the ability of a material to resist strain or rupture induced by external forces. See also Compressive Strength, Fatigue Strength, Flexural Strength, Shear Strength, Splitting Tensile Strength, Tensile Strength, Ultimate Strength, and Yield Strength.

Stress   Intensity of internal force (i.e., force per unit area) exerted by either of two adjacent parts of a body on the other across an imagined plane of separation; when the forces are parallel to the plane, the stress is called shear stress; when the forces are normal to the plane, the stress is called normal stress; when the normal stress is directed toward the part on which it acts it is called compressive stress; when it is directed away from the part on which it acts it is called tensile stress.

Strikeoff   To remove concrete in excess of that required to fill the form evenly or bring the surface to grade; performed with a straightedged piece of wood or metal by means of a forward sawing movement or by a power operated tool appropriate for this purpose; also the name applied to the tool. See also Screed and Screeding.

Structural Capacity   Expression of the ability of a pavement to carry traffic loads; Expressed as number of equivalent single axle loads in AASHTO design methodology.

Structure   An organized combination of connected units constructed to perform a function or functions requiring some measure of rigidity, stability and durability.

Styrene Butadiene Resins (SBR)   SBR resins are basically synthetic rubber in solution.

Subbase   A layer in a pavement system between the subgrade and base course or between the subgrade and a portland cement concrete pavement.

Subgrade   The soil prepared and compacted to support a structure or a pavement system.

Substructure   The part of a building or structure below the level of the plinth or adjoining ground.

Sulfate Attack   Chemical or physical reaction between certain constituents in cement and sulfates in the soil or groundwater; sufficient attack may disrupt concrete that is susceptible to it.

Sulfate Resistance	The ability of aggregate, cement paste, or mixtures thereof to withstand chemical attack by sulfate ion in solution.
Superplasticizer	See Water-Reducing Admixture (high range)
Superstructure	The part of a building or structure above level of the plinth or adjoining ground.
Supplementary Cementitious Material	Mineral admixtures consisting of powdered or pulverized materials which are added to concrete before or during mixing to improve or change some of the plastic or hardened properties of Portland cement concrete. Materials are generally natural or by-products of other manufacturing processes.
Surface Moisture	Water retained on surfaces of aggregates capable of mixing with portland cement in concrete; distinguished from absorbed moisture, which is contained inside the aggregate particles.
Surface Primer	A fluid material designed to 'wet' a substrate surface effectively and itself provide a surface onto which a primary adhesive, having lesser fluidity or 'wetting' properties, can bond efficiently
Surface Retarder	A retarder used by application to a form or to the surface of newly placed concrete to delay setting of the cement to facilitate construction joint cleanup or to facilitate production of exposed, aggregate finish.
Surface Tension	That property, due to molecular forces, that exists in the surface film of all liquids and tends to prevent the liquid from spreading.
Surface Texture	Degree of roughness or irregularity of the exterior surfaces of aggregate particles or hardened concrete.
Surface Vibrator	A vibrator used for consolidating concrete by application to the top surface of a mass of freshly mixed concrete; four principal types exist: vibrating screeds, pan vibrators, plate or grid vibratory tampers, and vibratory roller screeds.
Surface Voids	Cavities visible on the surface of a solid. See also Bug Holes.
Surface Water	See Surface Moisture
Suspended Floor	A floor that spans between supports.
Swelling	Increase in length or volume. See also Autogenous Volume Change,. Contraction, Expansion, and Volume Change.
Tamper	(1) An implement used to consolidate concrete or mortar in molds or forms. (2) A hand-operated device for compacting floor topping or other unformed concrete by impact from the dropped device in preparation for strikeoff and finishing; contact surface often consists of a screen or a grid of bars to force coarse aggregates below the surface to prevent interference with floating or troweling. See also Jitterbug.
Tamping	The operation of compacting freshly placed concrete by repeated blows or penetrations with a tamping device.
Temper	The addition of water and mixing of concrete or mortar as necessary to bring it initially to the desired consistency. See also Retempering.
Tendon	Steel rods or cables used in prestressing concrete units.
Tensile Strength	Maximum stress that a material is capable of resisting under axial tensile loading based on the cross-sectional area of the specimen before loading.

Texturing	The process of producing a special texture on either unhardened or hardened concrete.
Theories of Adhesion	(i) **Absorbtion:** If molecules can come closer together than 0.003 microns, they are automatically attracted to bond together (secondary force). Also explains why 'wetting' of surface is essential characteristic. Note: Primary force of bonding resulting from actual chemical reaction between adhesive and surface is very rare.
	(ii) **Diffusion:** It supposes actual diffusion or blending of surface materials together by migration of one into the other. Explains largely bond between two sheets. Which develops gradually over a period of time without added adhesive.
	(iii) **Electrostatic:** Russian theory based on observing sparks when self adhesive tapes are stripped rapidly.
	(iv) **Mechanical:** It supposes adhesive enters surface pores and forms a mechanical key to create bond. Rough surfaces may have fibers part-covering pores and DUST prevents bond to actual substrate surface and is the greatest enemy.
Thermal Expansion	Expansion caused by increase in temperature.
Thermal Movement	Change of dimension of concrete or masonry resulting from change of temperatures. See also Contraction and Expansion.
Thermal Shock	The subjection of newly hardened concrete to a rapid change in temperature which may be expected to have a potentially deleterious effect.
Throating	A groove under a sill, coping, or moulding.
Tie Bar	Bar at right angles to and tied to reinforcement to keep it in place; bar extending across a construction joint.
Tilting Concrete Mixer	See Mixer, Tilting.
Time of Haul	In production of ready-mixed concrete, the period from first contact between mixing water and cement until completion of discharge of the freshly mixed concrete.
Time of Set	Time required after addition of water to cement for cement paste, mortars, or concretes to attain a certain arbitrary degree of hardness or strength.
Time of Setting	See Initial Setting Time and Final Setting Time.
Tongue and Groove	A joint in which a protruding rib on the edge of one side fits into a groove in the edge of the other side, abbreviated "T&G." See also Keyway.
Topping	(1) A layer of high quality concrete placed to form a floor surface on a concrete base, or (2) a dry-shake application of a special material to produce particular surface characteristics.
Transit-mixed Concrete	Concrete, the mixing of which is wholly or principally accomplished in a truck mixer. (Same as truck mixed concrete.)
Transom	An intermediate horizontal member of a window frame, doorframe or similar structure.
Transverse Broom	Surface texture obtained using either a hand broom or mechanical broom that lightly drags the stiff bristles across the surface.
Transverse Crack	Crack that develops at a right angle to the long direction of the member.
Transverse Joint	A joint normal to the longitudinal dimension of a structure.

Transverse Reinforcement	See Reinforcement, Transverse.
Transverse Tine	Surface texture achieved by a hand held or mechanical device equipped with a rake-like tining head that moves laterally across the width of the paving surface.
Trial Batch	A batch of concrete used for establishing or checking proportions.
Trimming	Cutting/planning a member to size, trimmers and trimming joists or trimming rafters forming an opening.
Trowel	A flat, broad-bladed steel hand tool used in the final stages of finishing operations to impart a relatively smooth surface to concrete floors and other unformed concrete surfaces; also, a flat triangular-bladed tool used for applying mortar to masonry.
Truck Mixed Concrete	Concrete, the mixing of which is accomplished in a truck mixer.
Truck Mixer	A concrete mixer suitable for mounting on a truck chassis and capable of mixing concrete in transit. See also Horizontal-Axis Mixer, Inclined-Axis Mixer, and Agitator.
Tusk Tenon	A combined tenon and housed joint for bearing timbers.
Ultra-thin Whitetopping	Thin layer of new concrete (2-4 inches), usually high strength and fiber reinforced, placed over a prepared surface of distressed asphalt.
Unbonded Concrete Overlay	Overlay of new concrete placed onto distressed existing concrete pavement with a layer of asphalt or other medium between the new and old concrete surface to separate them.
Uncontrolled Crack	A crack that is located within a slab away from the sawed joints.
Under-Sanded	A concrete mixture that is deficient in sand content; a condition associated with poor workability or finishing characteristics.
Unit	Building material which is formed as a simple article complete in itself but which is intended to be part of a compound unit or building or structure. Examples are brick, block, tile and lintel
Unit Water Content	The quantity of water per unit volume of freshly mixed concrete, often expressed as pounds or gallons per cubic yard. It is the quantity of water on which the water-cement ratio is based and does not include water absorbed by the aggregate.
Unit Weight	See Bulk Density and Specific Gravity.
Unreinforced Concrete	See Plain Concrete.
Unsound Aggregate	An aggregate or individual particle of an aggregate capable of causing or contributing to deterioration or disintegration of concrete under anticipated conditions of service.
Uplift Beam	Beam-like movement detection device used to monitor slab lift during slab stabilization.
Urethanes	Polymers and copolymers produced by the reaction of isocyanates with polyols.
Veneer	A cover surface with thin coating of finer wood, thin outer coat of fine material.
Vibrated Concrete	Concrete compacted by vibration during and after placing.
Vibration	Energetic agitation of concrete produced by a mechanical oscillating device at moderately high frequency to assist consolidation and compaction.

Vibration Limit	That time at which fresh concrete has hardened sufficiently to prevent its becoming mobile when subject to vibration.
Vibration, External	External vibration employs vibrating devices attached at strategic positions on the forms and is particularly applicable to manufacture of precast items and for vibration of tunnel-lining forms; in manufacture of concrete products, external vibration or impact may be applied to a casting table.
Vibration, Internal	Internal vibration employs one or more vibrating elements that can be inserted into the concrete at selected locations, and is more generally applicable to in-place construction.
Vibration, Surface	Surface vibration employs a portable horizontal platform on which a vibrating element is mounted.
Vibrator	An oscillating machine used to agitate fresh concrete so as to eliminate gross voids, including entrapped air but no entrained air, and produce intimate contact with form surfaces and embedded materials.
Vibratory Plate Compactor	Motorized, one-man tool consisting of a vibrating square plate that transmits energy to compact granular materials.
Vinyl Resins	This is general term for substituted ethylenes and their copolymers like polytheylene, polystyrenes, etc., basically copolymers than homopolymers.
Vinylester	One of a group of synthetic resin produced by the reaction of acrylic with epoxy resin or Bisphenol A, and commonly prepared for application by mixing with vinyl group monomer and free radical catalyst at ambient temperature, and used as binders for resin mortars and concrete, and fiber laminates (mainly glass) adhesives.
Vitrification	The process of changing to a glassy substance through heat.
Volume Batching	The measuring of the constituent materials for mortar or concrete by volume.
Wall Hanger	A metal bracket suspended from a wall to support a member.
Wall Plate	A member built into or on a wall to distribute the load from a timber floor or roof.
Wash (or Flush) Water	Water carried on a truck mixer in a special tank for flushing the interior of the mixer after discharge of the concrete.
Water Bar	A metal strip in the seal of a door or window designed to prevent penetration of water.
Water Retentivity	Ability of a mortar to retain the mixing water for hydration purpose.
Water-Cement Ratio	The ratio of the amount of water, exclusive only of that absorbed by the aggregates, to the amount of portland cement in a concrete or mortar mixture; preferably stated as a decimal by weight.
Water-Cementitious Ratio	The ratio of the amount of water, exclusive only of that absorbed by the aggregates, to the amount of portland cement and other cementitious material (flay ash, pozzolan, etc.) in a concrete or mortar mixture; preferably stated as a decimal by weight.
Water-Gain	See Bleeding.
Water-Reducing Admixture	A material that either increases slump of freshly mixed mortar or concrete without increasing water content or maintains a workability with a reduçed amount of water, the effect being due to factors other than air entrainment; also known as water reducer.

Water-Reducing Admixture (High Range) A water-reducing admixture capable of producing large water reduction or great flowability without causing undue set retardation or entrainment of air in mortar or concrete.

Weather Moulding A moulded projection fixed to the bottom rail of an external door or sash to divert water from the sill or threshold. (A weather board is similar but un-moulded)

Weathering Changes in colour, texture, strength, chemical composition or other properties of a natural or artificial material due to the action of the weather.

Weight Batching Measuring the constituent materials for mortar or concrete by weight.

Welded-Wire Fabric Reinforcement Welded-wire fabric in either sheets or rolls, used to reinforce concrete.

Well-Graded Aggregate Aggregate having a particle size distribution that will produce maximum density; i.e., minimum void space.

Wet Covered with visible free moisture; not dry. See also Damp and Moist.

Wet Process In the manufacture of cement, the process in which the raw materials are ground, blended, mixed, and pumped while mixed with water; the wet process is chosen where raw materials are extremely wet and sticky, which would make drying before crushing and grinding difficult.

Whitetopping Concrete overlay pavement placed on an existing asphalt pavement.

Whitetopping, Conventional Overlay of new concrete, greater than 4 inches thick, placed onto existing asphalt pavement with no particular steps taken to ensure bonding or debonding.

Whitetopping, Ultra-thin See "Ultra-thin Whitetopping."

Wiggle Bolt Two-piece threaded bolt system used for tying lanes of concrete pavement; usually consists of a female section that is cast into a vertical slab face, and an angled male end which screws into the female coupler.

Wind Load The force exerted by wind on a structure or part of a structure.

Winning Mining or removal of clay.

Wire Mesh See Welded Wire Fabric.

Workability That property of freshly mixed concrete or mortar which determines the ease and homogeneity with which it can be mixed, placed, compacted, and finished.

Working Crack A crack in a concrete pavement slab that undergoes significant deflection and thermal opening and closing movements; Typically oriented transverse to the pavement centerline and near a non-functioning transverse contraction joint.

Yield The volume of fresh concrete produced from a known quantity of ingredients; the total weight of ingredients divided by the unit weight of the freshly mixed concrete.

Zero-Slump Concrete Concrete of stiff or extremely dry consistency showing no measurable slump after removal of the slump cone. See also Slump and No-Slump Concrete.

INDEX